Optical
Communication
Receiver Design

SPIE Tutorial Texts Series

Developed from SPIE short courses, this series comprises stand-alone tutorials covering fundamental topics in optical science and technology at the introductory and intermediate levels. Since 1989 the Tutorial Texts have provided an introduction to those seeking to understand foundation and new technologies. SPIE has expanded the series beyond topics covered by the short course program to encompass contributions from experts in their field who can write with authority and clarity at a pedagogical level. The emphasis is always on the tutorial nature of the text.

Series Editor:
Donald C. O'Shea, Georgia Institute of Technology, USA

IEE Telecommunications Series

The IEE Telecommunications Series, now containing 37 titles, has become an internationally respected library for professional and graduate-level books across the communications field. Aimed at the communications and/or engineering professional, many volumes contain substantial tutorial content. For many of the core titles, new editions are released on a regular basis to ensure up-to-date coverage of these fast-developing areas. The Series Editors and Publishers welcome comments on the Series and will always welcome and consider new ideas for publications.

Series Editors:
Professor C. J. Hughes, University of Essex, UK
Professor J. D. Parsons, University of Liverpool, UK
Dr. G. White, Company Consultant, British Telecom, UK

Optical Communication Receiver Design

Stephen B. Alexander
Ciena Corporation

SPIE OPTICAL ENGINEERING PRESS

A Publication of SPIE—The International Society for Optical Engineering
Bellingham, Washington USA

IEE

INSTITUTION OF ELECTRICAL ENGINEERS
London, UK

Library of Congress Cataloging-in-Publication Data

Alexander, Stephen B.
    Optical communication receiver design / Stephen B. Alexander.
      p. cm. – (Tutorial texts in optical engineering ; v. TT22)
    Includes bibliographical references and index.
    ISBN 0-8194-2023-9 (softcover)
    1. Optical communications—Equipment and supplies. 2. Optical
detectors—Design and construction. I. Title. II. Series.
    TK5103.59.A53   1997
621.382'7—dc20   95-46526
                                                  CIP

Copublished by

SPIE—The International Society for Optical Engineering
P.O. Box 10
Bellingham, Washington 98227-0010
Phone: 360/676-3290
Fax: 360/647-1445
Email: spie@spie.org
WWW: http://www.spie.org
SPIE Vol. TT22: ISBN 0-8194-2023-9

The Institution of Electrical Engineers
Michael Faraday House
Six Hills Way, Stevenage, Herts.
SG1 2AY United Kingdom
Phone: +44 (0)1438 313311
Fax: +44 (0)1438 360079
Email: books@iee.org.uk
WWW: http://www.iee.org.uk
IEE Telecommunications Series Vol. 37: ISBN 0-85296-900-7

Printed in the United States of America.

Second Printing

Contents

Preface

We are surrounded by an ongoing revolution in optical communication. Fiber-optic networks carrying gigabits per second span oceans and continents, and devices such as optical amplifiers, which were once regarded only as laboratory curiosities, are now commonplace. The vast capacity of optical communication systems has enabled the development of information infrastructures of both national and global extent. Optical communication techniques are not restricted to fiber-optics. Free-space optical communication offers the possibility of high-data-rate links among satellites and the Earth, allowing even greater flexibility in terms of network connectivity and access.

This text provides an overview of the design principles for receivers used in optical communication systems. The technology and techniques that are discussed are similar to those used in conventional microwave communication receivers; however, there are also significant differences because of the unique characteristics of the photodetection process. The text grew out of the notes for a short course in receiver design. The level of the material is targeted at the practicing engineer and the text contains some 500 references to provide a reader with pointers to the wide variety of work that is available in the open literature.

The material is organized into seven chapters, with Chapter 1 providing a brief review of the technologies used to construct optical communication links. Following the technology introduction, Chapter 2 illustrates the flow of system performance specifications into receiver requirements and is illustrated by the use of system link and receiver sensitivity budgets. Chapter 3 introduces the fundamentals of photodetection and the associated statistics. Semi-classical techniques are used, with appropriate references to quantum mechanical considerations as needed. The signal-to-noise ratio for both direct and coherent detection receivers is derived and the concept of a shot-noise-limited receiver is introduced. The characteristics and performance of photodetectors are reviewed in Chapter 4. The *p-i-n*, avalanche photodiode, and metal-semiconductor-metal photodetectors are covered in detail and a series of equivalent circuit models are developed so that the impact of device characteristics on achievable receiver performance can be determined.

The circuit analysis techniques used with electrical noise are omitted in many engineering curricula, and Chapter 5 provides a quick tutorial on the general subject of noise analysis and also serves to describe the specific analysis techniques needed to model optical receivers. In particular, we illustrate the concept of an equivalent input current-noise model for the receiver. Chapter 6 reviews the design of the receiver front end, covering the resistor terminated voltage amplifier, high-impedance amplifier, and transimpedance amplifier. Chapter 7 concludes the text with examples of receiver performance analysis. Direct detection, coherent detection, and optically preamplified receivers are discussed, as well as analog systems. Particular attention is given to the detection statistics associated with the various photodetectors and receiver structures.

Stephen B. Alexander
Savage, Maryland

Acknowledgments

A tutorial text of this kind is clearly not the result of the efforts of a single individual. We are all indebted to the pioneers in this field. The works of Personick, Smith, Goell, Hullett, and Muoi form the foundation for much of what is contained in this text.

The author would like to gratefully acknowledge Don O'Shea at Georgia Tech, the editor of the SPIE Tutorial Text Series. He invited me to prepare this manuscript hoping to get a final product within one year but then had the limitless patience necessary to tolerate my seemingly random, interrupt-driven schedule that prolonged the project *much* longer than either of us had intended. He reviewed each chapter as it was written and his comments and suggestions proved invaluable. Dixie Cheek at SPIE edited the text and managed to make grammatical sense out of the manuscript. She may even have finally taught me punctuation in the process!

Thanks to Vincent Chan of MIT for providing me with the initial opportunity to contribute to this field and to Pat Nettles, Steve Chaddick, and Dave Huber of Ciena Corporation for bringing me into the commercial world. More thanks to the past and present members of the MIT Lincoln Laboratory Optical Communication Technology Group, including Steve Bernstein, Roy Bondurant, Don Boroson, Ed Bucher, Dan Castagnozzi, Claudia Fennelly Roe Hemenway, Lori Jeromin, Bob Kingston, John Kaufmann, Emily Kintzer, Jeff Livas, Jean Mead, Mark Stevens, Eric Swanson, Fred Walther, and David Welford. Their 10+ years of demonstrated success in developing advanced optical communication systems contributed greatly to the content of this text. A special thanks to Al Tidd – a rare person who can build or fix just about anything.

The biggest thanks of all go to Edythe Alexander. Without her constant support, encouragement, sacrifice, patience, and understanding, none of this would have been possible.

Glossary

Symbols

α	absorption length of a material - usually in cm^{-1}
α_{dB}	attenuation constant of fiber - usually in dB/km
ε_o	permittivity of vacuum (8.85×10^{-12} F/m)
ε_r	relative permittivity
$\gamma(\omega)$	complex correlation coefficient
η	quantum efficiency
η_{AC}	AC quantum efficiency
η_{DC}	DC quantum efficiency
λ	wavelength
$\lambda(t)$	photon rate parameter - photons per second as a function of time
τ	transit-time
τ_a	transit-time for electrons in the absorbing region
τ_h	transit-time for holes
A	area
$A_v(\omega)$	voltage gain transfer function
B	bandwidth (Hz)
c	speed of light in a vacuum (3×10^8 m/s)
C_d	detector capacitance
C_j	junction capacitance

C_b	bonding-pad capacitance
C_p	package capacitance
C_t	total capacitance
E	energy (J)
$E(t)$	electric field
E_b	energy per bit
f	electrical frequency (Hz)
f_a	amplifier bandwidth
F	noise factor
$F(M)$	excess noise factor for multiplication gain
$F(\omega)$	noise figure (or factor) as a function of frequency
G_a	antenna (or aperture) gain
h	Planck's constant (6.626×10^{-34} J/s)
$\bar{i}$	time averaged DC photocurrent
i_{du}	unmultiplied dark current
i_{dm}	multiplied dark current
$i_{dark}\ i_{dk}$	dark current
i_{elec}	current-noise density of receiver electronics-noise
i_n or $i_n(\omega)$	current-noise density
$i_{n\text{-}d}$	dark current noise
$i_{n_{eq}}$ or $i_{n_{eq}}(\omega)$	equivalent input current-noise density $i_{n_{eq}}^2 = i_{shot}^2 + i_{rcvr}^2$
i_{rcvr}	current-noise density of receiver excluding signal shot-noise
$i_s(\omega)$ or $i_s(t)$	input signal current as a function of frequency or time
i_{shot}	current-noise density from photocurrent shot-noise
I_1, I_2, I_3, I_f	normalized noise bandwidth integrals
I_n	total current-noise
$I_{n_{eq}}$	total equivalent input current-noise
I_{shot}	total current-noise from shot-noise
I_a	length of absorbing region
L_b	bond-wire inductance

l_d	length of depletion region
L_{max}	maximum length
$L_N\{x\}$	Laguerre polynomial
L_p	loss during propagation
L_T	total loss
M	multiplication gain
M_e	multiplication gain for electrons
M_h	multiplication gain for holes
v	optical frequency (Hz)
n	number of photons, index of refraction
$n(f)$	noise spectral density
n_b	number of background photons
n_p	number of photons per bit
n_s	number of signal photons
N	total noise
N_o	noise density (W/Hz)
P	power (W)
P_{in}	input power
P_{rcvd}	received optical power
$P_{sig} \, P_s$	detected signal power
$P_{noise} \, P_n$	noise power
P_{out}	output power
$P(x)$	probability of occurring
$P(x|y)$	probability of occurring given that has occurred
r	photon arrival rate in photons per second
R	responsivity, reflectivity, resistance, data rate
R_j	junction resistance
R_l	load resistance
R_s	series resistance or source resistance
RC	resistor-capacitor

V_n	total current-noise
v_n or $v_n(\omega)$	voltage-noise density
$v_o(\omega)$ or $v_o(t)$	output voltage as a function of frequency or time
v_s	saturation velocity
v_e	velocity of an electron
v_h	velocity of a hole
ω	electrical frequency (rad/s)
ω_c	cutoff frequency (usually 3-dB point)
ω_p	frequency of a pole
ω_z	frequency of a zero
X_s	source reactance
$Y_c(\omega)$	correlation admittance
$Y_s(\omega)$	source admittance
z	propagation distance
$Z_{in}(\omega)$	input impedance
Z_o	characteristic impedance
Z_t or $Z_t(\omega)$	transimpedance gain

Abbreviations

AGC	automatic gain control
AFC	automatic frequency control
APD	avalanche photodiode
ASE	amplified spontaneous emission
ASE $\times$ ASE	ASE-cross-ASE noise
ASK	amplitude shift keying
AWGN	additive white Gaussian noise
BER	bit-error rate
BLIP	background limited IR photodetection
BJT	bipolar junction transistor
CB	common base
CC	common collector

CCD	charge-coupled device
CE	common emitter
CNR	carrier-to-noise ratio
CNDR	carrier-to-noise density ratio
CPFSK	continuous-phase FSK
CSO	composite second order
CTB	composite triple beat
CW	continuous wave
dBc	decibels relative to the carrier power
dBi	decibels relative to an isotropic radiator
dBm	decibels relative to 1.0 mW
DBR	distributed Bragg reflector
dBW	decibels relative to 1.0 W
DFB	distributed feedback
DPSK	differential phase shift keying
EDFA	erbium-doped fiber amplifier
FET	field-effect transistor
fF	femto-Farads
FM	frequency modulation
FP	Fabry-Perot
FSK	frequency shift keying
GBW	gain bandwidth
GEO	geosynchronous-Earth orbit
G-R	generation-recombination
HBT	heterojunction bipolar transistor
HEMT	high-electron-mobility transistor
HEO	high-Earth orbit
HFET	heterojunction FET
IF	intermediate frequency
ISI	intersymbol interference
JFET	junction FET
lasercom	laser communication
LED	light-emitting diode

LEO	low-Earth orbit
LO	local oscillator
MAP	maximum a posteriori
MESFET	metal-semiconductor FET
ML	maximum likelihood
MOPA	master-oscillator power-amplifier
MOSFET	metal-oxide-semiconductor FET
MMIC	monolithic microwave integrated circuit
MSM	metal-semiconductor-metal
NEB	noise equivalent bandwidth
NF	noise figure
NRZ	nonreturn to zero
OEIC	optoelectronic integrated circuit
OOK	on-off keying
PDFA	praseodymium doped fiber amplifier
pF	pico-Farads
PLL	phase-locked loop
PM	polarization maintaining
PMD	polarization mode dispersion
PMT	photomultiplier tube
PPM	pulse position modulation
PSD	power spectral density
PSK	phase shift keying
QE	quantum efficiency
QOS	quality of service
QPSK	quadrature phase shift keying
RIN	relative intensity noise
RAPD	reach-through APD
RC	resistance-capacitance
RF	radio frequency
RMS	root mean square
RZ	return to zero
$S \times ASE$	signal-cross-ASE noise

SAM	separate absorption and multiplication
SAGM	separate absorption and grading multiplication
SAW	surface acoustic wave
SLA	semiconductor laser amplifier
SL-APD	superlattice APD
SMSR	side-mode suppression ratio
SNR	signal-to-noise ratio
SNDR	signal-to-noise density ratio
TDRSS	Tracking and Data Relay Satellite System
TE	transverse electric
TM	transverse magnetic
WDM	wavelength-division multiplexing

Optical Communication Receiver Design

Chapter 1

OPTICAL COMMUNICATIONS

1.1 Introduction

Light has been used for communication since signal fires were first used to send messages. Paul Revere's lanterns, manually operated signal lanterns on ships, and signal flares are other examples of early optical communication systems. A revolution in technology occurred in 1880 when Alexander Graham Bell's photophone was used to send intensity modulated sunlight over a distance of a few hundred feet. From the late 1880's to the early 1960's low-capacity short-range links represented the state-of-the-art in optical communication.

The 1960's invention of the laser, the 1970's development of low-loss optical fiber, the 1980's demonstration of long-lived semiconductor laser diodes, and the 1990's development of practical optical amplifiers have ushered in a new era for optical communication. It is now possible for optical fiber systems to cross oceans and continents, while free-space systems provide high data-rate communication links between satellites at geosynchronous distances. Optical communications, in combination with microwave and wireless technologies, are enabling the construction of high-capacity networks with global connectivity.

The most common wavelengths used for optical communication fall between 0.83 and 1.55 microns. Other wavelengths are also used but this range encompasses the most popular applications. A wavelength of 1 micron corresponds to a frequency of 300 THz (300,000 GHz). This is a considerably higher frequency than those associated with conventional radio, microwave, or millimeter-wave communication systems. A high center frequency implies that extremely high-speed modulations should, in principle, be possible. A one-percent bandwidth, easily achieved in conventional microwave systems, would imply over a THz modulation bandwidth for optical carriers. This is far more modulation bandwidth than a microwave carrier can provide. Although wide bandwidth is important, it is not the only reason optical communications technology is causing such revolutionary changes in our ability to transfer information. To gain a broader understanding, we need to examine the advantages light has in both guided and unguided modes of transmission.

Optical fiber is an exceedingly low-loss propagating medium. Losses can approach 0.15 dB/km and, with modest amounts of optical amplification, multi-Gbps data streams can be transmitted over tens of thousands of kilometers without electronic regeneration. Conventional electronic systems are orders of magnitude less capable of providing such raw communications transport capacity. A good quality optical fiber loses half of the optical power in the propagating field in 15-20 km. In comparison, conventional coaxial cables will lose half the electrical power in just a few hundred meters! In addition to significantly lower loss, packaging and volume benefits are also realized when fiber is used. Low-loss coaxial cables frequently have diameters of a centimeter or more. An optical fiber has a diameter of a few hundred microns. This substantial reduction in cross-sectional area is particularly significant when fiber is used in crowded conduits.

In free-space systems, optical communication again has advantages. Smaller diameter transmitter and receiver apertures are needed to establish high data-rate communication links and, unlike the congested microwave bands, there is plenty of available spectrum. In addition, the narrow beam divergence of an optical system can be used to provide additional security by lowering the probability of detecting and intercepting the transmission. Free-space optical communication's chief drawback is that atmospheric effects such as wind, smoke, fog, and clouds have serious consequences if not taken into account in the overall system design.

1.2 Fiber-Optic Systems

The simplest form of a fiber-optic system is illustrated in Figure 1.1. As with any communication system, there is a transmitter, a receiver, and a channel to convey energy from transmitter to receiver. At the transmitter, the information is combined with the drive signals needed to operate a laser. The laser output is coupled into an optical fiber through which it propagates to the receiver. The light leaving the fiber is collected on a photodetector that generates an electrical signal in response to the optical excitation. The electrical signal is typically low-level and requires amplification and signal processing for the information to be recovered.

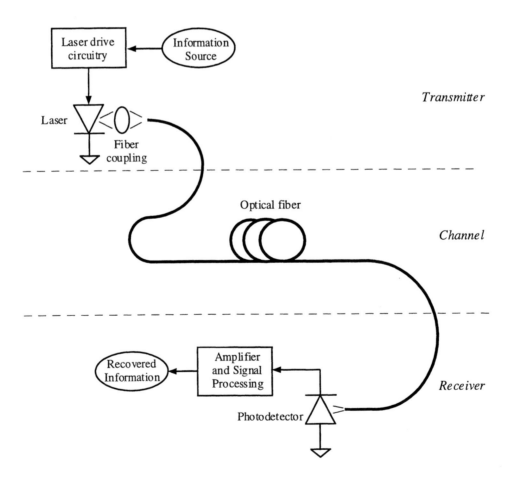

Figure 1.1 Block diagram of a fiber system.

In addition to the simple point-to-point link of Figure 1.1, optical fiber can be used to implement a variety of systems, as shown in Figure 1.2. A series of point-to-point links can be used to form long-haul repeatered systems spanning thousands of kilometers. Optical amplifiers can be used to build repeaterless systems. Fiber splitters can be used to form broadcast distribution systems, and wavelength selective elements can be used to construct wavelength division multiplexed (WDM) systems.

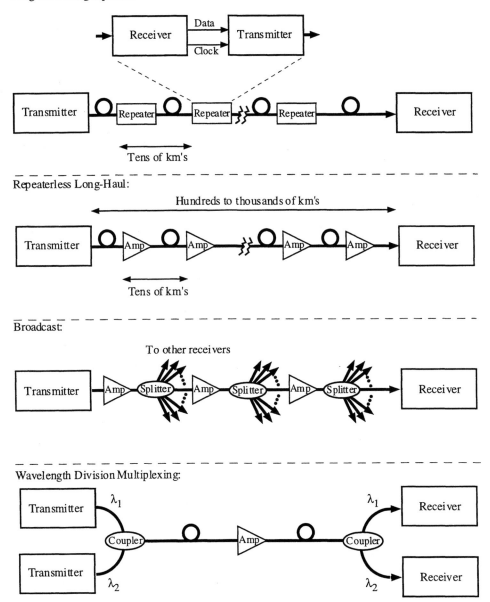

Figure 1.2 Example fiber based systems.

1.2.1 Optical Fiber

Optical fiber is a cylindrical dielectric waveguide that is most commonly fabricated using high-purity, low-loss materials such as silica glass [1]. Figure 1.3 illustrates the typical geometry of both single-mode and multimode optical fibers.

The term *mode* refers to the number of possible solutions to the electromagnetic wave propagation equation for the fiber. A single-mode fiber, carrying light at the wavelength for which it has been designed to be single-mode, will only allow one solution (i.e. one type of spatial distribution) to propagate in the fiber. A multimode fiber will support multiple spatial distributions. The various spatial modes in a multimode fiber propagate at different velocities, causing adjacent pulses in a modulated stream to rapidly disperse as they propagate. This modal-dispersion causes inter-symbol-interference, which severely limits the use of multimode fibers in high data-rate systems. Modal dispersion can be reduced by using graded index multimode fibers but cannot be eliminated.

Single-mode fibers operating at their designed wavelength do not exhibit modal-dispersion. However, a single-mode fiber does support two different configurations for the electric and magnetic fields corresponding to the two orthogonal polarization states for the single propagating mode. Any differences in the velocities at which the two polarizations propagate can give rise to polarization mode dispersion (PMD). Although PMD is typically small, it can cause significant degradation in long-haul high bit-rate systems.

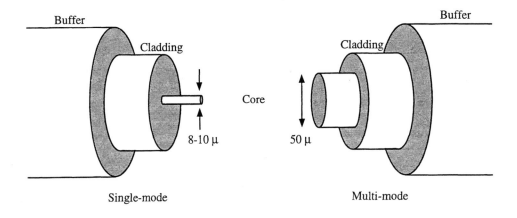

Figure 1.3 Typical geometry of optical fiber.

Most optical fiber does not maintain the polarization of the light originally launched into the fiber. This fiber is termed non-polarization-maintaining or non-PM. Fiber can be made to be polarization maintaining (PM) by varying the fabrication techniques. Fiber can also be fabricated with special characteristics such as the ability to polarize, or filter, light [2].

Light propagating in an optical fiber is subject to attenuation caused by absorption and scattering. The attenuation varies as a function of wavelength, as shown in

Figure 1.4. In the vicinity of 1550 nm there is a low-loss window where attenuation falls to approximately 0.15 dB/km. If we define the fiber bandwidth to correspond to the points at which the attenuation per km doubles to 0.3 dB/km, a bandwidth of about 200 nm, extending from 1450 nm to 1650 nm, is obtained. At 1550 nm, 200 nm corresponds to over 25 THz of bandwidth. Taking advantage of even 1% (250 GHz) of the bandwidth available in a single fiber is challenging. Most of today's systems efficiently utilize only a few tens of GHz of bandwidth.

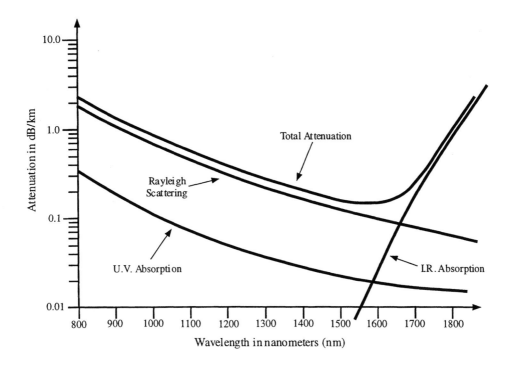

Figure 1.4 Intrinsic attenuation in silica glass fiber.

Figure 1.4 shows that three principle components contribute to the total intrinsic attenuation in an optical fiber. At long wavelengths there is an infrared (IR) absorption band. At short wavelengths the absorption is dominated by ultraviolet (UV) absorption and Rayleigh scattering. Rayleigh scattering is a commonly observed phenomena. When light propagates through the atmosphere it scatters off airborne molecules. The intensity of the scattered light is proportional to $(frequency)^4$ and thus short wavelength light is scattered the strongest. When the spectrum of the scattered light is combined with the human eye's response, the sky appears to be blue. Rayleigh scattering occurs in fiber because of random variations in the density and compositional uniformity of the glass lattice. These variations are frozen in when the fiber cools after being drawn and cause a random inhomogeneity in the refractive index of the fiber that acts as a collection of tiny scattering centers for the propagating light.

In addition to intrinsic attenuation, impurities in the glass may introduce local absorption peaks. For example, impurities associated with water vapor dissolved in the glass are known to introduce absorption peaks in the 1400 nm region. Metallic ions and other impurities can also cause significant absorption peaks if their concentrations are not controlled.

There are also nonlinear mechanisms that are observable when high power densities propagate in the fiber [3]. Stimulated Brillouin scattering is a relatively narrow bandwidth process that occurs at launched power levels of several mWs to a few Watts depending on fiber length and source bandwidth. Stimulated Raman scattering is a wider bandwidth effect with a higher power threshold that can come into play when many wavelengths propagate on a single fiber.

Single-mode fiber also exhibits both waveguide dispersion and material dispersion. Dispersion is a consequence of the velocity of propagation being different for different wavelengths of light. The combined effects of waveguide and material dispersion for three types of fiber are shown in Figure 1.5.

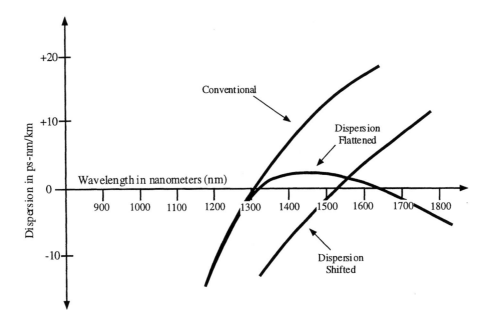

Figure 1.5 Dispersion in silica glass fibers.

Whenever a pulse travels through a dispersive media, the pulse tends to spread out in time. Dispersion in fibers is typically quoted in terms of the pulse spread in picosecond per nanometer of pulse bandwidth per kilometer of fiber traveled (ps/nm-km). Signals that spread energy over a wide bandwidth will consequently suffer from dispersion effects more than narrow band signals. Since a short pulse inherently has a wide bandwidth, dispersion significantly influences the ability of high bit-rate signals to propagate over long distances. Conventional fiber has a zero value of dispersion in the 1300 nm region. At 1550 nm this fiber has a dispersion of approximately 17 ps/nm-km. Note that even though the 1300 nm region has a dispersion zero, the value is zero only for one particular wavelength. Any signal with a substantial bandwidth will still suffer from dispersion.

Fiber can also be fabricated so that the dispersion zero occurs near 1550 nm, corresponding to the window of minimum attenuation. Dispersion reduced fiber has a few ps/nm-km of dispersion and is useful for WDM systems where four-wave mixing effects may limit performance [4, 5]. Dispersion compensating fiber and dispersion flattened fiber can also be fabricated. It is also possible to take advantage of fiber dispersion to build high-rate systems. Under certain circumstances, a relatively narrow pulse of light known as a soliton can travel through a fiber largely undisturbed by dispersion. This is because a soliton balances dispersion against self-phase modulation so that the pulse shape is preserved during long distance propagation. Self-phase modulation refers to a pulse's ability to modulate its own phase by varying its velocity of propagation. Self-phase modulation occurs in fiber because the refractive index of glass is weakly dependent on the light intensity. If the light is sufficiently intense the pulse is able to alter the local index of refraction. While solitons propagate largely unaffected by dispersion, they are still subject to attenuation. Optical amplifiers are required in order to maintain the soliton pulse energy at a sufficiently high level. There are also upper limits on the bandwidth a soliton can occupy before higher order dispersion and optical nonlinearity terms come into play. Even with these limitations, systems with rates of over 100 Gbps have been demonstrated in the laboratory [6].

Fiber dispersion also influences the type of optical source used in the transmitter. In general, a semiconductor laser diode is preferred over a light-emitting diode (LED) as a source in an optical fiber system. The typical LED has a broad spectral output, occupying several nanometers of optical bandwidth, even for low rate modulations. A semiconductor laser usually has a much narrower spectral output. When the laser is directly current modulated, the laser both intensity modulates and frequency modulates. The laser parameter known as chirp, which is usually given in nm/mA, relates the frequency change to the intensity change. Chirp can be minimized by using an external modulator or by altering the laser's design to minimize the amount of frequency modulation.

A semiconductor laser will also exhibit a noise termed *mode-partition* noise. Mode-partition noise arises because the output of a semiconductor laser is actually spread among a multitude of longitudinal modes. The main mode dominates all others, and when all the modes are observed as an ensemble, the laser output appears to be at a virtually constant power. In reality, small amounts of power are constantly being exchanged among the side modes and the main mode. As the various laser modes propagate down the fiber, dispersion causes the various longitudinal mode wavelengths to separate in time. This causes the power at the receiver to acquire a random time varying characteristic that is a function of the amount of dispersion present. Even though mode-partition noise is caused by interactions between the transmitter, channel, and receiver, it is a noise that is first observed at the receiver and is usually accounted for as a receiver degradation [7, 8]. Mode-partition noise is generally worse in systems using Fabry-Perot (FP) lasers than those using either distributed feedback (DFB) or distributed Bragg-reflector (DBR) structures. This is because both DBR and DFB lasers have a better side-mode suppression ratio (SMSR) than do most FP lasers. SMSR is a measure of the ratio of the laser's main-mode to the largest amplitude side-mode. Lasers with side-mode suppression ratios in excess of 30 dB are often considered to be free of mode partition noise in all but the most demanding systems.

Table 1.1 summarizes the characteristics of the three most popular wavelengths used with optical fibers. 850 nm is considered the first "window" of operation for optical fiber and corresponds to the wavelengths where GaAs devices operate. Even though loss and dispersion are both relatively high in the 850 nm region, this region remains popular for cost sensitive applications. 850 nm was the region where semiconductor lasers were first commercially available and is still attractive because of the wide availability of laser sources. The photodetector material is usually common silicon, further reducing fabrication costs. Some fiber systems have even been designed to use the very low-cost 780 nm laser diodes that are used in compact disc players.

Table 1.1 Candidate Wavelengths for Fiber Systems.

Operating Wavelength	Benefits	Drawbacks
850 nm	Low Cost Components	Highest Loss Highest Dispersion
1300 nm	Low Dispersion	Higher Loss
1550 nm	Lowest Loss region Fiber Amplifiers	Higher Dispersion unless dispersion shifted or dispersion reduced fiber is used

The 1300 nm region was the next "window" developed and many of today's installed systems operate in this region. The devices that operate in this region are commonly fabricated using the indium gallium arsenide phosphide (InGaAsP) material system. This region has the benefit of containing the dispersion zero for normal silica based single-mode fiber. Even though the loss is higher than the 1550 nm region, the smaller dispersion is beneficial for high-rate point-to-point systems. Fiber amplifiers for 1300 nm have proven harder to construct that those in the 1500 nm region. The semiconductor laser amplifier (SLA) is used with some 1300 nm systems when optical amplification is required.

The 1550 nm region is the lowest loss and has the advantage of simple erbium doped fiber amplifiers. The indium gallium arsenide (InGaAs) material system is now used to fabricate devices. Dispersion shifted fiber is available that reduces the dispersion to the level present in 1300 nm systems. The chief drawback to operating in the 1500 nm region remains the comparatively higher cost of the optoelectronic components.

In addition to these three "windows" for silica based fiber, there is interest in certain exotic fiber materials. Losses in these materials are projected to be below 0.01 dB/km. This would be a sufficiently low-loss channel that it would be possible to span thousands of km's without any intervening amplifiers or repeaters.

1.2.2 Fiber Coupling Considerations

Coupling of the transmitter laser's output power into a fiber is a key feature in many fiber optic systems. The optical power must be efficiently collected while any optical reflections back into the laser cavity that may destabilize the output are minimized. The output of a high-quality index-guided laser diode is usually a good fit to a Gaussian beam. Several schemes for coupling a Gaussian beam into a fiber have been developed [9]. Figure 1.6 illustrates four of the most common coupling schemes. Some form of an optical isolator is often required to minimize reflections back into the laser. The coupling efficiencies achieved are frequently better than 3 dB, with less than 1 dB having been reported using a laser machined fiber lens [10].

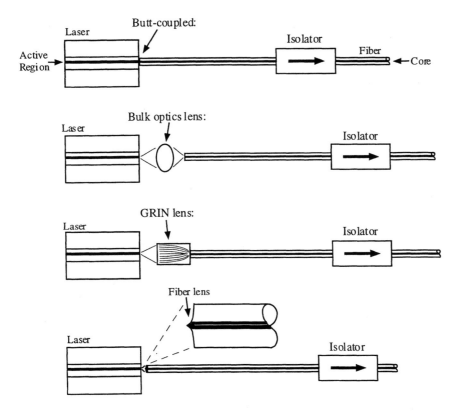

Figure 1.6 Fiber coupling.

1.3 Free-Space Systems

The high data-rate capability of optical communication systems also make them particularly attractive for use in free-space systems. As the amount of data available continues to grow, there will be increasing need to efficiently distribute the information for further processing, analysis, and presentation. Rapidly reconfigurable communication, treaty verification, natural resource determination, atmospheric sensing, weather forecasting, and environmental monitoring are functions that are often best performed from satellites. Any single satellite could directly transmit data down to the Earth's surface, but the costs and difficulties involved in maintaining a large number of individual ground stations can be prohibitive. A space-based communication network that allowed satellites placed in a variety of orbits such as low-earth-orbit (LEO), high-earth-orbit (HEO) or geosynchronous-earth-orbit (GEO) to communicate with each other would provide increases in flexibility and data availability [11].

An example of a global satellite network is shown in Figure 1.7. LEO platforms are connected to a constellation of GEO satellites which are themselves interconnected. The first steps to these types of systems have already been taken with efforts such as Europe's Silex and Japan's ETS-VI programs [12]. Similar intersatellite communication scenarios are anticipated for some of the advanced world-wide personal communication systems that are currently under development.

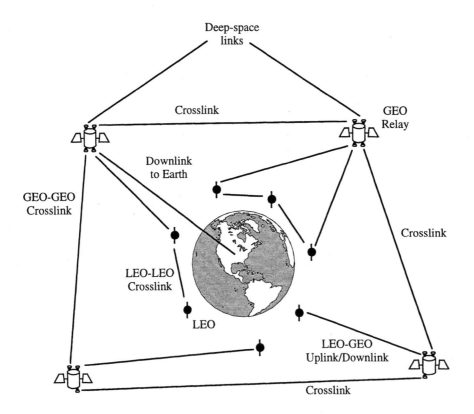

Figure 1.7 Network of interconnected satellites.

Some backbone satellites may be further connected to deep space sensors and probes. The up-down links between LEO and GEO are dynamic, changing rather frequently as satellites rise and set during their orbits. The cross-orbit links between GEO satellites are quasi-static and would typically change only to accommodate specific network routing and service provisioning functions [13].

Data rates on individual intersatellite links may range from a few kbps for command and telemetry functions, to the tens to hundreds of Mbps for data communications, to a Gbps or so for multispectral imaging of natural resources or for aggregating multiple lower rate channels. These comparatively high data-rates are expected even with the rapid advances in data compression techniques. Compression can dramatically reduce the required bandwidths in distribution systems where a human is the end user, with orders of magnitude reduction in data rate possible. In systems requiring computations to be performed on the information for analysis purposes, low-loss compression techniques are often required, which provide significantly smaller reductions in data rate.

The establishment of a communication link between two satellites has historically been accomplished via conventional microwave technology [14]. An example of an operational microwave system is the NASA Tracking and Data Relay Satellite System (TDRSS), which can support up to 300 Mbps data-rates using various frequency bands between 13 and 15 GHz. A conventional high data-rate communications package weighs several hundred pounds and requires an antenna several meters in diameter. A satellite serving as a high orbit node in the network architecture shown in Figure 1.7 would require several communication packages. Each package typically requires a dedicated receiver, a dedicated transmitter with an associated high-power traveling-wave-tube microwave amplifier, and one or more antennas. The combined effects of supporting several high data-rate microwave communications packages results in comparatively bulky and expensive relay satellites.

The many recent advances in optical communication technology have made the use of laser based intersatellite communication (lasercom) systems an attractive alternative to conventional microwave systems [12]. The maturing of semiconductor and solid-state laser technology and the widespread application of fiber-optic communications have resulted in the creation of a broad technology base from which to draw.

The extremely high center frequency present in an optical communication system results in very narrow beamwidths with high antenna gains for small size apertures. It is not unusual for an optical crosslink to achieve antenna (telescope) gains in excess of 110 dBi (dB relative to an isotropic radiator) with apertures of 20 cm (8 in.). This high gain translates directly into a significant reduction in the required transmitter power.

Using currently available technology, optical crosslinks can be designed that support data-rates in excess of 1.0 Gbps with apertures of under 25 cm (10 in.) and total package weights of under 90 kg (200 lb.). These types of lasercom packages make it possible to provide generic high data-rate communications capability to many satellites, facilitating the construction of a space network.

Optical techniques will not replace microwave systems in all applications, however. Low data-rate systems, and applications requiring Earth coverage from a single antenna, are often better served by microwaves. The optimum role for optical communications is most likely to be in providing the high data-rate (100+ Mbps) trunk lines.

With the high data-rates circulating in the spaceborne network, the problem of transmitting the information back to earth arises. Optical communication can also provide connectivity between satellites and Earth if ground station site diversity is employed. It is desirable to place the various ground sites far enough apart to guarantee that they are located in uncorrelated weather systems and that there is a high probability that at least one of the sites will have a clear, unobstructed view to the satellite. The ground sites would then be interconnected via conventional high data-rate optical fiber technology [15].

Figure 1.8 illustrates the major subsystems of a free-space optical communication link. The telescopes, spatial acquisition, spatial tracking, and point-ahead subsystems are used to establish and maintain a stable line-of-sight between the platforms that wish to communicate.

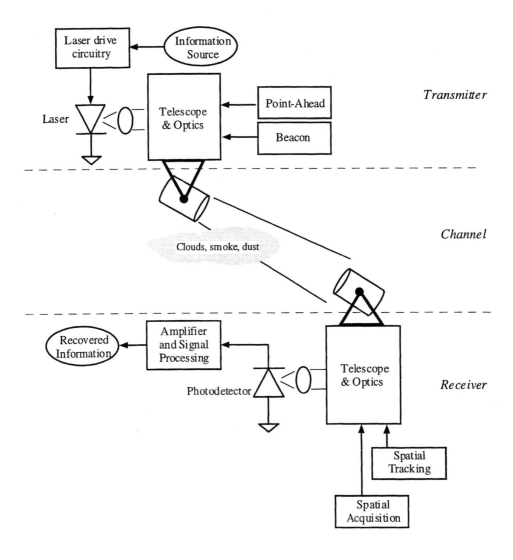

Figure 1.8 Block diagram of a free-space system.

Telescopes are frequently mounted in gimbals so that they can point to a variety of locations such as cross-orbit, down at the Earth, or up towards a relay satellite. The spatial tracking system utilizes the telescope's gimbal to accomplish coarse pointing and uses small, high-bandwidth, movable mirrors to accomplish fine tracking.

1.3.1 Choice of Laser Technology

Laser selection plays a significant role in a free-space system design in that it dictates the operating wavelength and usually determines the modulation format that is used. These in turn determine the type of photodetection and demodulation mechanisms employed by the receiver. There are several types of lasers that have been proposed for use in free-space systems. Characteristics of some of these lasers are listed in Table 1.2.

Semiconductor lasers have several desirable characteristics when used in free-space systems [16, 17]. They are small, rugged, and very power efficient, requiring only a few hundred milliwatts of prime power to provide tens of milliwatts of optical output, and they can be directly modulated by varying their bias current. Their principle drawback is that, although power levels are continuously being improved, they currently provide only a hundred milliwatts or so of usable optical power.

Table 1.2 Candidate Lasers for Free-Space Systems.

Laser Type	*Examples*	*Operating Wavelengths*	*Drawbacks*
Semiconductor	AlGaAs InGaAsP	830 nm 980 nm 1550 nm	Low powers
Solid State	Nd:YAG	1060 nm	Separate Modulator
Gas	CO_2 HeNe	10,600 nm 530 nm	Complex system Lifetime issues
Fiber	Erbium doped fiber	1550 nm	Unproven in a space environment

The solid-state laser is best illustrated by the Nd:YAG [18]. One or more 0.8 μ high power AlGaAs diode arrays are typically used to optically pump Nd:YAG, which then emits at 1.06 μ. The principle benefit to the Nd:YAG is that it provides as much as a few watts of usable continuous-wave (CW) output power at the expense of less efficient operation. Typically, a separate modulation element capable of modulating the high-power optical signal is required. This introduces additional optical alignment requirements and optical losses. The Nd:YAG may also be operated in a pulsed mode via either cavity-dumping or Q-switching.

The carbon dioxide (CO_2) as well as the helium neon (HeNe) gas lasers were some of the first lasers proposed for use in optical crosslinks [19]. They both suffer from being optomechanically cumbersome. Sealed enclosures and RF or electric discharge pumping is required. The lifetime of these lasers in a space environment has also been questionable because of outgassing in vacuum.

Fiber lasers are a recent development intended for use in the telecommunications industry [20]. They are typically fabricated using erbium doped fiber that is pumped at either 980 nm or 1480 nm, although examples of codoped fiber lasers that are pumped by Nd:YAG are also available. Output powers range from a few μW to several tens of mW. Fiber lasers can exhibit narrow linewidths with good frequency stability, and since the optical signal is already contained in a single-mode fiber, the spatial mode quality is excellent. Polarization control remains an issue however. High-power fiber amplifiers can also play a role in space-based systems. In a fiber amplifier a relatively short length of erbium doped fiber is pumped so as to provide gain at the 1550 nm region [21]. Fiber laser and high-power fiber amplifier technology is a relatively recent development [22, 23] and the lifetime of the high-power pump sources needs to be established.

A technique for increasing the available transmitter power that has recently become popular in the optical communication field is the use of a master-oscillator power-amplifier configuration, or MOPA, as shown in Figure 1.9 [24, 25]. This technique has been utilized very successfully in high-power microwave transmitters for decades and involves partitioning the transmitter into oscillator, modulator, and power amplifier subsystems.

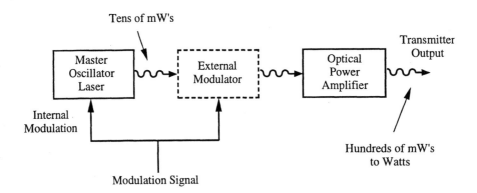

Figure 1.9 A master-oscillator power-amplifier (MOPA) transmitter.

By dividing transmitter functions among several components, it is often possible to achieve significantly higher performance than if a single component or device was required to perform all of the functions. In particular, the modulator need only be capable of handling a few tens to hundreds of milliwatts of optical power instead of the watt or so that might be required if the amplifier output was modulated.

The selection of a suitable laser system for an intersatellite link is a complex, time consuming task. Only a complete, system level, top-down approach is appropriate. It is easy to be misled into picking what at first appears to be the best transmitter solution only to discover that a suitable receiver is either extraordinarily expensive or unobtainable.

1.3.2 Spatial Pointing and Tracking Considerations

The high antenna gains and narrow beamwidths afforded by optical frequencies result in stringent spatial pointing and spatial tracking requirements. The transmitter laser typically generates a beam with a Gaussian intensity profile. The beam of waist r_w exits the transmitter telescope aperture and diffracts as it travels into the far-field, as shown in Figure 1.10.

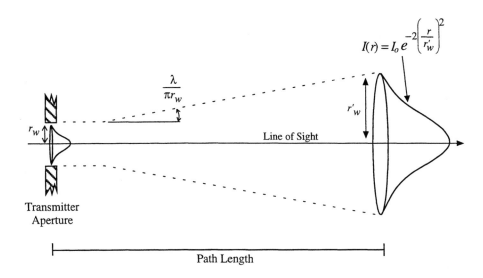

$$I(r) = I_o \, e^{-2\left(\frac{r}{r'_w}\right)^2}$$

Figure 1.10 Transmitter beam diffraction.

At the end of the propagation path the intensity profile is still Gaussian but the waist has now significantly expanded. The amount of beam diffraction scales directly with the wavelength of the transmitted field and hence an optical beam will have a much narrower beam divergence than a microwave beam exiting from the same size aperture. An optical system operating at 1 micron through a 10 cm aperture will have a beam divergence of only 6.4 microradians if the optics are of diffraction limited quality. This compares to a divergence of several milliradians for the same size aperture at microwave frequencies. The 6.4 microradian beam divergence translates into a main lobe spot size of just over 500 meters at a distance 80,000 km away from the satellite. Ideally the receiver would be located at the intensity peak of the transmitter's far field diffraction pattern so as to maximize the on-axis received signal.

It is necessary but not sufficient for the transmitter to illuminate the receiver aperture. The receiver must also be pointed so that the incoming optical field correctly illuminates the photodetector. Depending on the detector field-of-view and the instantaneous pointing error, an incoming off-axis signal field may either fully illuminate, partially illuminate, or even fail to illuminate the photodetector. All photodetectors obviously have a finite size, and even when the receiver is correctly pointed, the prescription for the receiver optical train must guarantee that the spot size of the received signal field be such that an acceptably small portion of the received signal field falls outside the detector area.

The extremely narrow beamwidths can cause errors in spatial pointing and tracking that are sufficient to seriously affect the received signal level. Before the communications link can be established, a spatial acquisition system must be used to guarantee that the transmitter is correctly illuminating the receiver aperture and that the incoming received field is efficiently illuminating the receiver photodetector. Knowledge of each spacecraft's position and relative motion allows initial pointing vectors to be established for each satellite, within an accuracy of a few milliradians. Once initial pointing is determined there are several possible acquisition system designs and algorithms that can be used to further reduce the pointing uncertainty [26-29]. An example of one is shown in Figure 1.11. The transmitting platform generates a beacon signal for the receiver to look for. The acquisition optics in combination with a charge-coupled-device (CCD) is used to image the entire pointing uncertainty region and find the beacon signal.

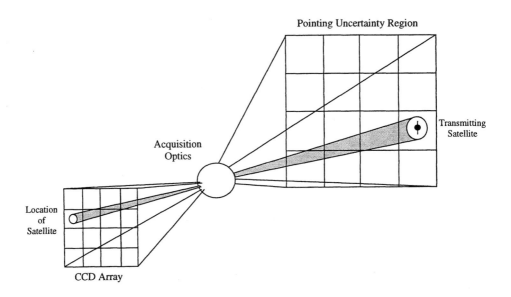

Figure 1.11 Spatial acquisition using a CCD.

Once the beacon is found at the receiver, spatial acquisition for the other platform must occur. Again there are several possible approaches to establishing a two-way link. An example of a cooperative acquisition using beacons between a satellite in geosynchronous Earth orbit and one in low Earth orbit is shown in Figure 1.12.

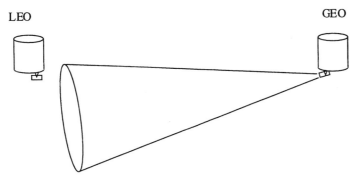

GEO illuminates LEO with a broad beacon

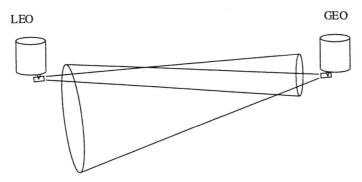

LEO acquires GEO beacon and returns a narrow beacon

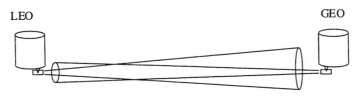

GEO tracks narrow LEO beacon and returns communications signal

LEO tracks received communications signal and returns narrow communications signal

Figure 1.12 An example of a spatial acquisition between satellites.

The transmitting GEO platform initially directs a widely diverging beacon down along its best estimate of the direction vector that points towards the LEO receiver. At the LEO receiver, a wide field of view CCD images the entire uncertainty region and looks for the presence of the beacon. Once the beacon position is found it is translated into a pointing vector that the LEO uses to direct a narrow beacon back up toward the GEO satellite. The GEO satellite acquires the narrow LEO beacon, begins a coarse spatial track to keep the LEO beacon centered in its field of view, and directs a very narrow (nominally diffraction limited) communications beam down toward the LEO. The LEO satellite senses the increase in received power from the GEO satellite, begins spatial tracking, and directs its own narrow communications beam up to the GEO. The GEO continues to spatial track on the LEO's communication beam and now both GEO and LEO track each other's motion.

The extremely narrow optical beamwidths and the relative motions of the satellites also require that either the transmitter be pointed ahead or that the receiver look behind the apparent line-of-sight by up to several tens of microradians so that the incoming optical signal field and the receiving platform arrive at the same location simultaneously. This problem is analogous to a hunter leading a moving target and is illustrated in Figure 1.13.

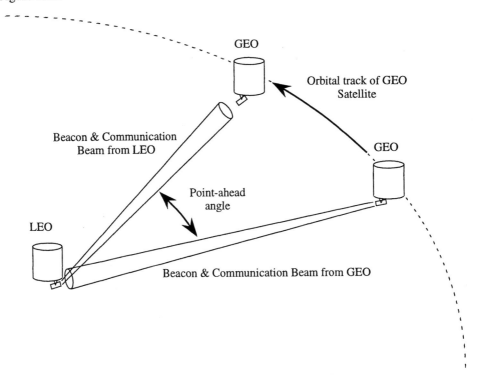

Figure 1.13 Transmitter point-ahead.

Spatial tracking information continuously drives the coarse and fine pointing systems so that the receiver line-of-sight is accurately pointed directly at the incoming received signal. A continuous tracking system is required since most satellites have inherent jitter levels that are far greater than the beamwidth of an optical communication link. It is not uncommon for three-axis-stabilized spacecraft to exhibit several hundred microradians of jitter. Solar-array stepper motors, momentum wheels, and attitude control systems all contribute to spacecraft micro-motion. Spin stabilized spacecraft require that the entire telescope assembly be despun to accommodate pointing, which introduces additional line-of-sight disturbances.

Spatial tracking systems frequently consist of either quadrant detector arrays or high-frame-rate CCD devices that provide azimuth and elevation error information. One example of a spatial tracking system is shown in Figure 1.14. The system uses coarse pointing optics (typically a gimbaled telescope) in combination with fast steering optics (typically a two-axis mirror) to achieve a wide bandwidth simultaneously with high dynamic range and substantial jitter rejection [30]. The relay optics convert incoming angle errors into position errors in the focal plane, which are sensed by a quadrant photodetector or "quad-cell." The quad-cell output is combined to provide azimuth and elevation error information that is then filtered and used to drive the fast steering optics.

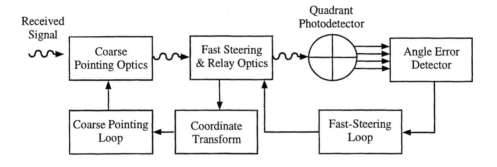

Figure 1.14 Spatial tracking loop.

Position information (or a portion of the error information) from the fast steering optics are sent to a coordinate transformer, which maps the local pointing vectors into the coordinate space needed by the coarse pointing optics. The transformed values are filtered and applied to the coarse pointing optics so that the fast steering optics are kept in the middle of their operating range. This serves to "unload" any long-term bias that builds up on the fast steering mechanisms [31-33].

The overall impact of uncorrected pointing jitter on communication link performance has been studied in detail and many approaches to spatial tracking system design have been proposed [34-36]. Spacecraft motion is subject to a wide variety of disturbances, not all of which will be known before launch, and not all of which will be constant over the life of the spacecraft. It is wise to view the spatial acquisition and tracking systems as the most critical components of an optical intersatellite link, and consequently they should be designed to be the most robust systems of all.

In spite of milliradians of initial pointing uncertainty and hundreds of microradians of residual spacecraft jitter that must be tracked out, a well designed spatial acquisition and tracking system can provide acquisition times of 1-10 seconds and ultimate tracking accuracy of better than 0.1 beamwidths. This typically reduces the degradation in the communications link due to uncorrected pointing errors to less than 1 dB.

1.4 References

1.   J.M. Senior, *Optical Fiber Communications: Principles and Practice,* 1985, Prentice-Hall: Englewood Cliffs, NJ, *Chap. 2 - Optical Fiber Waveguides.*

2.   D.N. Payne. "Special Fibers and Their Uses." in Conference on Optical Fiber Communication, OFC. 1987, Optical Society of America.

3.   G.P. Agarwal, *Nonlinear Fiber Optics,* 1989, Academic Press: New York, NY, *Section 1.3 - Fiber Nonlinearities.*

4.   A. Chraplyvy, "Limitations on Lightwave Communications Imposed by Optical-Fiber Nonlinearities," Journal of Lightwave Technology, 1990, vol. 8, no. 10, pp. 1548-1557.

5.   R.W. Tkach. "Tutorial on Combating Fiber Nonlinearities in Lightwave Systems." in Conference on Optical Fiber Communication, OFC. 1995, San Diego, Optical Society of America.

6.   S. Kawanishi, *et al.* "100 Gbit/s 50 km Optical Transmission Employing All-Optical Multi/Demultiplexing and PLL Timing Extraction." in Conference on Optical Fiber Communication, OFC. 1993, Optical Society of America.

7.   P.L. Liu, *Coherence, Amplification and Quantum Effects in Semiconductor Lasers,* Y. Yamamoto, Editor. 1991, John Wiley and Sons, Inc.: New York, *Chap. 10 - Photon Statistics and Mode Partition Noise of Semiconductor Lasers.*

8.   K. Petermann, *Laser Diode Modulation and Noise,* 1988, Kluwer Academic Publishers: Boston, *Chap. 7 - Noise Characteristics of Solitary Laser Diodes.*

9.   M. Saruwatari and K. Nawata, "Semiconductor Laser to Single-mode Fiber Coupler," Applied Optics, June 1979, vol. 18, no. 11, pp. 1847-1856.

10.   H. Presby. "Near 100% Efficient Fiber Microlenses." in Conference on Optical Fiber Communication, OFC. 1992, Optical Society of America.

11.   V.W.S. Chan, "Space Coherent Optical Communication Systems - An Introduction," Journal of Lightwave Technology, April 1987, vol. LT-5, no. 4, pp. 633-637.

12.   D. Begley, *Free-Space Laser Communications*, SPIE Milestone Series, Vol. MS 30, 1991, SPIE Optical Engineering Press: Bellingham, WA.

13.   B.N. Agrawal, *Design of Geosynchronous Spacecraft,* 1986, Prentice-Hall: Englewood Cliffs, NJ, *Chap. 2 - Orbit Dynamics.*

14. M.A. King, *Laser Satellite Communications,* M. Katzman, Editor. 1987, Prentice-Hall: Englewood Cliffs, NJ, *A Brief Outline of an RF Crosslink System Design*, pp. 214-231.

15. W. Chapman and M. Fitzmaurice. "Optical Space-to-Ground Link Availability Assessment and Diversity Requirements." in Free-Space Laser Communication Technologies III. 1991, Los Angeles, CA, SPIE.

16. W.S. Streifer, *et al.* "Characteristics of High Power GaAlAs Laser Diodes Useful for Space Applications." in Optical Space Communication. 1989, SPIE.

17. V.W.S. Chan, *et al.* "Heterodyne Lasercom Systems Using GaAs Lasers for ISL Applications." in IEEE International Conference on Communications, ICC-83. 1983, Boston, MA, IEEE Press.

18. M. Ross, *et al.*, "Space Optical Communications with the Nd:YAG Laser," Proceedings of the IEEE, March 1978, vol. 66, pp. 327-338.

19. J.H. McElroy, *et al.*, "CO_2 Laser Communication Systems for Near Earth Space Applications," Proceedings of the IEEE, February 1977, vol. 65, no. 2, pp. 221-251.

20. G.A. Ball and W.W. Morey. "Narrow-linewidth Fiber Laser with Integrated Master Oscillator Power Amplifier." in Proceedings of Optical Fiber Communications Conference, OFC. 1992, Optical Society of America.

21. "Special Issue on Optical Amplifiers," Journal of Lightwave Technology, February 1991, vol. 9, no. 1.

22. J.C. Livas, *et al.* "Gbps-Class Optical Communications Systems for Free-Space Applications." in Free-Space Laser Communication Technologies V. 1993, SPIE.

23. J.C. Livas, *et al.* "High Power Erbium-Doped Fiber Amplifier Pumped at 980 nm." in Conference on Lasers and Electro-Optics (CLEO). 1995, Baltimore MD, OSA.

24. J.C. Livas, *et al.*, "1 Gbit/s Injection-Locked DPSK Communication Experiments for Space Applications," Electronics Letters, 1991, vol. 27, no. 12.

25. S.B. Alexander, *et al.* "1 Gbps, 1 Watt Free-Space Coherent Optical Communication System." in Annual Meeting of the IEEE Lasers and Electro-Optics Society LEOS-92. 1992,

26. K. Komatu, *et al.* "Laser Beam Acquisition and Tracking System for ETS-VI Laser Communication Equipment." in Free-Space Laser Communication Technologies II. 1990, SPIE.

27. J.C. Boutemy. "Use of CCD Arrays for Optical Link Acquisition and Tracking." in Optical Systems for Space Applications. 1987, SPIE.

28.   P. Van Hove and V.W.S. Chan, "Spatial Acquisition Algorithms and Systems for Intersatellite Optical Communication Links," 27 November 1984, Report Number 667 (DTIC AD-A150794), MIT Lincoln Laboratory.

29.   J.M. Lopez and K. Yong, *Laser Satellite Communications,* M. Katzman, Editor 1987, Prentice-Hall: Englewood Cliffs, NJ, *Chap. 6 - Laser Beam Pointing Control, Acquisition, and Tracking Systems.*

30.   W. Stoelzner, *Laser Satellite Communications,* M. Katzman, Editor. 1987, Prentice-Hall: Englewood Cliffs, NJ, *Chap. 5 - Optical Configuration and System Design.*

31.   G.C. Loney. "Design of a High-Bandwidth Steering Mirror for Space-Based Optical Communications." in Active and Adaptive Optical Components. 1991, SPIE.

32.   R.W. Cochran and R.H. Vassar. "Fast Steering Mirrors in Optical Control Systems." in Advances in Optical Structure Systems. 1990, SPIE, Vol. 1303.

33.   L.R. Hedding and R.A. Lewis. "Fast Steering Mirror Design and Performance for Stabilization and Single Axis Scanning." in Acquisition Tracking and Pointing IV. 1990, SPIE.

34.   C.C. Chen and C.S. Gardner, "Impact of Random Pointing and Tracking Errors on the Design of Coherent and Incoherent Optical Intersatellite Communication Links," IEEE Transactions on Communications, 1989, vol. 37, no. 3.

35.   E.A. Swanson and V.W.S. Chan, "Heterodyne Spatial Tracking Systems for Optical Space Communication," IEEE Transactions on Communications, 1986, vol. 34, pp. 118-126.

36.   E.A. Swanson and J.K. Roberge, "Design Considerations and Experimental Results for Direct Detection Tracking Systems," Optical Engineering, 1989, vol. 28, no. 6, pp. 659-666.

Chapter 2

SYSTEM PERFORMANCE

A generalized optical communication link is illustrated in Fig. 2.1. The information to be transmitted to the receiver is assumed to exist initially in an electrical form. The information source modulates the field generated by the optical source. The modulated optical field then propagates through a transmission channel such as an optical fiber or a free-space path before arriving at the receiver.

The receiver may perform optical processing on the incoming signal. The optical processing may correspond to a simple optical filter or it may involve interferometers, the introduction of additional optical fields, or the use of an optical amplifier. Once the received field is optically processed it is detected. The photodetection process generates an electrical signal that varies in response to the modulations present in the received optical field. Electrical signal processing is then used to finish recovering the information that is being transmitted.

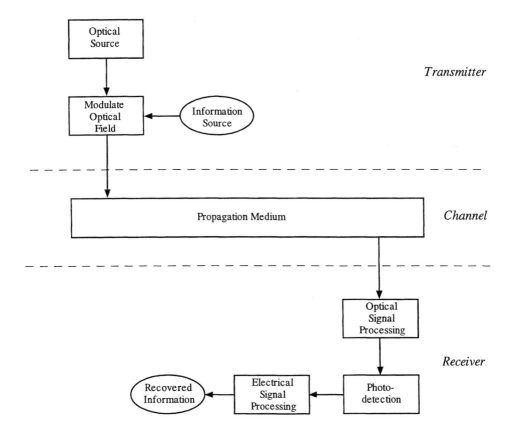

Figure 2.1 A generalized optical communication link.

2.1 Analog and Digital Optical Communications

Optical communication links can be further divided into two broad classes: analog systems and digital systems. Figure 2.2 illustrates the distinction between the two classes.

Analog communication systems transfer continuous-time waveforms from transmitter to receiver and are designed to minimize the noise and distortion present in the recovered waveform [1-3]. They are used in cable-TV distribution systems, antenna remoting, and phased-array feed applications. They can have significant cost and performance advantages over digital systems when the information to be conveyed is wide-band or high-frequency waveforms and accurate analog-to-digital conversion would be prohibitive.

Digital systems are used where the information exists as a stream of binary digits (bits) and are designed to minimize the number of bit-errors in the output data stream [4-7]. Digital systems are used for telephone systems and computer interconnects and form the backbone data-links of world-wide networks such as the Internet. Digital systems specify data rate, and an associated probability of bit error, or bit-error-rate (BER), such as 10^{-8}, is then used as a measure of performance. A BER specification of 10^{-8} means that, on average, the digital link would experience one error in every one hundred million bits. Whether this is sufficient performance depends strongly on the needs of the users of the communication link. If this link were carrying a non-compressed digitized version of conventional quality television, a viewer watching the reconstructed video could easily tolerate a single bit in error every second and so 10^{-8} would be considered an acceptable error rate for this class of user. Other users of this link who might wish to transfer financial information or computer files would demand significantly better error performance, on the order of 10^{-12} to 10^{-15}.

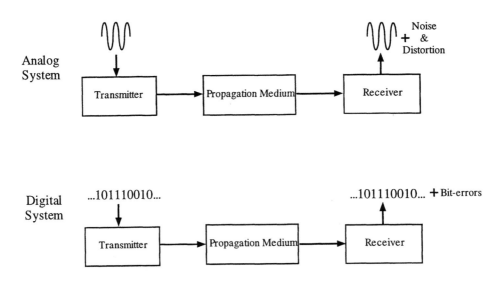

Figure 2.2 Analog and digital optical communication systems.

Some free-space satellite systems make use of end-to-end error correction schemes and can utilize communication channels operating with error rates as high as 10^{-3} to 10^{-6} while still providing end users with the desired error rates. Submarine lightwave systems have also benefited from the application of forward error correction coding [8]. The penalty paid for using error correction coding is in the form of bandwidth expansion [9, 10]. A higher channel rate is needed to accommodate the additional parity bits coding requires. A rate 1/2 code requires two channel bits for each information bit.

A communication system's error rate quoted in terms of a single probability of bit error is naturally a time averaged quantity and may not adequately specify the error performance of a system. Optical communication systems that propagate through the atmosphere experience time varying signal fading due to atmospheric turbulence and scattering. Free-space systems that have marginal spatial tracking systems also experience significant fades in received signal power during periods of severe angular disturbance. Fiber systems sensitive to polarization, optical feedback effects, or lightning induced transients can also exhibit fading of the received signal. Systems subject to signal fading have time varying error rates and are prone to having errors occur in bursts, and their performance should be described in terms of a statistical distribution of error rates.

Fundamental physical effects, imperfections in the transmitter and receiver, and the characteristics of the channel cause the information recovered at the receiver to be only an *estimate* of the information originally transmitted. The estimate may be exceedingly accurate, with no readily observable differences, or the estimate may be poor, with significant errors introduced. Some communication link users may find a certain amount of observable error tolerable while others may require virtually ideal performance. The *quality-of-service* (QOS) that the users of the communication link demand will be the primary determinant of what amount of error is acceptable.

In addition to specifying the error in the information recovered, QOS can have many other dimensions. Some users may require a certain guarantee on the availability of the link. This is particularly important for free-space systems where link availability is influenced by weather and in fiber systems where link availability is influenced by cable cuts, lightning strikes, or power failures. Other users may require that the link be error-free for a specified period of time. Still others may want guarantees on how rapidly the link will be restored if it does fail.

Whether the system is analog or digital, we can define a set of receiver subsystems as illustrated in Fig. 2.3. The received signal may be optically processed before being photodetected. Once converted into an electronic signal by the photodetector, conventional signal processing can be used to extract the desired information.

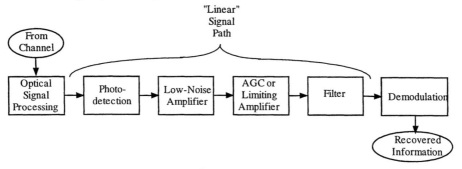

Figure 2.3 Receiver subsystems.

2.2 The Link Budget

The overall system level performance of an optical communication link is easily quantified in terms of a system link budget. This is the same technique that is used in microwave communication links. A link budget provides a convenient tally sheet where a system's performance constraints may be accounted for. Finite transmitter power, modulation imperfections, optical gains and losses, receiver sensitivity, propagation losses, electronic imperfections, etc., are all conveniently expressed in terms of their impact on the ultimate communications performance of the link.

All link budgets will have at least three entries, one for the transmitter, one for the channel, and one for the receiver. An elementary link budget is illustrated in Table 2.1. This example uses an average transmitter power of 1.0 milliwatts and a propagation loss of 20 dB. A 20 dB loss means that 1% of the transmitted power actually arrives at the receiver in a useful form. The rest is lost during propagation. The receiver sensitivity is -30 dBm (dBm is dB relative to a 1.0 mW power level). This sensitivity is the amount of optical signal power that must be present at the receiver to meet the link user's quality-of-service requirements.

Entries in link budgets are usually expressed in terms of decibels so that values can be conveniently added instead of multiplied. The link margin is computed by determining the amount of power actually present at the receiver and then subtracting the receiver sensitivity. In the example shown in Table 2.1 a positive 10 dB margin is obtained. This means that ten-times the power needed to meet the user's requirements is available at the receiver. If a negative margin had been obtained, the desired performance could not be obtained with the specified combination of transmitter, channel, and receiver.

Table 2.1 An elementary link budget.

Item	Value	dB value
Transmitter:		
1) Average Transmitter Power	1.0 mW	0.0 dBm
Channel:		
2) Propagation Losses	1% Transmission	-20.0 dB
Receiver:		
3a) Signal Power at Receiver		-20.0 dBm
3b) Receiver Sensitivity		-30.0 dBm
Link Margin (also called power margin). Equals 3a) minus 3b)		+10.0 dB

Any communication system that desires to maintain a specified QOS will require that a minimum signal-to-noise ratio (SNR) be maintained within the receiver. The receiver sensitivity specification is essentially the amount of optical signal power needed to obtain the SNR required to provide the desired quality-of-service. The exact SNR needed is dependent on the receiver noise statistics, type of modulation, detection, and demodulation employed, the imperfections in the modulator, photodetector, and demodulator, optical nonlinearities, and any error correcting coding techniques that may be used.

The exact amount of signal-to-noise ratio required is usually determined and then an additional amount known as "link-margin" or "received-power-margin" is purposely added to account for uncertainties in estimating the performance of the individual subsystems, any additional degradation that may be expected as components age, and unforeseen events. Thus if a 10 dB signal-to-noise ratio is required to achieve the desired QOS, a 13, 16, or even 20 dB SNR, corresponding to 3, 6, or 10 dB of margin, will be used as the design goal. The exact amount of margin used will depend upon the confidence in the individual performance estimates, any time varying characteristics of the channel, and the aging stability of the components.

The graphical technique illustrated in Fig. 2.3 is frequently used to provide additional insight into the various contributions to a link budget. The x-axis is representative of the propagation distance along the channel but is usually not drawn to scale. Instead, each contributor to the link budget is allocated an equal amount of space along the x-axis.

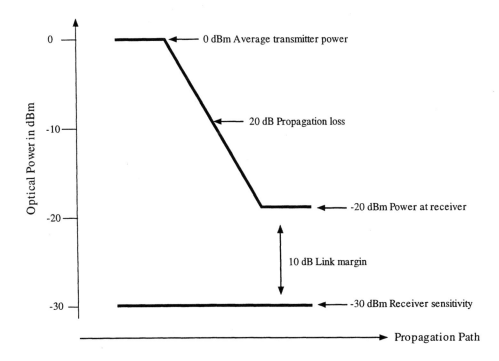

Figure 2.3 Graphical representation of the link budget of Table 2.1.

The y-axis corresponds to the amount of optical power present in the link. At least two curves are drawn. One corresponds to the receiver sensitivity. This is usually assumed to be essentially constant regardless of where the receiver is located. The second curve is the optical power level as a function of location within the link. For our example, the power curve starts at 0 dBm, decreases linearly as we propagate through the channel and ends at -20 dBm. The -30 dBm receiver sensitivity curve is also drawn. The 10 dB link margin is clearly evident.

The graphical technique can also represent known limitations in a systems performance. For example, if it was known that there was an interfering background signal of -33 dBm and that the receiver would overload for received signals above -6 dBm, the graphical technique produces the plot shown in Fig. 2.5. The goal is to maintain at least a 10 dB SNR where the noise level is actually set by the background.

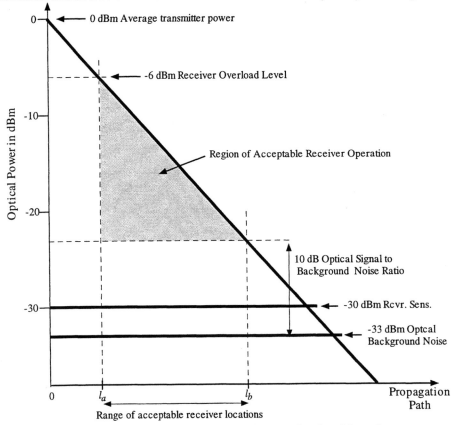

Figure 2.5 Graphical link budget with overload and interference.

Care must be used when *only* a SNR specification is used and information about the noise statistics is not known. Most SNR calculations are done assuming the noise exhibits additive-white-Gaussian noise (AWGN) statistics [11-14]. This is a result of the fact that in most cases in microwave and RF systems, AWGN is an accurate description of the statistics of the noise present in the receiver. We will see in later chapters that assuming the noise statistics in an optical receiver to be those of AWGN can lead to inaccurate performance specifications for the receiver.

2.2.1 Link Budget for the Fiber-Optic Channel

A more elaborate example of a link budget is that for a fiber link. Figure 2.6 illustrates the principal entries in the link budget for a fiber-optic communication link. The entries are 1) average transmitter laser power, 2) modulation losses, 3) losses incurred in coupling the transmitter to the fiber, 4) other transmitter optical losses such as the insertion loss in an inline isolator or any reflection induced losses, 5) fiber attenuation, 6) fiber dispersion, 7) fiber nonlinearity, 8) losses due to incomplete collection of the light at the fiber output, and 9) the receiver sensitivity.

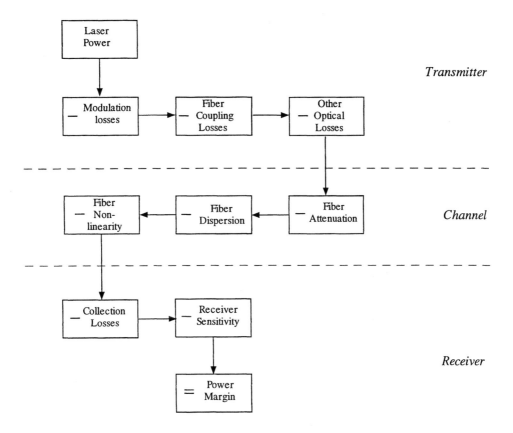

Figure 2.6 Components of a fiber system link budget.

The entries shown in Fig. 2.6 can be further broken down, often to the individual component level, resulting in a link budget with potentially dozens of individual contributors. To illustrate the components in a fiber link-budget we will use the example fiber link illustrated in Fig. 2.7 The transmitter has an average output power of 1.0 mW. This is obtained from a laser that has a 3 mW average output. The internal optical isolator is known to have a 1.1 dB insertion loss. The remaining loss is due to imperfect coupling of the laser output to the optical fiber.

The transmitter output is fusion spliced to a 1:2 fiber splitter. The splice has a 0.1 dB loss. The splitter has a 3 dB splitting loss and a 0.5 dB excess loss. The output from the splitter is fusion spliced to a 20 km length of fiber with a loss of 0.21 dB per km at the wavelength of operation. This 20 km span is connected to a 50 km length using a mechanical splice with 0.25 dB loss. The 50 km span has a loss constant of 0.23 dB per km.

The 50 km path is fusion spliced and then split four ways using a 4:4 star coupler with 0.8 dB excess loss. One of the outputs of the star is mechanically coupled to the input fiber to the receiver, which has a -31 dBm sensitivity at our data rate and wavelength of operation.

Although the other terminations are not shown explicitly in Fig. 2.7, we will assume that all fiber ends are properly terminated in low reflection loads. Even with the isolator in the transmitter, reflections from unterminated fiber ends can lead to optical interference effects that can cause noise and instability in the system. For simplicity we will also assume that fiber dispersion and nonlinearity are insignificant. These are reasonable assumptions given the relatively low mW class power levels used, a low chirp source, and modest data rates.

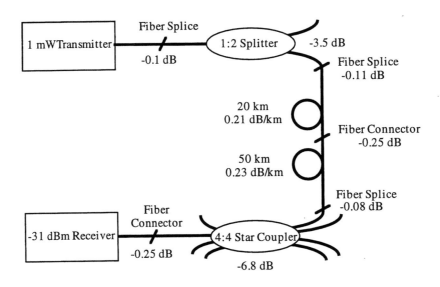

Figure 2.7 An example of a fiber link.

The link budget for this system is shown in Table 2.2. The 1.0 mW effective transmitter power is obtained by including the effects of the fiber coupling losses and inline optical isolator losses on the 3 mW laser output power. The losses in the fiber splices, connectors, splitters, and star couplers are included in the channel losses. The combination of transmitter and channel results in -26.79 dBm of optical signal power at the receiver. We have assumed that the effects of fiber dispersion and nonlinearity are negligible for this system. This is a valid assumption for a low chirp optical source, relatively low optical powers, moderate bandwidths, and moderate propagation distances.

Table 2.2 An example of a fiber system link budget.

Item	*Value*	*dB value*
Transmitter:		
1) Average Transmitter Power	3 mW	4.8 dBm
2) Modulation Losses	Included in Receiver Sensitivity	0.0 dB
3) Fiber Coupling Losses		-3.7 dB
4) Other Optical Losses (Isolator)		-1.1 dB
Effective Transmitter Power	1.0 mW	0.0 dBm
Channel:		
5a) Splitting Losses	3.5 + 6.8	-10.3 dB
5b) Splice & Connector Losses	0.1 + 0.11 + 0.25 + 0.08 + 0.25	-0.79 dB
5c) Fiber Attenuation		-15.7 dB
6) Fiber Dispersion	Negligible	0 dB
7) Fiber Nonlinearity	Negligible	0 dB
Receiver:		
8) Collection Losses	Included in Receiver Sensitivity	0.0 dB
Signal Power at Receiver		-26.79 dBm
9) Receiver Sensitivity		-31.0 dBm
Link Margin		+4.21 dB

The receiver sensitivity of -31.0 dBm was measured at the receiver's input fiber and consequently accounts for any internal receiver losses such as imperfect collection of the light exiting from the fiber. The link exhibits just over a 4 dB margin, meaning that there is 2.5 times the amount of optical power required to provide the QOS associated with the -31.0 dBm receiver sensitivity.

Whether this is adequate margin depends on the aging stability of the link and any margin that must be reserved for future events. Assume that it was known that over the design lifetime of the link, the laser output would drop by 50% (a -3.0 dB loss) and that four accidental cable cuts could be expected; with each repair introducing 0.25 dB of additional loss. The bare minimum margin that would be appropriate at the beginning of life for this link would be 4.0 dB. In order to accommodate unforeseen events and additional degradation in link components, a margin of 6-8 dB would be more appropriate. A graphical representation of the link budget from Table 2.2 is shown in Fig. 2.8.

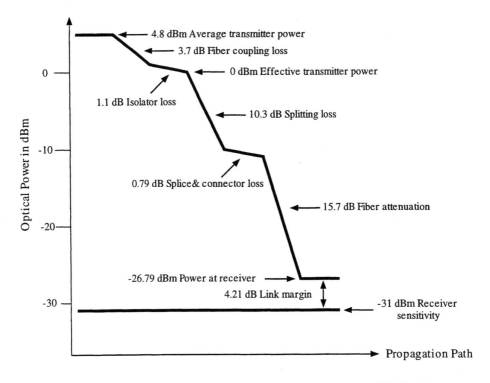

Figure 2.8 Graphical representation of the link budget of Table 2.2.

An important parameter in any fiber link budget is the loss during propagation in the fiber. If P_{in} is the amount of optical signal power launched at the input of a fiber of length L, the amount of optical power at the fiber output P_{out} is given by

$$P_{out} = P_{in} \exp(-\alpha L), \tag{2.1}$$

where α is the attenuation constant for the fiber. The attenuation constant is usually expressed in the units of cm^{-1} or "inverse centimeters." Centimeters are appropriate for semiconductor waveguide devices but not for long lengths of fibers. In fibers it has become customary to express the attenuation constant in the more practical units of dB/km by using the relation

$$\alpha_{dB} = -\frac{10}{L} \log_{10} \frac{P_{out}}{P_{in}} = 4.343\alpha, \tag{2.2}$$

where α is now in km^{-1} (10^5 cm^{-1}).

Note that a loss of 1 dB/km corresponds to an attenuation constant of 2.3×10^{-6} cm^{-1}, an extremely low value compared to the 0.9 - 0.1 cm^{-1} obtained in the low-loss semiconductor waveguide materials used in opto-electronic devices.

Using the expression for loss in dB/km allows Eq. 2.1 to be rewritten as

$$P_{out} = P_{in} \exp\left(\frac{-\alpha_{dB}L}{10}\right), \tag{2.3}$$

where L is the fiber length in kilometers.

We can rearrange Eq. 2.3 to solve for the maximum length that a fiber system can have by assuming 1) a zero dB link margin, 2) that all of the transmitter power is used as input to the fiber P_{in}, 3) that the receiver sensitivity is equal to the power out of the fiber P_{out}, and 4) that the fiber attenuation is α dB/km.

This approach will serve to provide an upper bound on a fiber system's maximum length. The maximum length is given by

$$L_{max} = \frac{10}{\alpha_{dB}} \log_{10} \frac{P_{in}}{P_{out}}. \tag{2.4}$$

We see that the maximum distance L_{max} has a direct inverse dependence on the fiber attenuation but is only logarithmically dependent on the amount of power launched or the receiver sensitivity. The first goal of a long-span fiber system is therefore to use the lowest loss fiber practical. A factor of ten reduction in fiber loss yields a factor of ten increase in link length. A factor of ten increase in transmitter power or a factor of ten improvement in receiver sensitivity will yield only a 33% increase in link length. This relationship is illustrated in Table 2.3.

Table 2.3 Fiber loss, transmitter power, and receiver sensitivity.

Maximum length (km)		Fiber loss (dB/km)	Transmitter power (dBm)	Receiver Sensitivity (dBm)
15.0	Baseline	2.0	0.0	-30.0
20.0	10 times higher power	2.0	**10.0**	-30.0
20.0	10 times more sensitive	2.0	0.0	**-40.0**
150.0	10 times lower loss	**0.2**	0.0	-30.0

The tradeoff between fiber loss, transmitter power, and receiver sensitivity in the guided-wave propagation model of a fiber system is important to understand. It is a fundamentally different tradeoff from the free-space propagation case, where signal attenuation varies with the inverse-square of the path length and a factor of ten improvement in either transmitter power or receiver sensitivity will more than triple the maximum link length.

2.2.2 Link Budget for the Free-Space Channel

Figure 2.9 illustrates the principal entries in the link budget for a free-space optical communication system. The entries are 1) average transmitter power, 2) modulation losses, 3) transmitter antenna gain, 4) transmitter optical loss, 5) space propagation loss, 6) losses in the channel from scattering attenuation, etc., 7) spatial pointing or tracking loss, 8) receiver antenna gain, 9) receiver optical loss, and 10) receiver sensitivity. Just as in the fiber system example, these ten entries can be broken down even further, resulting in link budgets with dozens of individual contributors.

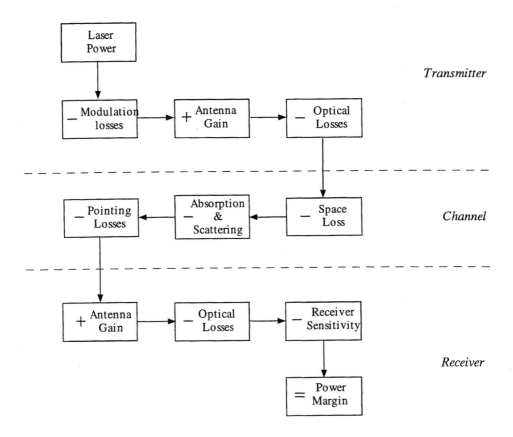

Figure 2.9 Components of a free-space system link budget.

Although fiber systems frequently operate with over a 10 dB margin, it is not unusual for free-space systems to be specified to operate in a "power-starved" regime where only a 3 or 6 dB link margin is maintained. Any additional available power margin would generally be used to reduce the required telescope apertures to save on spacecraft weight and launch costs.

Figure 2.10 shows examples of the geometry of some orbital scenarios that occur in free-space systems. The longest propagation path typically considered for systems that are not concerned with deep-space communications occurs between two geosynchronous satellites positioned over opposite hemispheres and corresponds to a distance of approximately 84,000 km, depending on whether the link enters the atmosphere and on the exact satellite positions. GEO to Earth links have nominal path lengths of 37,500 km while LEO to GEO links have nominal propagation paths of 44,500 km [15]. The actual distances depend on the exact orbits of the satellites and the altitudes of the downlink receivers.

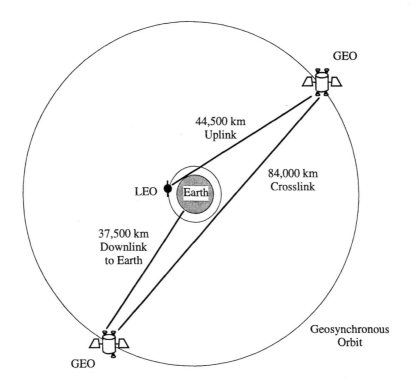

Figure 2.10 Example free-space link scenarios.

An example of a link budget for a free-space system is shown in Table 2.3. We are interested in establishing communications between two platforms. One will be at LEO, while the other is at GEO. The link will be designed for a maximum separation of 40,000 km, or approximately one geosynchronous orbit distance away. The source is to be a semiconductor laser emitting 100 mW of average power at 830 nm. In this example, the optics in the transmitter path are assumed to introduce 6.0 dB of loss due to obscuration in the telescope, beam truncation, imperfect coatings, beam wavefront aberrations, surface reflections, and bulk absorption losses. The telescope aperture is taken to be 25 cm, which produces an ideal antenna gain of 119.5 dB at 830 nm.

The receiving platform is assumed to use a telescope identical to the transmitter, with an optical train that introduces similar losses. At the receiver there is also an additional 1.0 dB loss from imperfect spatial pointing between the platforms and a 1.0 dB loss due to picking off a portion of the received field to operate an independent spatial tracking receiver. For the crosslink scenario assumed, the receiver will have -80.6 dBW or just under 9 nW of optical power to work with. The receiver is known to require an average input signal level of -54.0 dBm to produce the desired quality-of-service. This results in just over a 3 dB margin, meaning there is initially over twice the power needed to close the link. Once the effects of aging in the system components are included, the margin may be reduced by several dB at the end of the design life of the system.

Table 2.4 An example of a free-space system link budget.

Item	*Value*	*dB value*
Transmitter:		
1) Average Transmitter Power	100 mW @ 830 nm	-10.0 dBW
2) Transmitter Antenna Gain	25 cm @ 830 nm	+119.5 dBi
3) Transmitter Optical Losses		-6.0 dB
Effective Radiated Power	103.5 dBW	
Channel:		
4) Space Loss	40,000 km	-295.6 dB
5) Losses in a Vacuum		0 dB
6) Spatial Pointing Loss		-1.0 dB
Receiver:		
7) Receiver Antenna Gain	25 cm @ 830 nm	+119.5 dBi
8a) Receiver Optical Losses		-6.0 dB
8b) Spatial Tracking Pick-off		-1.0 dB
Power at Communications Photodetector	8.71 nW	-80.6 dBW
9) Rcvr sensitivity at 100 Mbps and 10^{-6}	-54.0 dBm	-84.0 dBW
Link Margin		+3.4 dB

The antenna gains in this budget are computed using the formula for the ideal gain of a uniformly illuminated unobscured circular aperture. The gain is quoted in dB relative to an isotropic radiator (dBi) and is given by [16, 17]

$$G_a = 10 \log_{10}\left(\frac{\pi D}{\lambda}\right)^2 \quad \text{(dBi)}, \tag{2.5}$$

where D is the aperture diameter in meters and λ is the wavelength in meters.

In practice this ideal gain is never realized. The transmitter aperture is not uniformly illuminated but is instead typically illuminated by a TEM_{00} Gaussian mode beam that is aberrated by the imperfect optics, truncated by the finite diameter of the telescope primary mirror, and often partially obscured by the telescope secondary mirror and supporting struts [18]. These imperfections in the transmitter telescope distort the far-field pattern and transfer energy from the main-lobe into the side-lobes. This results in a reduction of the achievable on-axis antenna gain.

Since the receiver is assumed to be in the transmitter's far-field and the channel is truly "free-space," with no absorption, scattering, or aberrations, the receiver aperture will be uniformly illuminated by a plane-wave with a flat phase-front and a uniform amplitude distribution. Again, the achievable antenna gain will be reduced due to imperfections in the telescope and supporting optics and truncation due to finite detector size, and often obscuration due to the secondary mirror and its supports will further reduce performance [17-19]. It is not unusual for 3-6 dB of loss to occur in a well engineered telescope system.

The loss due to beam spreading in free-space propagation is given by

$$L_p = 10 \log_{10}\left(\frac{\lambda}{4\pi z}\right)^2 \quad \text{(dB)}, \tag{2.6}$$

where z is the propagation path length in meters. Combining Eq. 2.5 and Eq. 2.6 allows the overall transmission loss to be calculated as

$$L_T = 10 \log_{10}\left(\frac{\pi D_t D_r}{4\pi z}\right)^2 \quad \text{(dB)}, \tag{2.7}$$

where
    D_t is the diameter of the transmitter aperture in meters,
    D_r is the diameter of the receiver aperture in meters,
    z is the path length in meters.

2.3   Specifying an Analog Receiver's Performance

Analog systems employ continuous modulations of the optical field. The transmitter in an analog system will modulate the optical field's amplitude, frequency, phase, or polarization in direct proportion to the input waveform. The receiver recovers information from the continuous electrical signal generated in response to the incoming optical signal field. An analog receiver's goal is to accurately recover a continuous-time waveform. The four primary measures of performance are 1) the signal-to-noise ratio of the recovered waveform, 2) the amplitude of the recovered waveform, 3) the amount of distortion present in the recovered waveform, and 4) the amount of variation in optical power that is tolerable before the user's quality-of-service requirements are violated. This last measure is commonly called the system dynamic range. The exact requirements will vary depending on the intended application.

Analog systems can be broadly divided into the two groups illustrated in Fig. 2.11. In those that use baseband modulation, the laser field varies directly with the analog information. In systems based on subcarrier schemes, a microwave carrier is modulated by the information. The resulting modulated carrier is then used to modulate the laser. Several subcarriers can be combined on a single laser. This is one technique that allows tens to hundreds of cable-TV channels to be transported by a single transmitter laser [20].

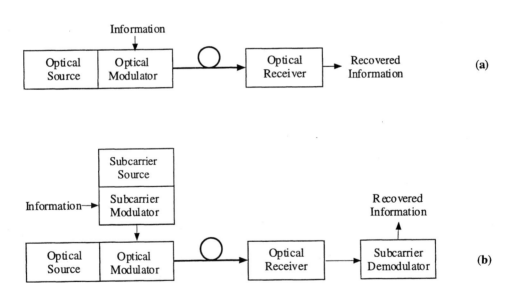

Figure 2.11 Analog receivers. (a) Baseband modulation. (b) Subcarrier modulation.

In cable-TV systems, we are particularly concerned with SNR and distortion. The SNR is usually specified as a carrier-to-noise ratio (CNR). CNR is defined as the amount of power present in the TV video carrier divided by the amount of noise present in the TV channel bandwidth. This is usually specified in dB relative to the carrier, or dBc.

In a cable-TV system using an optical feed, the output of the optical receiver would be a comb of RF video carriers, with each carrier corresponding to a single TV-channel.

CNR values below 40 dBc will generally result in an unacceptable quality-of-service because of objectionable amounts of noise in the picture. Many cable-TV systems use between 45 dBc and 55 dBc CNR as a target value. An analog receiver's CNR performance may be summarized in the type of a plot illustrated in Fig. 2.12.

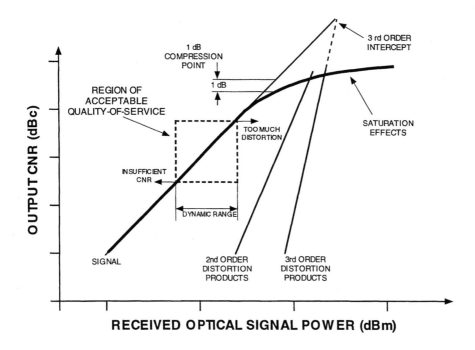

Figure 2.12 Example of an analog receiver CNR performance plot.

In an analog receiver CNR performance plot, the amount of carrier-to-noise ratio at the receiver output is plotted as a function of the amount of optical signal power received. The output CNR improves with increasing optical power until the receiver's output begins to distort due to nonlinear effects. When there is a 1 dB difference between the actual output CNR and the expected output CNR, the 1 dB compression point can be defined. The point at which the distortion products cross the signal power curve determines the nth order intercept point.

The desired region of operation is typically bounded at the low-end by insufficient output CNR and at the high-end by increasingly evident distortion products. Distortion is inevitable for high enough optical power levels since ultimately, no receiver is linear over all potential values of input optical power. Distortion is particularly important in analog systems because even small amounts of distortion present in the recovered waveforms can significantly influence the quality-of-service the link provides.

Neglecting phase related effects, the output voltage from an analog receiver can be represented as a power series of the form

$$v_{out}(t) = A_0 + A_1 P_{rcvd}(t) + A_2 P_{rcvd}^2(t) + A_3 P_{rcvd}^3(t) + \cdots + A_i P_{rcvd}^i \qquad (2.8)$$

In a perfectly linear receiver all of the coefficients A_i are zero for $i \geq 2$. Note that the A_0 term corresponds to an offset voltage. Nonlinearities result in the production of three principle types of distortion, harmonic distortion, resulting in harmonics of the input signal; intermodulation distortion, resulting in the mixing between input signals; and crossmodulation distortion, resulting in modulations on one input signal interfering with a second input signal.

Consider a TV system where each channel has three primary carriers, the video carrier, the audio carrier, and the color subcarrier. Many cable systems have upwards of 50 channels, and consequently over 150 individual RF subcarriers may be present on a single optical carrier used in a fiber-optic trunk distribution cable. Any nonlinearity in the system results in mixing between the various subcarriers. If the mixing products are larger than the video and color information present in the TV picture, interference lines and spurious colors become annoyingly visible. Dominant distortions are termed composite triple beat (CTB) and composite second order (CSO) [20]. To avoid these effects, cable-TV system nonlinear products are typically kept at levels below -60 dBc.

The second commonly used technique for specifying analog receiver performance is with a plot of input-output signal powers. Figure 2.13 illustrates a simple model for the entire analog link. We are interested in plotting the output signal and noise as a function of the input signal and noise. We assume that the optical power and the type and amount of analog modulation are fixed and that our only variable is the amount of RF power present at the input to the link.

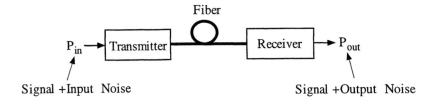

Figure 2.13 Link model for determining input-output signal powers.

If in the link illustrated in Fig. 2.13 we vary the input signal power over a complete range from zero to maximum, a plot of the type illustrated in Fig. 2.14 is obtained. With no input power, there is still noise present in the receiver output. Ideally this would just be whatever noise was originally present at the input of the link, such as the thermal noise of an RF source.

As the input signal is increased, an output signal begins to appear. At the point where there is equal output signal power and output noise power, the noise figure of the link can be defined. As the input signal continues to increase the output signal increases in direct proportion. At a high enough input signal level nonlinearities appear that begin degrading the accuracy of the recovered information. Since we are using a complete link model, these nonlinearities could be from the transmitter, channel, or receiver.

We note that this approach *does* couple receiver performance to transmitter and channel characteristics, making less obvious the receiver's contribution. In order to keep receiver effects separate, the input-output signal power approach is often used in combination with the CNR performance plot of Fig. 2.12.

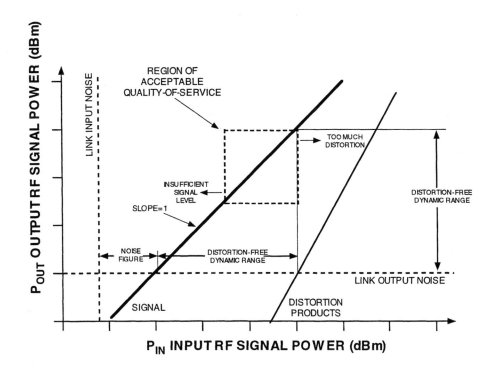

Figure 2.14 Example of an analog receiver signal power plot.

The distortion-free dynamic range is simply the span in signal power between 1) the signal being equal to the link output noise and 2) the distortion products being equal to the link output noise. It is again possible to superimpose the region of acceptable QOS. The lower bound will be set by insufficient input RF signal. The upper bound will be set by too high a level of input RF, causing an intolerable amount of nonlinear distortion. If we continue to increase the input RF power we will eventually begin to saturate the link, and we could define a compression point just as was done in Fig. 2.12.

The overall implication of this discussion on analog receiver performance is that for an analog link to provide an adequate quality-of-service, relatively large amounts of signal-to-noise ratio are required and only small amounts of distortion are tolerable. It is not unusual for the SNR in an analog system to be measured in many tens of dB. The nonlinear distortion products are frequently kept 50 to 60 dB below the desired signal level. In spite of these apparently stringent requirements, analog fiber optic systems can provide unprecedented signal quality, bandwidth, and reliability.

2.4   Specifying a Digital Receiver's Performance

Digital systems use discrete modulations of the optical field. Depending on the data to be transmitted, the optical field's amplitude, frequency, phase, or polarization is varied among a number of well-defined discrete values. The receiver recovers a sequence of binary digits (bits) from the incoming optical signal field. The technique used to specify a digital receiver's sensitivity is different from that used with an analog receiver. The primary measure of a digital link's performance is the probability that the receiver will make an incorrect decision, resulting in a bit-error. In a digital receiver, the amount of optical signal power needed to obtain the desired bit-error-rate is specified. If one varies the amount of received optical power and measures BER at the same time, a plot similar to the one illustrated in Fig. 2.15 is obtained.

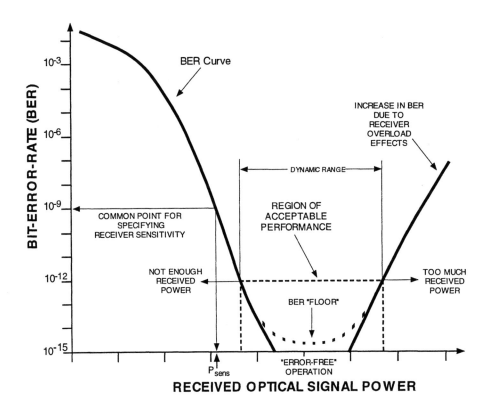

Figure 2.15 Example of a digital receiver performance plot.

At low levels of received power the error rate will be relatively high. As the received power level increases, the probability of making a bit-error decreases. In a correctly designed link, a point will be reached that satisfies the user's quality-of-service requirements for bit-error-rate. For the case illustrated in Fig. 2.15, this is a BER below 10^{-12}. At higher received power levels the bit-error-rate will usually continue to decrease and the system will become error-free. However, it is possible for a BER "floor" to be observed in some systems. These are usually caused by subtle receiver degradations that are independent of the amount of received signal power. Some forms of interference, clock jitter, data pattern dependence, and setup and hold-time violations in digital circuit designs can cause BER floors to be observed.

Ultimately, if the received power continues to increase, the receiver will eventually overload and the error rate will rise. The difference between the point at which there is just enough received power to meet the desired BER QOS and the point at which receiver overload causes the BER to rise to unacceptable levels is the receiver dynamic range.

The amount of optical signal power that produces a receiver BER of 10^{-9} is a widely used measure of receiver performance and is frequently termed the receiver *sensitivity*. Receiver sensitivity is generally a function of data rate and wavelength and can be specified two ways. The receiver sensitivities used in Tables 2.1, 2.2, and 2.3 were quoted in terms of the average optical power level in dBm. This is the most common technique for specifying a sensitivity. The other technique quotes the number of photons required per bit of information and is sometimes preferred since it is both wavelength and data rate independent. Photons per bit is also a direct measure of the energy-per-bit to noise-density ratio (Eb/No) that is used to predict the theoretical performance of digital communication systems [11, 12, 21].

The relationship between average power and photons-per-bit is simple to obtain. The energy in a lightwave is known to be carried by photons with discreet values of energy, given by

$$E = h\nu \quad \text{(Joules)},$$

(2.9)

where
$\quad$ h is Planck's constant (6.626×10^{-34}),
$\quad$ ν is the frequency of the light $= c/\lambda = (3 \times 10^8)/\lambda$.

The relationship between average power at a given wavelength and the photon arrival rate is given by

$$r = \frac{P}{h\nu} \quad \text{(photons per second)},$$

(2.10)

where P is the average received optical power in watts. Photons per bit are calculated by dividing the photon arrival rate in photons per second by the data rate

$$\text{photons per bit} = n_p = \frac{r}{R} = \frac{P}{h\nu R},$$

(2.11)

where R is the data rate in bits per second.

Regardless of the form in which receiver sensitivity is quoted, the proper context must also be given. A digital receiver's sensitivity always has a bit error rate associated with it. For terrestrial fiber telecommunication systems the most common convention is to use 10^{-9}. Free-space systems will frequently use 10^{-6}. For a given data rate, specifying receiver sensitivity in photons per bit without the corresponding BER obtained is misleading, just as specifying sensitivity in terms of dBm without also quoting both wavelength and BER is misleading. Using Eq. 2.11, the expression of the receiver sensitivity used in Table 2.4 is 166 photons per bit at a data rate of 100 Mbps and a 10^{-6} BER.

2.5 The Receiver Sensitivity Budget

A receiver performs two essential functions in a communications system. It detects the received signal field and then extracts the information contained in the detected signal. The ultimate performance achievable by a receiver, as summarized by a simple output CNR or sensitivity specification, is the result of the combined performance of several receiver subsystems.

The contribution of each subsystem to the ultimately achievable receiver sensitivity can be tabulated in a receiver sensitivity budget. The budget typically has entries for the photodetector, the amplifier chain, and the demodulator. The exact entries will be determined by the detailed architecture and circuit design of the receiver.

An example of a receiver sensitivity budget for a digital receiver is illustrated in Table 2.5. The table starts with the theoretical maximum performance achievable using an optimum (i.e. maximum likelihood) demodulator for the modulation format utilized in the system. In this example this is taken to be 40 photons per bit for an error rate of 10^{-9}. Not all of the photons illuminating the detector will be converted into an electrical signal. Some will be reflected off the detector surface and some will not be absorbed, and consequently there is an efficiency loss due to non-ideal photodetection. The photodetector is 85% efficient in our example, corresponding to a loss of 0.71 dB.

The photodetector, being an imperfect physical device, will also introduce some extraneous excess noise into the system, and the detected signal field may be further corrupted either by noise from a strong optical background or from noise already present in the transmitter laser. In this example there is a stray illumination from an interfering optical source, causing 1.0 dB of degradation, in addition to 1.25 dB of loss from the excess noise in the photodetector.

The receiver amplifier chain introduces additional electronics noise and distortion into the system. Normally the first amplifier in a chain dominates the other noise sources in a receiver. The other stages still have a small effect, however, and in our example the combined effects from the noise in the following stages amount to 1.0 dB. There is also some nonlinear distortion, resulting in the recovered waveform being slightly different from the waveform expected at the demodulator. The effect of the distortion is 0.25 dB.

The amplifier chain output also contains the modulation imperfections that were present at the transmitter. Even though it is really a transmitter modulation loss, the effects of imperfect modulation often first appear at the demodulator and therefore are sometimes accounted for at the receiver. No modulation scheme will be perfect and some losses will inevitably occur. Within the demodulator and clock recovery subsystems there will be additional degradation due to imperfect filter bandwidths, jitter in the recovered clock signal, and decision aperture imperfections

In our example the modulation was determined to be 1.0 dB from ideal. The filters in the receiver are not quite ideal and introduced an additional 0.6 dB loss. The time at which the demodulator sampled the recovered waveforms was off by a small amount, causing a 0.4 dB loss. The combined effect of all of these individual contributions sum to an overall degradation of 6.21 dB. Instead of the ideal 40 photons-per-bit receiver sensitivity, we have a sensitivity of 166 photons-per-bit. An early system level evaluation of the effects of receiver, channel, and transmitter imperfections is critical in determining the ultimate performance achievable by a communication link. The locations of significant degradations must be identified. If the modulator implementation is introducing several dB of system degradation it is unlikely that any amount of heroics in receiver design will make up the loss.

Table 2.5 A digital receiver sensitivity budget.

Item	Value	dB Value
Ideal Performance for 10^{-9} BER	40 Photons per bit	16.0 dB
Photodetection losses		
Efficiency	85%	0.71 dB
Excess Detector Noise		1.25 dB
Background Noise		1.0 dB
Amplifier losses		
Amplifier Noise		1.0 dB
Amplifier Distortion		0.25 dB
Demodulator losses		
Imperfect Modulation		1.0 dB
Non-ideal Filter Bandwidths		0.6 dB
Sample Time Jitter and Offset		0.4 dB
Actual Performance	166 Photons per bit	22.21 dB
Receiver Degradation from Ideal		6.21 dB

2.6 References

1.   J. Gowar, *Sec. 17.4 - Analogue Systems* in *Optical Communication Systems,* 1984, Prentice-Hall: Englewood Cliffs, NJ.

2.   M.F. Mesiya, *Chap. 15 - Design of Multichannel Analog Fiber-Optic Transmission Systems* in *Optical-Fiber Transmission,* E. Basch, Editor. 1986, Howard W. Sams Division of Macmillan Inc.: Indianapolis, IN.

3.   J.M. Senior, *Sec. 10.7 - Analog Systems* in *Optical Fiber Communications: Principles and Practice,* 1985, Prentice-Hall: Englewood Cliffs, NJ.

4.   H. Carnes, *et al., Chap. 14 - Digital Optical System Design* in *Optical-Fiber Transmission,* E. Basch, Editor. 1986, Howard W. Sams Division of Macmillan Inc.: Indianapolis, IN.

5.   J.M. Senior, *Sec. 10.5 - Digital Systems* in *Optical Fiber Communications: Principles and Practice,* 1985, Prentice-Hall: Englewood Cliffs, NJ.

6.   R.M. Gagliardi and S. Karp, *Chap. 7 - Digital Communications - Binary Systems* in *Optical Communications,* 1976, John Wiley & Sons: New York.

7.   J. Gowar, *Chap. 15 - The Regeneration of Digital Signals* in *Optical Communication Systems,* 1984, Prentice-Hall: Englewood Cliffs, NJ.

8.   S. Yamamoto, *et al.,* "BER Performance Improvement by Forward Error Correction Code in 5 Gbit/s 9000 km EDFA Transmission System," Electronics Letters, April 1994, vol. 30, no. 9, pp. 718-719.

9.   J.J. Spilker, *Chap. 15 - Viterbi Decoding of Convolution Codes* in *Digital Communications by Satellite,* 1977, Prentice-Hall Information Theory Series: Englewood Cliffs, New Jersey.

10.   J.M. Wozencraft and I.M. Jacobs, *Chap. 6 - Implementation of Coded Systems* in *Principles of Communication Engineering,* 1965, John Wiley & Sons: New York.

11.   R.E. Ziemer and W.H. Tranter, *Chap. 6 - Noise in Modulation Systems* in *Principles of Communications,* 1976, Houghton Mifflin Company: Boston.

12.   J.M. Wozencraft and I.M. Jacobs, *Chap. 2 - Probability Theory* in *Principles of Communication Engineering,* 1965, John Wiley & Sons: New York.

13.   R.E. Ziemer and W.H. Tranter, *Appendix A - Physical Noise Sources and Noise Calculations in Communication Systems* in *Principles of Communications,* 1976, Houghton Mifflin Company: Boston.

14.   W.B. Davenport and W.L. Root, *Chap. 6 - Spectral Analysis* in *An Introduction to the Theory of Random Signals and Noise,* 1958, McGraw-Hill: New York.

15. B.N. Agrawal, *Chap. 2 - Orbit Dynamics* in *Design of Geosynchronous Spacecraft,* 1986, Prentice-Hall: Englewood Cliffs, NJ.

16. B. Klein and J. Degnan, "Optical Antenna Gain 1: Transmitting Antennas," Applied Optics, 1974, vol. 13, no. 9, pp. 2134-2141.

17. W. Stoelzner, *Chap. 5 - Optical Configuration and System Design* in *Laser Satellite Communications,* M. Katzman, Editor. 1987, Prentice-Hall: Englewood Cliffs, NJ.

18. B. Klein and J. Degnan, "Optical Antenna Gain 3: The Effect of Secondary Element Support Struts on Transmitter Gain," Applied Optics, 1976, vol. 15, no. 4, pp. 977-979.

19. J. Degnan and B. Klein, "Optical Antenna Gain 2: Receiving Antennas," Applied Optics, 1974, vol. 13, no. 10, pp. 2397-2401.

20. "Special Issue on Broad-Band Lightwave Video Transmission," Journal of Lightwave Technology, January 1993, vol. 11, no. 1.

21. R.E. Ziemer and W.H. Tranter, *Chap. 7 - Digital Data Transmission* in *Principles of Communications,* 1976, Houghton Mifflin Company: Boston.

Chapter 3

PHOTODETECTION

A variety of theoretical techniques have been developed to describe the generation, propagation, and detection of light. Geometrical-optics, also known as ray-optics, is primarily concerned with image formation from lenses and mirrors. Wave-optics describes light as a scalar sinusoidal lightwave and is adequate to explain many interference and diffraction effects. Electromagnetic-optics introduces a vector form for the lightwave, which allows the explanation of polarization effects and propagation in dielectric media. Quantum-optics is the most complete and fundamental technique and allows the prediction of virtually all observed phenomena, including the details of interactions between lightwaves and atoms.

A complete treatment of quantum-optics is beyond the scope of this text. Fortunately, a perfectly adequate theory of photodetection can be developed by combining some of the simpler concepts from quantum-optics with conventional electromagnetic and wave-optic descriptions of light. This "semi-classical" approach takes advantage of what has been termed the wave-particle duality of light and will be the technique used in this text. Light will be modeled both as an electromagnetic lightwave and as a stream of incident "particles" known as photons.

3.1 The Detection of Optical Signals

The ability to respond to light is a fundamental requirement of all optical receivers. Several methods to detect the presence of an optical signal have been developed, with photographic film probably being the most widely manufactured "detector." In communication applications, the photodetection method employed must convert the received optical signal into an electrical signal that is then processed by conventional electronics to recover the information being transmitted. Table 3.1 lists the techniques that are those most often associated with the detection of optical signals [1].

Table 3.1 Photodetection techniques

Thermal Effects	Wave Interaction Effects	Photon Effects
Thermoelectric Effect Pyromagnetic Effect Pyroelectric Effect Liquid Crystals Bolometers	Parametric Down-conversion Parametric Up-conversion Parametric Amplifiers	Photoconductors Photoemissives Photovoltaics

Thermal effects involve energy being absorbed from the received optical signal so that the photodetector's temperature is altered. The change in detector temperature in turn alters some other device parameter that can then be externally sensed. Examples of thermal effect detectors are bolometers, which change their electrical resistance when illuminated, and pyroelectric detectors, which change their capacitive charge. Thermal effect detectors typically have slow response times when used in high-sensitivity applications and are generally not used in communication systems.

A second, somewhat unconventional technique uses the interaction between lightwaves and a nonlinear material to form sum or difference frequency lightwaves. These wave interactions are the basis for the construction of optical parametric amplifiers and frequency doublers. Unfortunately, efficient optical parametric amplifiers are relatively complex to implement and have not been widely used in communication applications.

In the third and most popular technique, the photodetector absorbs photons from the incident lightwave through atomic interactions in the photodetector material. These interactions produce photo-excited electrical carriers or "photocarriers." The generation of a photocarrier corresponds to the formation of an electron-hole pair in the photodetector material. When these photocarriers transport charge they form an electrical photocurrent that can be processed using conventional electronics. Excellent results have been obtained using this technique and virtually all practical communication systems use it. The remainder of this text will focus on this technique. It is sometimes referred to as "photon effect" photodetection.

3.2 Photon Counting

An optical signal arriving at a receiver can be modeled as a stream of incident photons. For a monochromatic (single frequency) lightwave, each photon has the same energy as all the others and the photon energy is *independent* of the power in the received optical signal. The rate at which the photons arrive at the detector *does* vary with the power in the optical signal. A strong signal corresponds to a high photon arrival rate. A weak signal would have a low photon arrival rate. From a receiver's point-of-view, the fundamental parameter of interest is then the number of photons received during a specified time interval. This is essentially an event counting process known as *photon counting* and is illustrated in Figure 3.1.

Figure 3.1(a) illustrates an ideal photon-counter. A constant amplitude monochromatic lightwave with photons of energy $E = hv$ illuminates an ideal photon-to-photocarrier converter. The photons arrive at the converter at discrete points in time and form what is known as a *point process*, as shown in Fig. 3.1(b). Point processes are random processes in which the observable events are the "points" in time at which the events occur.

The ideal photon-to-photocarrier converter is an infinite bandwidth device that converts each and every incident photon into a photoelectron. The photocarriers generated in the photodetector would then be "counted" by an ideal electrical counter. If the counter is reset at regular time intervals, the number of photons detected by the receiver within each time interval is obtained, as illustrated in Fig. 3.1(c). If we purposely vary the number of photons available for counting by modulating the transmitter's photon source and we observe and correctly interpret the variation in the receiver's photon count, we can transfer information from transmitter to receiver.

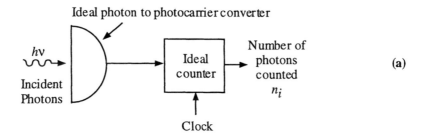

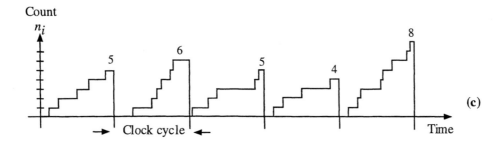

Figure 3.1 Photon counting. (a) Ideal photon detector and counter. (b) Photon arrivals. (c) Resulting photon count.

The generation of photons in an optical transmitter is known to be a random process [2, 3]. Consequently, the point process that describes their arrival times at the photodetector is also random. For a given observation-time T, the number of photocarrier events generated in an ideal photodetector illuminated by constant-amplitude, monochromatic light, is described by the Poisson distribution [2, 4, 5], given by

$$P(n|T) = \frac{(rT)^n e^{-rT}}{n!},$$ (3.1)

where

$P(n|T)$ = probability of counting n photocarriers in a T second observation time,

r = mean photon arrival rate in photons per second.

If we fix the observation-time T and the mean photon arrival rate r we obtain

$$P(n) = \frac{(n_s)^n e^{-n_s}}{n!},$$ (3.2)

where
$P(n)$ = probability of counting n photon events given

$n_s = rT$ = mean number of signal photons counted per observation.

    Three examples of the Poisson distribution are shown in Fig. 3.2. The distribution is skewed to the right because n must always be greater than or equal to zero. In Fig. 3.2, the distribution corresponding to a mean photon arrival rate of 5 photons per observation-time shows the most obvious skew. The distribution for 60 photons per observation-time shows the least.

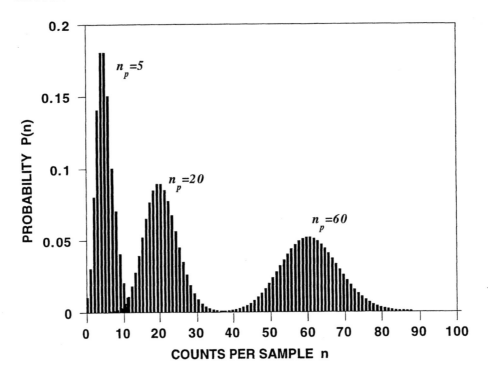

Figure 3.2 Examples of the Poisson distribution.

    Poisson distributions are accurate descriptions of many processes that appear as a series of random independent events. Calls at a telephone exchange, arrivals of customers at a business, and cars passing a specific point on a roadway are all examples of random processes that are often modeled as Poisson point processes.

    We emphasize that the use of a Poisson distribution to describe photon counting is correct for a photodetector illuminated by coherent (i.e. monochromatic) light such as is generated by a laser. There are many optical sources that do not generate coherent light. Light from fluorescent and incandescent light bulbs, the Sun, and the amplified spontaneous emission from optical amplifiers are all examples of incoherent light. Incoherent light of this type is also called thermal or Gaussian light. Sources of background illumination are often characterized this way.

The distribution that describes the counting statistics when the photodetector is illuminated by narrowband incoherent light is known as the Bose-Einstein distribution [6, 7] given by

$$P(n) = \left(\frac{1}{1+n_b}\right)\left(\frac{n_b}{1+n_b}\right)^n, \tag{3.3}$$

where

$P(n)$ = probability of counting n photon events given

n_b = mean number of background photons from incoherent source $= \dfrac{N_o}{h\nu_o}$,

N_o = spectral density of incoherent source in watts per Hz $= \dfrac{P_{opt}}{B_o}$,

P_{opt} = total optical power from incoherent source in Watts,

B_o = optical bandwidth of incoherent source in Hz,

T = observation time in seconds, and

$T \le \dfrac{1}{B_o}$ to meet narrowband requirement.

If the received light has both a coherent signal and a narrowband incoherent cbackground omponent the photon counting is described by a Laguerre distribution [6, 7], given by

$$P(n) = \left(\frac{1}{1+n_b}\right)\left(\frac{n_b}{1+n_b}\right)^n \exp\left\{\frac{n_s}{1+n_b}\right\} L_N\left\{\frac{-n_s}{n_b(n_b+1)}\right\}, \tag{3.4}$$

where

$P(n)$ = probability of counting n photon events,

n_b = mean number of photons from incoherent background source $= \dfrac{N_o}{h\nu_o}$,

n_s = mean number of photons from coherent signal source $= \dfrac{P_{sig}}{h\nu_{sig}}$,

P_{sig} = optical signal power in watts,

$L_N\{x\}$ = Laguerre polynomial.

3.3 Modeling Photodetection

The ideal photon counter of Fig. 3.1 is impossible to realize in actual practice. A more realistic model of the photodetection process is shown in Fig. 3.3. Here the ideal photon-to-photocarrier converter is replaced by one with a finite conversion efficiency, and an electrical low-pass filter is included to account for the finite response time of a practical photodetector.

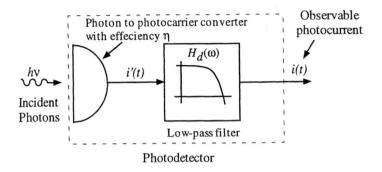

Figure 3.3   Model of a photodetector.

Figure 3.4(a) illustrates the current output from the photon-to-photocarrier converter when the number of photocarriers per observation-time is small. The photocurrent generated is a series of current impulses. Each impulse carries a charge of q coulombs, corresponding to the charge of a single electron.

The effect of the electrical low-pass filter that accounts for a finite photodetector bandwidth is shown in Fig. 3.4(b). The pulses correspond to the impulse response of the photodetector filter and have a finite width τ that is a function of the filter bandwidth. Each pulse still carries a charge of q coulombs. In the example illustrated in Fig. 3.4(b), we are using a filter whose impulse response is given by a simple decaying exponential. As with any linear system, the impulse-response of the filter and the transfer function of the filter form a Fourier-transform pair:

$$H_d(\omega) = \mathrm{FT}\{h_d(t)\}. \tag{3.5}$$

The filter transfer function is unrestricted beyond the requirement that its response must go to zero as ω goes to ∞. The filter's passband can be lowpass or bandpass and it may have constant gain, loss, or be randomly time varying. The filter usually represents the combined effects of photodetector transit-time, junction capacitance, and parasitic reactance.

As the number of photons per observation-time increases further, a continuous photocurrent as illustrated in Fig. 3.4(c) is obtained. Associated with this photocurrent will be a probability density function $P(i)$ and a mean DC photocurrent I_{DC} as illustrated in Fig. 3.4(d).

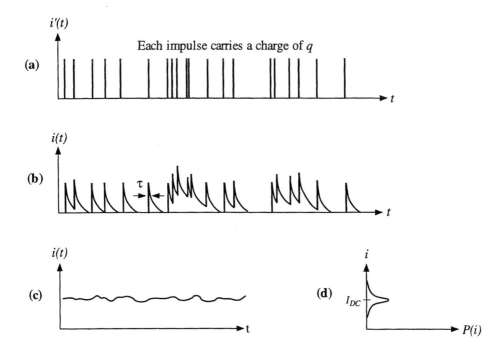

Figure 3.4   **Characteristics of photodetector currents (adapted from Fig. 2.1 Kingston [8]). (a) Ideal photocurrent consisting of a series of current impulses. (b) The observable photocurrent for a low photon arrival rate. (c) Photocurrent for a high photon arrival rate. (d) Current probability density function.**

We can model the output current from the photodetector illustrated in Fig. 3.4(b) using a summation of individual photodetector impulse responses as [9-11]

$$i(t) = \sum_{j=1}^{N} h_d(t - \tau_j),$$

(3.6)

where

$h_d(t)$ = impulse response of the photodetector,

   N = total number of electron - hole pairs generated,

   τ_j = random time at which the jth photocarrier is generated.

The output of a photodetector as described by Eq. 3.6 is known as a filtered Poisson process. Each impulse carries a charge of q since

$$q = \int_{-\infty}^{+\infty} h_d(t) dt.$$

(3.7)

The number of carriers generated during a given observation-time depends on the power in the received optical signal during the observation-time. If we include a time-varying power due to information modulating the transmitter's power, we can define a rate parameter $\lambda(t)$ for the photocarriers produced by the photodetector that is a function of time:

$$\lambda(t) = \frac{\eta}{h\nu} P_{rcvd}(t) \quad \text{(photocarriers per second)}, \tag{3.8}$$

where $P_{rcvd}(t)$ = received power in watts, $h\nu$ = photon energy, and we have included the inherent inefficiency in the conversion from photons to photocarriers in any realizable detector by an efficiency term η, called the *quantum efficiency*. Quantum efficiency (QE) is defined as

$$\eta = \frac{\text{number of photocarriers produced}}{\text{number of incident photons}} \tag{3.9}$$

$$0 \leq \eta \leq 1.$$

For a photodetector with an 85% quantum efficiency, there will be an average of 85 photocarriers generated for every 100 photons that are incident on the photodetector.

We now want to relate the received power to the strength of the received optical field. For a lightwave propagating in free-space the electric field can be expressed as

$$E = E(\vec{p}, t) = \text{Re}\left[E(\vec{p}, t) e^{j2\pi\nu t + \phi} \right] \quad \left(\frac{\text{volts}}{\text{unit length}} \right) \tag{3.10}$$

$$= E(\vec{p}, t) \cos(2\pi\nu t + \phi),$$

where $\vec{p}$ = position of the observation point and ν = frequency of the lightwave = c/λ. The instantaneous intensity of the field at the point $\vec{p}$ is given by

$$I(\vec{p}, t) = \frac{1}{Z_o} \left| E(\vec{p}, t) \right|^2 \quad \left(\frac{\text{watts}}{\text{unit area}} \right), \tag{3.11}$$

where Z_o is the impedance of the medium. This is 377 Ohms for free-space.

The amount of power incident on a photodetector of area A is then just the integral of the intensity function over the detector's area:

$$P(t) = \int_A I(\vec{p}, t) dA \quad \text{(watts)}. \tag{3.12}$$

Equation 3.8 can then be written in terms of received electric-field intercepted by the detector's surface as

$$\lambda(t) = \frac{\eta}{h\nu} \frac{|E(t)|^2}{Z_0} \quad \text{(photocarriers per second).} \tag{3.13}$$

Since the photocarrier generation rate is proportional to incident power, and power is proportional to the magnitude of the electric field squared, a photodetector is essentially a square-law detector for electric fields.

The total number of photocarriers generated during an observation-time is obtained by integrating the rate parameter during the observation-time as

$$N_j = \int_0^T \lambda_j(\tau) d\tau = \text{photoelectrons counted during the } j^{th} \text{ observation time.} \tag{3.14}$$

The Poisson distribution from Eq. 3.2 that describes the photon counting during the jth observation interval becomes

$$P(N_j = N) = \frac{\left(\int_0^T \lambda_j(\tau) d\tau\right)^N}{N!} \exp\left[-\int_0^T \lambda_j(\tau) d\tau\right]. \tag{3.15}$$

This is called a conditional inhomogeneous Poisson process with rate parameter $\lambda(t)$. It is conditional on the amount of received optical power and is inhomogeneous because the received power will generally vary with time. If the received power were constant the process would be called homogeneous.

For a photodetector that generates N photocarriers during a time interval of T seconds, the rate parameter and the observed photocurrent are then

$$\lambda(t) = \frac{N}{T}, \quad i(t) = \lambda(t)q, \tag{3.16}$$

where q = charge of an electron $= 1.6 \times 10^{-19}$ C. If we normalize to an observation-time of one second, the basic unit of amperes is the measure of the current produced during that second.

Let us assume that the photodetector is illuminated by a coherent light with a constant amount of power P_{rcvd}. The resulting average photocarrier generation rate will then also be a constant. Using Eqs. 3.8 and 3.16, the photocurrent produced by the photodetector is given by

$$i(t) = \lambda(t)q = \frac{\eta q}{h\nu} P_{rcvd}(t) \quad \text{(amps).} \tag{3.17}$$

The term $\eta q/h\nu$ is known as the responsivity of the detector:

$$R = \frac{\eta q}{h\nu} \quad (\frac{\text{amps}}{\text{watt}}). \tag{3.18}$$

Responsivity varies with wavelength, as shown in Fig. 3.5. Responsivity increases with wavelength because there are more photons per watt at long wavelengths than there are at short wavelengths. This is a direct result of the photon energy decreasing with wavelength. Since the amount of photocurrent is determined by the number of photons, not the energy of the photons, longer wavelengths generate more photocurrent per watt than do the short wavelengths.

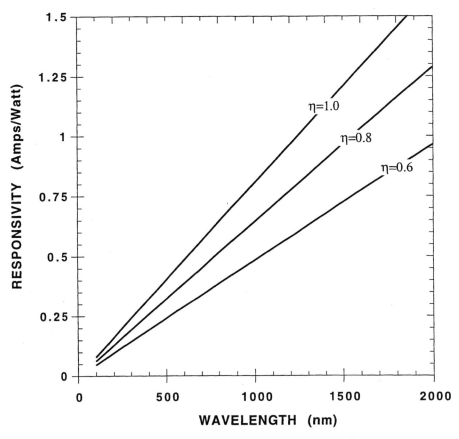

Figure 3.5 Responsivity versus wavelength.

3.4 Photocurrent Statistics

We have seen that photodetection is inherently a statistical process and that for coherent or nearly coherent light, the photocurrent is essentially a Poisson process. In order to predict the performance of a receiver we need to know the statistics of the photocurrent. In particular we are interested in the mean and the variance of the photocurrent. The mean is of interest because it corresponds to the signal that we are interested in recovering information from. The variance is of interest because it represents the amount of noise power present in the photodetector's output and is a measure of how distorted the recovered signal will be.

Let us define a "noise" current as the difference between the instantaneous photocurrent $i(t)$ and the average DC photocurrent as

$$i_{noise}(t) = i(t) - I_{DC} \quad \text{(amps).} \tag{3.19}$$

The noise current of Eq. 3.19 can be thought of as a source of current-noise internal to the photodetector that is in parallel with a noise free DC photocurrent. A current-noise arising from random current pulses such as are produced from photodetection is termed *shot-noise*. Shot-noise is also found in vacuum tubes and in *p-n* junctions in semiconductor diodes and transistors. Shot-noise that arises during photodetection due to the quantum mechanical nature of light is known as *quantum shot-noise*. A fundamental derivation of the origin of quantum shot-noise requires a complete quantum mechanical treatment and interested readers are referred to refs. 3,12, and 13.

To understand the impact of photocurrent shot-noise on the detection of a received signal we need to determine both the signal i_s and noise i_n present in the photodetector's output. For a constant power illumination, the rate-parameter $\lambda(t)$ is a constant and the "signal" is simply the mean DC photocurrent. The noise corresponds to the variance in the photocurrent. The results are well known. For a filtered, homogeneous Poisson process, the mean and variance are given by [9,11]

$$i_s(t) = \overline{i(t)} = \frac{nq}{h\nu} P_{rcvd} \int_0^t h_d(\tau)d\tau \quad \text{(amps),}$$

$$\tag{3.20}$$

$$i_n^2(t) = \text{var}\{i(t)\} = \frac{nq}{h\nu} P_{rcvd} \int_0^t h_d^2(\tau)d\tau \quad \text{(amps}^2\text{).}$$

In Eq. 3.20, we have made the assumption that the response of the photodetector is causal, that is, $h_d(t) = 0$ for $t < 0$. As long as we are observing for a time longer than the impulse response of the photodetector, we can change the limits of integration to be from zero to ∞, and neither the mean nor variance will then be time varying.

Note that both the mean and the variance of the photocurrent are directly proportional to the received signal power. This implies that as the signal power increases, *both* the signal component in the photocurrent *and* the noise component of the photocurrent increase. This is fundamentally different from a microwave receiver, where an increase in received signal power increases only the recovered signal component. The microwave receiver's noise level, having been set by thermal noise, remains constant.

The spectral-density of quantum shot-noise is found by computing the autocorrelation of the noise and then Fourier transforming the result [14, 15]. The resulting power spectrum for the photocurrent associated with a filtered homogeneous Poisson process is illustrated in Fig. 3.6. At low frequencies the height of the spectrum is given by [11]

$$i_n^2(f) = 2qI_{DC} \quad (\frac{\text{amps}^2}{\text{Hz}}). \tag{3.21}$$

The total noise is then given by

$$I_n^2 = 2qI_{DC}B \quad (\text{amps}^2), \tag{3.22}$$

where B is the bandwidth that the noise is being observed in.

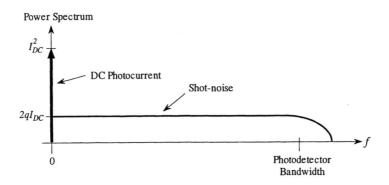

Figure 3.6 Power spectrum of the photocurrent from a photodetector under constant power illumination.

In most optical communications applications $P_{rcvd}(t)$ and consequently $\lambda(t)$ correspond to the information being transmitted and are expected to be time-varying. This makes the Poisson process inhomogeneous and conditional on the received signal power. We also allow the photodetector filter function $h_d(t)$ to have gain, loss, or time-varying characteristics. To account for a filter that may vary from photocarrier to photocarrier, we modify Eq. 3.6 as

$$i(t) = \sum_{j=1}^{N} G_j h_d(t - \tau_j), \tag{3.23}$$

where G_j = gain associated with the jth photocarrier, where $0 \le G_j \le G_{\text{max}}$.

The mean and variance of the photocurrent when the photodetector has a randomly varying gain and the rate-parameter is time varying are then [9, 11]

$$\overline{i(t)} = \overline{G} \int\limits_{-\infty}^{t} \lambda(\tau) h_d(t-\tau) d\tau \quad \text{(amps)},$$

$$\text{var}\{i(t)\} = \overline{G^2} \int\limits_{-\infty}^{t} \lambda(\tau) h_d^2(t-\tau) d\tau \quad \text{(amps}^2). \tag{3.24}$$

We are interested in recovering a signal $i_s(t)$ related to the information that is modulating the received signal. The signal corresponds to the photocurrent mean, while the noise corresponds to the photocurrent variance. Expressed in terms of the received signal power, the signal and noise components of the photocurrent are

$$i_s(t) = \overline{i(t)} = \frac{\eta q}{h\nu} \overline{G} \int\limits_{-\infty}^{t} P_{rcvd}(\tau) h_d'(t-\tau) d\tau \quad \text{(amps)},$$

$$i_n^2(t) = \text{var}\{i(t)\} = q \frac{\eta q}{h\nu} \overline{G^2} \int\limits_{-\infty}^{t} P_{rcvd}(\tau) h_d'^2(t-\tau) d\tau \quad \text{(amps}^2). \tag{3.25}$$

In Eq. 3.25, instead of q, we are using a normalized photodetector impulse response $h_d'(t)$ that has an area of unity so that we can explicitly show the charge associated with a photocarrier.

Let us assume that the modulating signal is a baseband signal, that is, it is centered at DC. For this case, the power spectrum of the photocurrent associated with a conditional Poisson process will be as illustrated in Fig. 3.7.

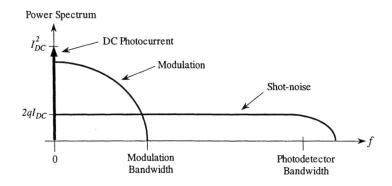

Figure 3.7 Power spectrum of the photocurrent from a photodetector illuminated by an optical signal carrying baseband intensity modulation.

When using a classical electromagnetic-wave formulation it is important to understand the limitations introduced by the Heisenberg uncertainty principle. The uncertainty principle governs the fundamental accuracy with which a measurement can be made and can be thought of as an irreducible "noise" in our measurement system.

Since reception and detection are essentially forms of measurement in which the current, frequency, power, or voltage present in a receiver is "measured," they are also governed by uncertainty. The uncertainty principle is usually expressed as a limitation in the ability to measure a particle's momentum and its position. It applies just as well to our ability to accurately measure an electromagnetic wave [12, 16]. Uncertainty implies that we cannot determine the energy in a wave or the time at which the wave possesses the energy to an accuracy better than that given by

$$\Delta E \Delta t \geq \frac{h}{4\pi},$$
(3.26)

where

ΔE is the uncertainty in energy
Δt is the uncertainty in time
h = Planck's constant.

Since energy in a lightwave is related to the number of photons, and time is related to phase, the Heisenberg uncertainty can also be expressed within the constraints of our semi-classical formulation as [17]

$$\Delta n \Delta \phi \geq \frac{1}{2},$$
(3.27)

where

Δn is the uncertainty in the number of photons n in the lightwave
$\Delta \phi$ is the uncertainty in the phase of the lightwave.

The uncertainty principle as expressed in Equations 3.26 and 3.27 has an important consequence. The more accurately we determine an electromagnetic field's energy, the worse our ability to measure its phase becomes. In principle one could construct a receiver that would measure the energy (i.e. photon number) to an arbitrary accuracy but only if one were willing to accept a corresponding increase in uncertainty with respect to phase.

This concept of trading photon number uncertainty against phase uncertainty gives rise to the concept of *squeezed light,* which can exhibit sub-Poisson statistics [9]. Although squeezed light has been successfully generated, there have been few useful applications and for most practical purposes, Poisson statistics and quantum shot noise form the fundamental limit to the achievable performance in an optical communications receiver. It is sufficient for our receiver design purposes to view the generation of quantum shot noise as an inherent part of the photodetection process that is an unavoidable result of the random arrival times of incident photons.

3.5 Ideal Direct Detection

Direct detection describes an optical detection technique that is closely related to photon counting [18]. Ideally, all of the energy in the received optical field is detected. A simple form of a direct detection receiver is shown in Figure 3.6. The photodetector converts the incident field into a photocurrent that is then processed electronically. For simplicity, the electronic processing is modeled as an ideal, noiseless, unity-gain electronic amplifier that presents a load of R_l ohms to the photodetector.

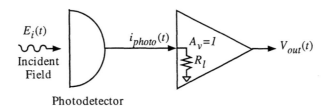

Figure 3.8 Direct detection.

Let us model the photodetector as being illuminated by an incident optical field:

$$E_i(t) = \sqrt{2P_s Z_0}\, \cos(\omega_s t + \phi) \quad (\frac{\text{volts}}{\text{meter}}). \tag{3.28}$$

The average power in this field is then

$$P_{avg} = \frac{\left(\dfrac{|E_i(t)|}{\sqrt{2}}\right)^2}{Z_0} = P_s \quad (\text{watts}). \tag{3.29}$$

The photodetector responds to the square of the field as given by Equation 3.19 and generates a photocurrent

$$\begin{aligned}
i_{photo}(t) &= \frac{\eta q}{h v} \frac{|E_i(t)|^2}{Z_0} \quad (\text{amps}), \\
&= \frac{\eta q}{h v} \frac{2 P_s Z_0}{Z_0} \cos^2(\omega_s t + \phi), \\
&= \frac{\eta q}{h v} 2 P_s \frac{1}{2} \left[1 + \cos 2(\omega_s t + \phi)\right].
\end{aligned} \tag{3.30}$$

where we have made use of the trigonometric identity

$$\cos^2 \theta = \frac{1}{2}(1 + \cos 2\theta). \tag{3.31}$$

The photodetector's low-pass nature prevents it from responding to the double frequency term and consequently the output is just the average DC term:

$$i_{photo}(t) = i_{dc} = \frac{\eta q}{h\nu} P_s \quad \text{(amps)}. \tag{3.32}$$

Thus the output of a direct detection receiver is directly proportional to the power in the received field. Strictly speaking, direct detection receivers respond only to fluctuations in the power in the received field. Phase, frequency, and polarization information about the received signal is "lost." However, we will see in Chapter 7 that by including additional optical signal processing elements such as interferometers or polarizers in the received signal path ahead of the photodetector, we can create a direct detection receiver that is sensitive to optical phase, frequency, or polarization as well as intensity.

We are interested in determining the ultimate signal-to-noise ratio achievable at the output of the direct detection receiver. If we define the "signal" to be the electrical power dissipated in the load resistor due to the DC photocurrent, then the signal power will be equal to the mean photocurrent squared times the load resistor:

$$P_{signal} = i_{dc}^2 R_l = \left(\frac{\eta q}{h\nu} P_s\right)^2 R_l \quad \text{(watts)}. \tag{3.33}$$

The "noise" is determined by the quantum shot noise that is inherent in the photodetection process. This was given by Equation 3.22. The electrical power in the shot noise is equal to the mean-square noise current times the load resistor:

$$P_{noise} = i_{noise}^2 R_l = 2q i_{dc} B R_l \quad \text{(watts)}. \tag{3.34}$$

Quantum shot noise forms the fundamental limit to the signal-to-noise ratio and consequently sets the ultimate sensitivity limit for an optical receiver. The maximum achievable signal-to-noise ratio in a direct detection optical receiver is given by the ratio of the signal power to noise power, or

$$SNR_{dd} = \left(\frac{P_{signal}}{P_{noise}}\right) = \frac{i_{signal}^2 R_l}{i_{noise}^2 R_l} = \frac{\left(\frac{\eta q}{h\nu} P_{signal}\right)^2}{2q \frac{\eta q}{h\nu} P_{signal} B} = \frac{1}{2} \frac{\eta P_{signal}}{h\nu B}. \tag{3.35}$$

Equation 3.35 represents the fundamental *shot-noise limit* for the signal-to-noise ratio achievable by a direct detection receiver. The SNR is limited only by the shot-noise that arises because of the quantum mechanical nature of light. We will see in later chapters that because of additional noise contributed by the receiver's electronics, obtaining near shot-noise limited performance can be challenging.

3.6 Ideal Coherent Detection

The term "coherent detection" as applied to optical communications is used to describe a receiver that exploits the coherence properties of light [19]. A single local oscillator (LO) laser that is spatially combined with the incoming signal field is typically employed. Absolute phase knowledge of the received signal may be used but is not required for the receiver to be termed "coherent." This is significantly different from classical radio frequency receivers where coherent detection implies the use of a phase synchronous local oscillator. A simple coherent detection receiver is shown in Fig. 3.9. A beamsplitter with an intensity reflection coefficient ε is used to spatially combine an optical field with amplitude E_s at optical frequency v_s and phase angle ϕ with a second optical field E_{lo} at optical frequency v_{lo}.

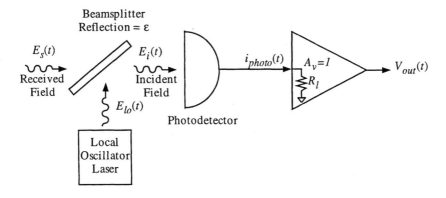

Figure 3.9 Coherent detection.

If we assume that the electric field vectors lie oriented in the same spatial plane and are of the same polarization and spatial phase and the beamsplitter is linear and lossless, then the resulting combination of fields incident on the photodetector is given by

$$E_i(t) = \sqrt{\varepsilon}E_{lo}(t) + \sqrt{1-\varepsilon}E_s(t),$$

$$= \sqrt{\varepsilon}\sqrt{2P_{lo}Z_0}\cos(2\pi v_{lo}t) + \sqrt{1-\varepsilon}\sqrt{2P_{rcvd}Z_0}\cos(2\pi v_s t + \phi).$$

(3.36)

The photodetector responds as a square law detector for the electric field as given in Equation 3.13 and generates a photocurrent:

$$i_{photo}(t) = \frac{\eta q}{hv} \frac{1}{Z_0} \left| \sqrt{2P_{lo}Z_0 \varepsilon} \cos(2\pi v_{lo} t) + \sqrt{2P_{rcvd}Z_0(1-\varepsilon)} \cos(2\pi v_s t + \phi) \right|^2. \quad (3.37)$$

We simplify Eq. 3.37 by apply Eq. 3.31 to expand the $\cos^2$ terms and by applying

$$\cos\alpha\cos\beta = \frac{1}{2}\cos(\alpha-\beta) + \frac{1}{2}\cos(\alpha+\beta) \quad (3.38)$$

to expand the product terms. We can ignore the double and sum optical frequency terms because they are well beyond the photodetector's electrical bandwidth. The resulting photocurrent is given by

$$i_{photo}(t) = \frac{\eta q}{hv} \left\{ \begin{array}{l} P_{lo}\varepsilon \\ +P_{rcvd}(1-\varepsilon) \\ +2\sqrt{P_{lo}P_{rcvd}\varepsilon(1-\varepsilon)} \cos[2\pi(v_s - v_{lo})t + \phi] \end{array} \right\}. \quad (3.39)$$

In communications applications, the received signal P_{rcvd} is much smaller than the local oscillator power P_{lo} and its contribution to the DC photocurrent can be ignored. The DC photocurrent is therefore

$$I_{DC} = \frac{\eta q}{hv} P_{lo}\varepsilon. \quad (3.40)$$

The mixing of the incident fields through the square-law nature of the photodetector gives rise to an AC photocurrent at the difference frequency between the signal laser and the local oscillator laser. This difference frequency term is called the intermediate frequency, or IF. As long as the IF is within the photodetector's electrical bandwidth the signal photocurrent will take the form

$$i_s(t) = \frac{\eta q}{hv} 2\sqrt{P_{lo}P_{rcvd}\varepsilon(1-\varepsilon)} \cos(\omega_{if} t + \phi), \quad (3.41)$$

where $\omega_{if} = 2\pi(v_{lo} - v_s)$. Note that unlike the direct detection receiver, the coherent receiver's signal term contains information about the frequency and phase of the received signal.

In signal-to-noise ratio calculations we are interested in the IF power available. In a heterodyne receiver the difference frequency ω_{if} is non-zero and the phase ϕ is not controlled. Obtaining the IF power simply requires the determination of the RMS magnitude of the sinusoidal AC photocurrent:

$$i_{rms} = \frac{i_{peak}}{\sqrt{2}} = \frac{\eta q}{hv}\sqrt{2P_{lo}P_{rcvd}\varepsilon(1-\varepsilon)} \quad \text{(Amps)}. \tag{3.42}$$

The signal-to-noise ratio for the coherent receiver can then be obtained using the same procedure as was used for the direct detection receiver. The noise power comes from the photocurrent's quantum shot noise caused by the LO laser as

$$P_{noise} = 2qI_{DC}BR_l = 2q\frac{\eta q}{hv}P_{lo}\varepsilon BR_l. \tag{3.43}$$

The signal power is just the electrical power contained in the IF signal:

$$P_{signal} = i_{rms}^2 R_l = \left(\frac{\eta q}{hv}\right)^2 2P_{lo}P_{rcvd}\varepsilon(1-\varepsilon)R_l. \tag{3.44}$$

The signal to noise ratio is then given by

$$SNR_{het} = \left(\frac{P_{signal}}{P_{noise}}\right) = \frac{\left(\frac{\eta q}{hv}\right)^2 2P_{lo}P_{rcvd}\varepsilon(1-\varepsilon)}{2q\frac{\eta q}{hv}P_{lo}\varepsilon B} = \frac{\eta(1-\varepsilon)P_{rcvd}}{hvB}. \tag{3.45}$$

In the limiting case of a strong local oscillator and a small beamsplitter reflectivity, the shot noise limited SNR for a heterodyne receiver becomes

$$\lim_{\substack{P_{lo}\to\infty \\ \varepsilon\to 0}} SNR_{het} = \frac{\eta P_{rcvd}}{hvB}. \tag{3.46}$$

Note that the shot noise limited SNR obtained in the heterodyne receiver is a factor of two (i.e. 3 dB) better than the SNR obtained in a direct detection receiver.

Recall that in deriving Eq. 3.46 we assumed that a heterodyne receiver was being used so that the difference frequency ω_{if} was non-zero and the phase ϕ was not controlled. Another type of coherent receiver known as a *homodyne* receiver uses phase tracking to force $\omega_{if} = 0.0$ and controls the phase ϕ. For a homodyne receiver, Equation 3.41 takes the form of a DC signal:

$$i_s = \frac{\eta q}{hv}2\sqrt{P_{lo}P_{rcvd}\varepsilon(1-\varepsilon)}\cos\phi . \tag{3.47}$$

For the case of perfect phase tracking, $\phi = 0$, and the signal-to-noise ratio becomes

$$SNR_{hom} = \left(\frac{P_{signal}}{P_{noise}}\right) = \frac{\left(\frac{\eta q}{hv}\right)^2 4P_{lo}P_{rcvd}\varepsilon(1-\varepsilon)}{2q\frac{\eta q}{hv}P_{lo}\varepsilon B} = \frac{\eta 2(1-\varepsilon)P_{rcvd}}{hvB}. \tag{3.48}$$

The shot-noise limit for a homodyne receiver is then

$$\underset{\substack{P_{lo}\to\infty \\ \varepsilon\to 0}}{limit} \quad SNR_{hom} = 2\frac{\eta P_{rcvd}}{hvB}. \tag{3.49}$$

Note that the shot-noise limited SNR obtained in a homodyne receiver is a factor of two better than that obtained in the heterodyne receiver and a factor of four (6 dB) better than the SNR of a direct detection receiver. We will see in later chapters that SNR is not the whole story, particularly when used to predict the performance of digital receivers.

We can summarize the photodetection process using the equivalent circuit illustrated in Fig. 3.10. The photocurrent is a function of the square of the incident electric field times the photodetector's responsivity, convolved with the impulse response of the photodetector. The quantum shot noise associated with the generation of a photocurrent is proportional to the DC photocurrent and the bandwidth of the photodetector.

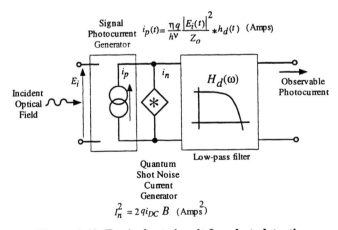

Figure 3.10 Equivalent circuit for photodetection.

3.7 References

1. R.J. Keyes, *Chapter 2* in *Optical and Infrared Detectors,* 1977, Springer-Verlag: New York.

2. A. Yariv, *Chap. 10 - Noise in Optical Detection and Generation* in *Optical Electronics,* 1985, Holt, Rinehart and Winston: New York.

3. H. Haken, *Chap. 1 - What is Light?* in *Light Volume 1 Waves, Photons, Atoms,* 1981, North-Holland Publishing Company: New York.

4. J.B. Kennedy and A.M. Neville, *Chap. 9 - Poisson Distribution* in *Basic Statistical Methods for Engineers and Scientists,* 1986, Harper & Row: New York, Chapter 10.

5. C.W. Helstrom, *Section 5.3 - Poisson and Affiliated Stochastic Processes* in *Probability and Stochastic Processes for Engineers,* 1984, Macmillan Publishing Company: New York.

6. R.M. Gagliardi and S. Karp, *Chap. 2 - Counting Statistics* in *Optical Communications,* 1976, John Wiley & Sons: New York.

7. S. Karp, *et al., Section 3.5 - Counting Statistics* in *Optical Channels,* 1988, Plenum Press: New York.

8. R.H. Kingston, *Chap. 2 - The Ideal Photon Detector* in *Detection of Optical and Infrared Radiation,* 1978, Springer-Verlag: New York.

9. E.V. Hoversten, *Chap. F8 - Section 2.3 - Detector Statistical Model* in *Laser Handbook,* F.T. Arecchi and E.O. Schulz-DuBois, Editors. 1972, American Elsevier Publishing Company: New York.

10. D.L. Snyder, *Chap. 4 - Filtered Poisson Processes* in *Random Point Processes,* 1975, John Wiley & Sons: New York.

11. R.M. Gagliardi and S. Karp, *Chap. 4 - The Optical Detector Response Process* in *Optical Communications,* 1976, John Wiley & Sons: New York.

12. J.P. Gordon, "Quantum Effects in Communications Systems," Proceedings of the IRE, 1962, vol. 50, pp. 1898-1908.

13. M.J. Buckingham, *Chap. 11 - Quantum Mechanics and Noise* in *Noise in Electronic Devices and Systems,* 1983, Halsted Press Division of John Wiley & Sons: New York.

14. M.J. Buckingham, *Chap. 2 - Mathematical Techniques* in *Noise in Electronic Devices and Systems,* 1983, Halsted Press Division of John Wiley & Sons: New York.

15.   W.B. Davenport and W.L. Root, *Chap. 6 - Spectral Analysis* in *An Introduction to the Theory of Random Signals and Noise,* 1958, McGraw-Hill: New York.

16.   B.M. Oliver, "Signal-to-Noise Ratios in Photoelectric Mixing," Proceedings of the IRE, 1961, vol. 49, pp. 1960-1961.

17.   P. Carruthers and M.M. Nieto, "Phase and Angle Variables in Quantum Mechanics," Reviews of Modern Physics, April 1968, vol. 40, no. 2, pp. 411-440.

18.   R.M. Gagliardi and S. Karp, *Chap. 5 - Noncoherent (Direct) Detection* in *Optical Communications,* 1976, John Wiley & Sons: New York.

19.   R.M. Gagliardi and S. Karp, *Chap. 6 - Coherent (Heterodyne) Detection* in *Optical Communications,* 1976, John Wiley & Sons: New York.

Chapter 4

PHOTODETECTORS

The heart of an optical communication receiver is the opto-electronic device that is used as the photodetector. Ideally, a photodetector would detect all incident photons, respond to the fastest changes in the incoming signal that were of interest, and not introduce additional noise beyond the inherent quantum shot-noise from the received signal. In most practical applications, additional desirable characteristics can be defined. The photodetector should be small, lightweight, rugged, reliable, and cost-effective, and its characteristics should remain unaffected by age and environment. Unfortunately, realistic photodetectors have limited bandwidths with finite response times. They introduce unwanted noise into the detection process, and the probability of detecting an individual photon is less than 100%. Some detector technologies are fragile and environmentally sensitive; others may have finite lifetimes or may degrade unacceptably as they age.

The photon-effect based photodetectors that are used in optical communication systems are those that directly generate photocurrents from interactions between photons and atoms in the detector material. When light penetrates into a photodetector material, the volume of material illuminated contains a tremendous number of atoms. The probability that a *specific* atom will absorb a photon and generate free carriers that form a photocurrent is quite small. However, since the number of atoms is huge, the probability that *some* atom will interact with an incident photon to form photo-excited free carriers can be quite high in a well designed detector. Photon-effect photodetectors can be constructed to simultaneously exhibit the high sensitivities and fast response times needed for high data-rate communications and are frequently grouped into one of four general categories; photomultipliers, photoconductors, photodiodes, and avalanche photodiodes.

4.1 Photomultipliers

A photomultiplier tube (PMT) is a form of vacuum tube that utilizes the photoelectric effect and the secondary emission of electrons to provide high current gains once a photon is initially detected. A schematic illustration of a photomultiplier tube is shown in Fig. 4.1. Incident photons fall on a photo-emissive photocathode consisting of a low work-function metal or semiconductor. If the photon energy exceeds the work function of the cathode material there is a reasonable probability that an electron will be ejected from the material [1]. The electron emitted from the cathode is accelerated by an applied electric field towards a series of additional electrodes called dynodes. The dynodes are coated with a material that is prone to secondary emission of electrons and are connected in series between resistors, causing a voltage gradient to form along the dynodes. The dynodes generate additional electrons via secondary emission, with the last electrode forming an anode that collects the emitted electrons. The first dynode may generate 5-10 additional electrons, the second 25-100, etc. This current multiplication process continues for each dynode in the tube, with current gains in excess of 10^6 possible.

The long-term average of the probability of detecting an incident photon is equal to the quantum yield or quantum efficiency of the photomultiplier. It is defined as the number of electrons released by the photocathode per incident photon. A typical high performance photocathode might have a QE of 30%. This photocathode would generate one electron for approximately every 3 incident photons. This also implies of course that 70% of the incident photons go undetected.

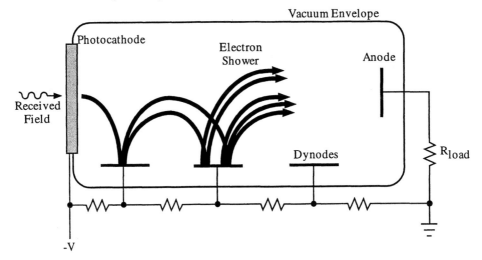

Figure 4.1 Photomultiplier Tube (PMT).

Photomultipliers are widely used in laboratory instruments and in some low-speed photodetection systems. The high current gains achievable in photomultipliers can partially compensate for their comparatively low quantum efficiencies. This feature has allowed PMTs to be used to form extremely sensitive optical detection systems. The application of PMTs in high data rate communication systems is limited by their bandwidths and physical characteristics. Conventional tubes have only a few hundred MHz bandwidth. Wider bandwidths require more complex traveling wave configurations [2]. Although they are still proposed for use in some specialized free-space optical communication systems [3], PMTs are physically large, require high voltage power supplies, and are considered mechanically fragile. In most communications applications, they have largely been displaced by solid-state photodetectors based on semiconductor materials.

The photomultiplier principle is routinely used in the construction of microchannels that are used in image intensifiers [4-6]. In a microchannel array image intensifier, a large number of coated glass capillaries are used to form a parallel array of continuous dynodes. The coatings provide both electrical resistance and support electron multiplication. The capillaries may be only a few tens of microns in diameter and a few millimeters long. Several million such capillaries may be used in a high resolution imager. The array of parallel capillaries is sandwiched between a photocathode plate and a phosphor plate and a high voltage is applied across the plates. This forms a voltage gradient along the interior walls of the capillaries. Weak light is imaged onto the photocathode, generating free-electrons. These free electrons cascade down the interior of the capillaries, causing electron multiplication, and a substantially larger number of electrons will strike the phosphor. An intensified image will then appear on the phosphor plate.

4.2 Photoconductors

Photoconductors are based on photon absorption in semiconductor materials. Fig. 4.2 illustrates the three principal absorption processes that can occur in semiconductors [4, 6, 7]. Intrinsic band-to-band absorption occurs when the photon energy is greater than the material bandgap energy E_g and a photon is able to excite an electron from the valence band up into the conduction band. The hole and electron each constitute a charge carrier. The application of an electric field to the semiconductor will cause the hole and electron to be transported through the material and into the external circuit, causing a photocurrent to flow. When the electron-hole pair makes its way to the external circuit their combined effect is to induce a current flow of q charge in the circuit [7, 8]. Since q is the charge of a single electron, the term *photoelectron* is frequently used interchangeably with photocarrier.

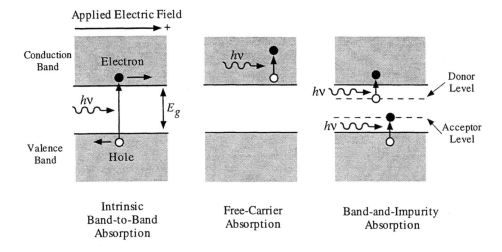

Figure 4.2 Photon absorption mechanisms.

Intrinsic band-to-band absorption is the dominant absorption mechanism in most semiconductors used for photodetection. The requirement that the photon energy $E = h\nu$ be sufficient to create an electron-hole pair can be expressed as

$$h\upsilon > E_g \quad \text{or} \quad h\frac{c}{\lambda} > E_g. \tag{4.1}$$

Rearranging to solve for the maximum allowable wavelength yields

$$\lambda_{max} = \frac{hc}{E_g} \quad \text{or} \quad \lambda_{max}(\text{nm}) = \frac{1240}{E_g(\text{eV})}, \tag{4.2}$$

where

$$E_g(\text{eV}) = \text{bandgap energy in electron - volts.}$$

The bandgap, the corresponding maximum usable wavelength, and typical operating wavelengths for several popular semiconductor materials are listed in Table 4.1.

Free-carrier absorption corresponds to the "heating" of the semiconductor material and is a secondary effect at the near infrared wavelengths used for optical communications. Band-to-impurity absorption is another secondary effect at near-IR wavelengths. It is used in the mid-infrared region to construct photodetectors responsive at mid-IR wavelengths as long as 30 microns.

Table 4.1   Characteristics of various semiconductors

Material	*Bandgap (eV)*	*Maximum Wavelength (nm)*	*Typical Operating Range (nm)*
Si	1.12	1110	500-900
Ge	0.67	1850	900-1300
GaAs	1.43	870	750-850
$In_xGa_{1-x}As_yP_{1-y}$	0.38 - 2.25	550-3260	1000-1600

The $In_xGa_{1-x}As_yP_{1-y}$ materials are particularly attractive because their bandgap can be "tuned" by varying the crystal composition. The particular composition x=0.70, y=0.64 yields $In_{0.70}Ga_{0.30}As_{0.64}P_{0.36}$, a material commonly called indium-gallium-arsenide-phosphide (InGaAsP), with properties appropriate for use at 1300 nm. A slightly different composition, x=0.53, y=1.0, is commonly referred to as indium-gallium-arsenide (InGaAs) and finds use at both 1310 nm and 1550 nm [7].

A signal whose photon energy is sufficient to generate photocarriers will continuously lose energy to the semiconductor crystal lattice as the optical field propagates through the semiconductor. As illustrated in Fig. 4.3, at the air-to-semiconductor interface there is a reflection loss due to the differences in index of refraction. The Fresnel reflectivity for an optical signal at normal incidence to an interface between two materials is given by

$$R = \frac{(n_1 - n_2)^2}{(n_1 + n_2)^2},$$

(4.3)

where

$n_1 =$ index or refraction of first material (for air $n_1 \cong 1$),

$n_2 =$ index of refraction of second material.

For many semiconductors used in the 800 nm to 1550 nm region, n is between 3.2 and 3.6 and the reflectivity can be as high as 33%. Reflection losses can be reduced with anti-reflection coatings to less than 1%.

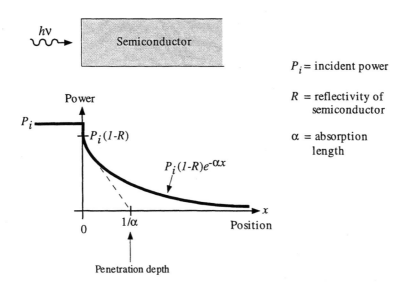

Figure 4.3 Optical absorption in a semiconductor.

Inside the semiconductor, the field decays away exponentially as energy is transferred to the semiconductor. The material can be characterized by an absorption length α and a penetration depth $1/\alpha$. Penetration depth is the point at which $1/e$ of the optical signal power remains.

The power in the optical field decays with distance. The amount of power absorbed in the semiconductor as a function of position within the material is then

$$P_{abs}(x) = P_i(1-R)(1 - e^{-\alpha x}). \tag{4.4}$$

The number of photons absorbed is the power in watts divided by the photon energy ($E = h\nu$). If each absorbed photon generates a photocarrier (i.e. an electron-hole pair), the number of photocarriers generated per number of incident photons for a specific semiconductor with reflectivity R and absorption length α is given by

$$\eta(x) = \frac{\text{number of photocarriers produced}}{\text{number of incident photons}} = (1-R)(1 - e^{-\alpha x}), \tag{4.5}$$

where $0 \le \eta(x) \le 1$. This has the exact same form as Eq. 3.9 and is an equivalent definition for the photodetector's quantum efficiency.

For a photodetector to have a high quantum efficiency, the surface reflection must be low and the length of the absorbing region must be at least one absorption length long. A long absorption region also implies that carriers will take more time reaching the external circuitry, which will tend to reduce the bandwidth of the photodetector. There is an inherent tradeoff between our desire to make the absorption region long to guarantee that all photons are absorbed and our desire to minimize the time between photocarrier generation and our ability to sense the resulting current. For a wide bandwidth photodetector, one tries to select a semiconductor material in which the absorption length is sufficiently large to guarantee that 80-90% of the photons will be absorbed in only a few microns of semiconductor material. The absorption length and corresponding penetration depth vary among the semiconductor materials and are a strong function of wavelength, as shown in Fig. 4.4 [4, 7, 8].

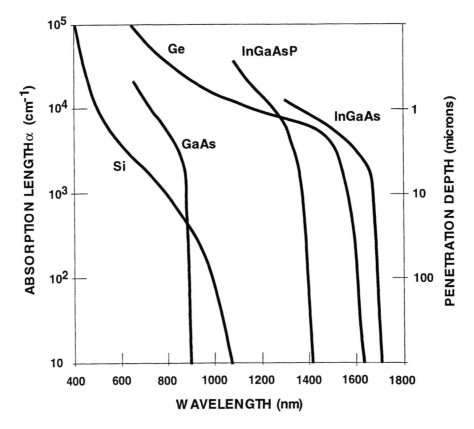

Figure 4.4 Optical absorption for various materials.

Near the material's bandgap, there is tremendous variation, with a hundred nanometers' change in wavelength causing a two to three order of magnitude variation in absorption. In the region of the material's maximum usable wavelength, the absorption rapidly decreases, the material becomes more transparent, and the photodetector quantum efficiency drops dramatically.

By absorbing photons, a photoconductive photodetector forms what is essentially an illumination-controlled variable resistor. The more photons absorbed, the greater the number of electrical carriers produced and the lower the electrical resistance of the semiconductor material. A photoconductive detector can be as simple as a slab of bulk semiconducting material with ohmic contacts at the ends, as illustrated in Fig. 4.5(a), or it may be an interdigitated structure with ohmic contacts on the top, as shown in Fig. 4.5(b). In use, the photoconductor is usually placed in a simple DC biasing circuit and the photocurrent i_{photo} is monitored.

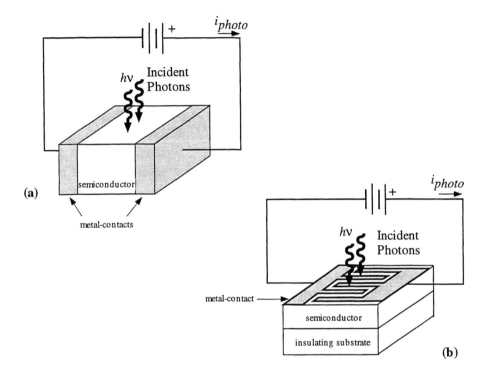

Figure 4.5 Photoconductive detectors. (a) Bulk semiconductor. (b) Semiconductor film with an interdigitated electrode structure.

Variations in the power in the incident optical field modulate the resistance of the photodetector, leading to a corresponding modulation of the current flow i_{photo} in the circuit. Once incident photons cause charge carriers to form in the valence and conduction bands, the applied electric field causes the carriers to move toward the electrical terminals. The net result is a reduction in the resistance of the material, which allows an increase in the current circulating in the detector circuit.

The photogenerated carriers do not last indefinitely. There is an associated lifetime that is a measure of the average time it takes for the carriers to recombine. The photocurrent generated in a photoconductor illuminated by an optical signal field of average power P_{rcvd} is influenced by both the carrier lifetime and the time the carriers require to make their way through the device. The photocurrent is given by [4, 9]

$$i_{photo}(t) = \frac{\eta q}{h\nu}\left(\frac{\tau_{carrier}}{\tau_{transit}}\right)P_{rcvd}(t) + i_{dark}, \qquad (4.6)$$

where

$\dfrac{\eta q}{h\nu} = R$ = responsivity of the semiconductor material,

$\tau_{carrier}$ = mean carrier lifetime,

$\tau_{transit}$ = transit time between the electrical contacts,

i_{dark} = dark current.

The dark-current flows regardless of the amount of illumination present and is a form of leakage current. Dark-current arises in essentially all photodetectors operating at room temperature because of random thermally generated electron-hole pairs and is typically a small contributor to the overall photocurrent. In a photoconductive detector, dark-current corresponds to the residual resistance of the photoconductor when it is not illuminated.

In Eq. 4.6 we can define a photoconductor gain term G given by

$$G = \frac{\tau_{carrier}}{\tau_{transit}}. \qquad (4.7)$$

The gain will be greater than one if the mean carrier lifetime exceeds the transit time in the device. Carrier lifetime is a strong function of the semiconductor materials used in the photoconductor. It is possible for the photoconductor gain to be much greater than one if the carrier's lifetime can be made to significantly exceed the transit time. This corresponds to a photogenerated carrier looping through the circuit many times before recombining. Each time a photocarrier makes a complete trip through the circuit it contributes a charge of $q = 1.6 \times 10^{-19}$ coulombs to the current flow in the circuit.

An example of the history of a single electron-hole pair in a semiconductor of length l_a is illustrated in Fig. 4.6. At the top of the figure an incident photon has been absorbed at time $t = 0$, resulting in the creation of an electron-hole pair. The electron and hole drift under the influence of the applied electric field, with drift velocities v_e for the electron and v_h for the hole. A carrier's velocity is determined by how mobile the carrier is in the specific semiconductor and the magnitude of the applied field. The carrier drift velocities are given by

$$v_e = \mu_e E \quad \text{and} \quad v_h = \mu_h E, \qquad (4.8)$$

where μ_e = mobility of electrons and μ_h = mobility of holes.

In many semiconductors, the mobility of an electron is greater than that of a hole. Figure 4.6 uses an example of $v_e = 4v_h$, which means that the electron velocity is also four times the hole velocity.

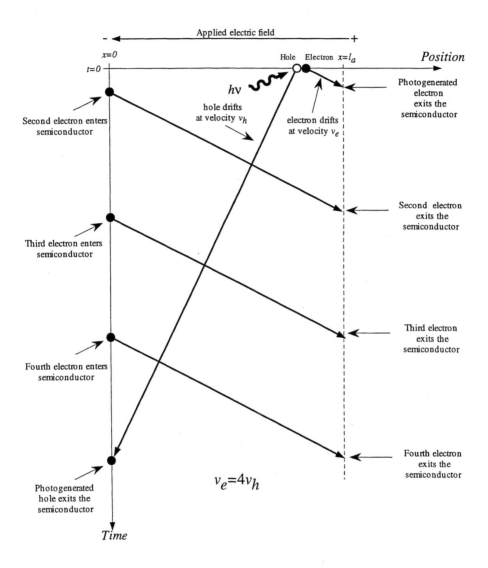

Figure 4.6 Carriers in a photoconductor.

The photogenerated electron drifts under the influence of the applied field and ultimately exits the semiconductor. The semiconductor then acquires a net positive charge because the photogenerated hole still exists. In order to maintain bulk charge neutrality, a second electron enters and begins drifting. The second electron also exits before the photogenerated hole exits and consequently a third electron enters to maintain neutrality. The process continues until either the hole recombines with an electron or the hole exits the device.

Eventually the photogenerated hole does exit the photoconductor at the negative contact and the process terminates. Note that four electrons traveled through the photoconductor during the time required for the hole to reach the device terminal and that consequently we may say that the photoconductor has a gain of four. Note that the *same* photocarrier does not live $\tau_{carrier}$ seconds and make multiple trips. Carrier lifetime actually refers to how long the electron-hole pair exists. The gain is then just the equivalent number of photocarriers generated per photon absorbed.

In the example illustrated in Fig. 4.6, the velocity of a hole is one-fourth that of an electron. The carrier lifetime used in Eq. 4.6 and Eq. 4.7 is therefore determined by the time that would be required for the hole to exit the device. The worst case occurs when a hole is created at the edge of the absorbing region farthest from the negative contact of the device. The time required is

$$\tau_{carrier} = \frac{l_a}{v_h}, \tag{4.9}$$

where l_a is the length of the photoconductor's absorbing region.

The transit-time is determined by the time required for an electron to travel through the absorbing region and is given by

$$\tau_{transit} = \frac{l_a}{v_e}. \tag{4.10}$$

Equation 4.8 indicates that a carrier's velocity is a linear function of electric field. This is true in most semiconductors for electric fields of low or moderate field-strengths. At high field-strengths, in excess of approximately 10^5 V/cm, a saturation velocity between $6{\times}10^6$ cm/s and 10^7 cm/s is observed. The application of even higher strength fields will not increase a carrier's velocity. In most semiconductors the saturation velocities for both holes and electrons are similar, with the hole generally being slightly slower, and a single value for saturation velocity v_s is used for both carriers.

In some semiconductors, such as GaAs and InP, the electron drift velocity exhibits a distinct peak for field-strengths in the vicinity of 10^4 V/cm. Electron velocities as high as $2{\times}10^7$ cm/s are obtainable [7]. Since the velocity of the faster carrier determines the transit-time in a photoconductor, high-gain photoconductors are sometimes operated at relatively low voltages.

Throughout our discussion of the example portrayed in Fig. 4.6 we have implicitly assumed that the mean hole recombination time in the semiconductor is long compared to the time required for the hole to be swept out of the photoconductor by the applied field. Recombination processes in a semiconductor are probabilistic in nature and a single quantity such as carrier lifetime represents a time-average. A large number of carriers are observed and their average lifetime becomes the definition of carrier lifetime. In reality there is an inherent statistical variation in the carrier lifetime. This means that the photoconductor gain will have an amount of randomness associated with it. This is equivalent to a noise source within the photoconductor and is known as generation-recombination (G-R) noise [4, 7, 9]. We will defer discussing the effects that G-R noise, as well as other photodetector noise processes, have on receiver performance until Chapter 5.

Carrier lifetime is not just a part of the photoconductor gain, it is also intimately connected with the bandwidth of the photoconductive detector. Since the duration of the photocurrent that is generated in response to the absorption of a photon depends on the lifetime of the longest lived carrier, any fluctuations in the received optical field faster than the carrier lifetime will not be clearly observed. This is an inherent bandwidth limitation that can be accounted for by modifying Eq. 4.6 to be

$$i_{photo}(\omega) = RG \frac{P_{rcvd}(\omega)}{\sqrt{1 + \left(\dfrac{\omega}{\omega_c}\right)^2}}, \tag{4.11}$$

where

$$\omega_c = \text{cutoff frequency} = \frac{1}{\tau_{carrier}} = \text{the bandwidth of the photoconductor,}$$

$P_{rcvd}(\omega) = $ power in the optical signal that is modulated at ω,

$\omega = $ electrical frequency,

$$G = \text{gain} = \left(\frac{\tau_{carrier}}{\tau_{transit}}\right),$$

$$R = \text{responsivity} = \frac{\eta q}{h v}.$$

Note that as the carrier lifetime increases, the bandwidth decreases. This is the opposite of what occurred with photoconductor gain, where an increase in carrier lifetime was beneficial. This allows us to draw an important conclusion. Photoconductor gain cannot be increased arbitrarily by increasing carrier lifetime: there is an inherent gain-bandwidth limitation

Given this constraint, one way to achieve high performance is to first engineer the semiconductor so that the carrier lifetime is as long as possible while still maintaining adequate bandwidth. The photoconductor gain is then increased by decreasing the transit-time as much as possible. Transit time is then minimized by keeping the path between the electrodes short.

The interdigitated electrode structure that was illustrated in Fig. 4.5(b) is a popular technique for reducing transit time while maintaining adequate carrier lifetime and responsivity. The electrodes can also be made optically transparent, further increasing responsivity. Interdigitated photoconductors based on InGaAs have been fabricated with gains of 50-100 and bandwidths of a few GHz. Silicon based photoconductors have been demonstrated with gains as high as 1000. The bandwidth in these high-gain devices is lower, typically a few MHz maximum [4, 7-9].

4.3 The *p-n* Photodiode

A semiconductor *p-n* junction can also be used as a photodetector. A *p-n* junction forms a diode, and consequently a junction used as a photodetector is frequently called a photodiode. The concepts that we applied to photoconductors also apply to photodiodes. The basic operation of a photodiode relies on the absorption of photons in a semiconductor material. The photogenerated carriers are separated by an applied electric field, and the resulting photocurrent is proportional to the incident optical power. The velocity at which the carriers move in the depletion region is related to the strength of the electric field across the region and the mobility of the carriers. Holes generally move more slowly than electrons unless the device is operating under a high electric field strength, at which point the saturation velocity of both carriers is approximately equal.

Figure 4.7 illustrates a photodiode. The *p-n* junction and an example bias circuit are shown in Fig. 4.7(a). The junction consists of a semiconductor with the doping profile shown in Fig. 4.7(b). The left side is a *p*-type semiconductor containing a surplus of holes. It has been doped with acceptor atoms to a density of N_{accept}. The right side is an *n*-type semiconductor with a surplus of electrons. It has been doped with donor atoms to a density of N_{donor}. We have assumed an abrupt junction with a greater doping of donors than acceptors, that is, $N_{donor} > N_{acceptor}$. We have also used a junction formed from a single semiconductor material such as Si, Ge, or GaAs. This is called a homojunction. A diode can also be formed between dissimilar semiconductors such as GaAs and AlGaAs. Junctions of dissimilar semiconductors are called heterojunctions.

When the junction is formed, electrons from the *n*-type material and holes from the *p*-type material immediately diffuse across the junction in an attempt to equalize the density of free-carriers throughout the semiconductor. Holes and electrons recombine as they cross the junction. The number of free-electrons in the *n*-type material decreases and the *n*-type material acquires a net positive charge that starts to repel holes flowing out of the *p*-type material. The number of free-holes in the *p*-type material is similarly reduced and the *p*-type material acquires a net negative charge that starts to repel the flow of electrons from the *n*-type material. Holes and electrons continue to flow until the charge buildup illustrated in Fig. 4.7(c) is sufficiently large to prevent any additional carriers from crossing the junction. The region in the vicinity of the junction in which this charge buildup occurs is known as the *space-charge* region. This region is also fully depleted of holes and electrons and is equivalently called the depletion region.

The energy diagram for the photodiode is shown in Fig. 4.7(d). An energy-band diagram illustrates the relationship between the top of the valence-band and the bottom of the conduction-band as a function of position within the junction. Also shown is the Fermi level that describes the energy level with a 50% probability of occupancy. The profile of the electric field in the depletion region is shown in Fig. 4.7(e). The lengths l_p and l_n are the lengths of the *p*-type and *n*-type depletion regions. The depletion region occurs between l_p and l_n and has a length $l_d = l_p + l_n$. Note that separation of charges occurs only in the space-charge region and consequently a *p-n* junction can support an electric field only within the space-charge region. The field reaches a peak value at the *p-n* interface and does not extend outside the space-charge region.

A photon that is absorbed by the semiconductor in the depletion region will cause the formation of an electron-hole pair. The hole and electron will be transported by the electric field to the edges of the depletion region. Once the carriers leave the depletion region they travel to the terminals of the device and form a photocurrent i_{photo} flowing in the external circuitry.

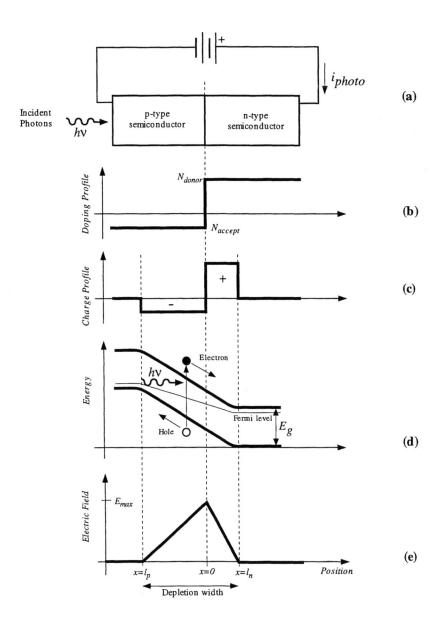

Figure 4.7 **The reverse-biased *p-n* diode photodetector. (a) Structure of an abrupt junction. (b) Doping profile. (c) Charge profile. (d) Energy-band diagram. (e) Electric field profile.**

The built-in electric field associated with the existence of the *p-n* junction is usually supplemented by the application of an external field that reverse biases the diode. The stronger the applied field, the wider the depletion region and the more rapidly any carriers generated within the depletion region will move to the edges of the region. The field cannot be increased arbitrarily. Eventually, the carrier saturation velocity is reached in the vicinity of the *p-n* interface and if the field continues to increase, the semiconductor material's breakdown voltage will eventually be reached.

The average time required for a photocarrier to cross the depletion region is called the *transit-time* and is equal to

$$\tau_t = \frac{l_d}{v_s},$$ (4.12)

where

l_d = length of depletion region,

v_s = average carrier saturation velocity.

Equation 4.12 assumes that both the photogenerated hole and electron travel at the same saturation velocity. This is a valid assumption in the high electric field region near the *p-n* junction since the saturation velocity for holes and electrons is similar in most popular semiconductors. However, as carriers travel away from the junction the electric field strength diminishes and the carrier velocity is reduced to below the saturation level. In most semiconductors, electrons move faster than holes and the disparity in charge velocity will affect the response time of the photodiode. The effect of a noticeable difference in the electron drift velocity and the hole drift velocity is illustrated in Fig. 4.8.

In the left half of Fig. 4.8(a) the received photon is absorbed just at the most negative end of a depletion region that is l_d long. The photogenerated electron-hole pair begins to drift apart because of the applied electric field, and since it is very near the negative end, the hole immediately exits the region. The space charge barrier prevents another hole from entering into the depletion region and consequently the multiple carrier passages that were observed in a photoconductor do not occur in a photodiode. The electron travels across the depletion region at a velocity v_e and reaches the positive end of the depletion region at time $t = l_d/v_e$. This generates an impulse of photocurrent as shown in the right half of Fig 4.8(a). The impulse lasts for as long as the electron was in motion and has an area q that is equal to the charge of the electron.

In Fig. 4.8(b) the received photon is absorbed just at the most positive end of the depletion region. The photogenerated electron-hole pair again begins drifting and this time the electron immediately exits the region. The holes now travel across the depletion region, traveling at velocity v_h and reaching the other end after $t = l_d/v_h$ seconds. We have assumed that $v_e > v_h$ and consequently the current pulse that is generated by the hole is longer in duration that the current pulse generated by the electron.

In Fig. 4.8(c) the received photon is absorbed in the middle of the depletion region. The photogenerated electron-hole pair drifts apart, with the electron traveling at velocity v_e and the hole traveling at velocity v_h. The resulting photocurrent pulse shape is shown in the right hand portion of Fig. 4.8(c). The photocurrent increases rapidly, based on the speed of the fastest carriers, but the photon induced event lasts for a duration governed by the speed of the slowest carrier. Thus the slowest carrier generally determines a photodiode's response time and consequently the bandwidth.

Figure 4.9 illustrates the case where a pulse of N photons is exponentially absorbed in a semiconductor according to Eq. 4.4. The photocurrent pulse now exhibits an exponential decay corresponding to more of the photons being absorbed near the negative end of the depletion region, thus minimizing the hole transit-time. The duration of the pulse is still determined by the slower holes traveling the entire depletion length.

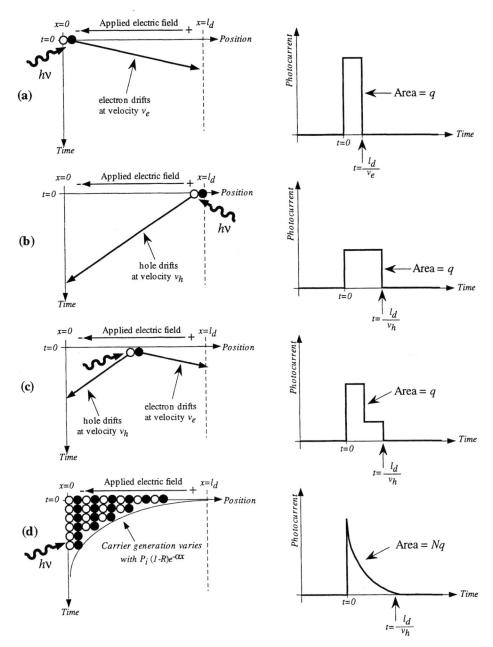

Figure 4.8 Photodiode responses when the hole and electron drift velocities differ. (a) A single photon is absorbed and carriers are generated near the negative end of the absorption region. (b) A single photon is absorbed near the positive end. (c) A single photon is absorbed in the middle of absorbing region. (d) A pulse of light containing N photons is absorbed in the photodiode. The carriers are generated in proportion to the exponentially decreasing optical signal field.

It is also possible for photons to be absorbed outside the depletion region. Carriers in a semiconductor can either randomly diffuse or they can drift under the influence of an applied electric field. Carriers generated outside the depletion region do not benefit from the applied electric field and consequently they move at their diffusion velocity instead of their drift velocity. Carrier diffusion-time is typically much longer than carrier transit-time. It can take several nanoseconds for a carrier to diffuse a micron. The same trip can require only a few tens of picoseconds (or less) of transit-time in a fully depleted diode with a high electric field. Since the bandwidth of a photodiode is inversely related to the time required for the photogenerated carriers to reach the external circuit, if a wide bandwidth is to be obtained, as many photons as possible should be absorbed in the depletion region. Carrier diffusion effects should be minimized.

A carrier's diffusion length is defined as the average distance the carrier will diffuse in a non-depleted semiconductor before it recombines. If a photon is absorbed more than one diffusion length away from the depletion region, there is a high probability that the resulting carriers will recombine before they reach the depletion region. These photons, even though absorbed in the photodetector, do not contribute to the photocurrent i_{photo}. Only photons absorbed *within* one diffusion-length of the depletion region contribute to the photocurrent.

If the photon is absorbed in the p-type material within an electron's diffusion length of the depletion region, the probability that the electron will diffuse to the depletion region is high. Once it reaches the depletion region it is rapidly transported by the electric field to the opposite end of the diode. A similar situation occurs for a photon absorbed within one diffusion length of the depletion region in the n-type, the difference being that the hole must now diffuse to the depletion region before being transported by the electric field.

The ratio of carriers generated in the depletion-region to those generated outside the depletion region is strongly dependent on the device dimensions, construction, and materials used. In some materials and photodiode structures, it is possible for a significant portion of the total photocurrent to come from diffusion related carriers. Such photodetectors can have a high quantum efficiency when measured with a constant amplitude optical signal field and an apparently wide bandwidth for rapidly modulated signals. There is however a noticeable difference in quantum efficiency, depending on the modulation frequency of the optical signal being detected [10]. Near DC the quantum efficiency may be high. At higher frequencies, the quantum efficiency can fall-off dramatically. This frequency dependence and the corresponding effect in the time-domain are illustrated in Fig. 4.9.

Figure 4.9(a) illustrates an idealized optical pulse that is received by a photodetector. An ideal photodiode would generate a pulse of photocurrent that would exactly match the shape of the incident optical pulse. A photodetector with a significant diffusion component will have a response similar to the one illustrated in Fig. 4.9(b). The leading edge of the pulse will exhibit some rounding-off and the trailing edge will have a noticeable "tail," corresponding to relatively long carrier diffusion times. In some materials where the diffusion length is long, this tail can persist for a comparatively long time after the optical pulse is actually received.

The effect of a diffusion component on the frequency response of the photodetector is illustrated in Fig. 4.9(c). At low frequencies, the quantum efficiency is essentially the same as the DC quantum efficiency that is obtained when the photodiode is subjected to a constant power optical signal.

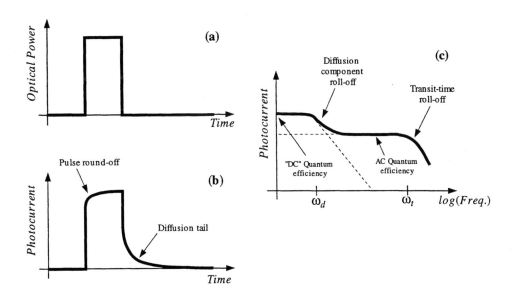

Figure 4.9   Effects of a carrier diffusion component in the photocurrent.
(a) Received optical signal. (b) Resulting photocurrent when diffusion
is significant. (c) Corresponding frequency response.

If we modulate the power of the received optical signal at a relatively high frequency (hundreds of MHz), a decrease in photocurrent will be observed. This corresponds to a reduction in the photodiode's ability to respond to fluctuations faster than the diffusion time of the carriers. The magnitude of the photocurrent in a photodiode with a significant diffusion component can be represented as the combination of a diffusion related term and a drift related term as

$$i_{photo}(\omega) = \frac{(\eta_{DC} - \eta_{AC})q}{hv} \frac{P_{rcvd}(\omega)}{\sqrt{1 + \left(\dfrac{\omega}{\omega_d}\right)^2}} + \frac{\eta_{AC}q}{hv} P_{rcvd}(\omega), \qquad (4.13)$$

where
$$\omega_d = \frac{1}{\tau_d},$$

τ_d = carrier diffusion time,

η_{DC} = DC quantum efficiency,

η_{AC} = AC quantum efficiency,

and ω_d is the frequency at which the photocurrent will have dropped to $1/\sqrt{2}$ of its DC value. This is the half-power point and is equivalent to the bandwidth of the diffusion component [10].

At frequencies higher than the diffusion roll-off there is usually a wide-plateau in the response that corresponds to the usable AC quantum efficiency. This plateau will often extend for at least a decade or so in frequency (e.g. 100 MHz to 1 GHz is one decade in frequency) before another roll-off is noticed. At very high frequencies the transit-time of carriers crossing the depletion region begins to dominate the response time and transit-time effects will cause another roll-off in the photodiode's response. The roll-off follows a $\sin(x)/x$ response curve [7]. The frequency corresponding to a $\sqrt{2}$ decrease in photocurrent due to transit-time effects is given by

$$\omega_t = \frac{2.78}{\tau_t},$$ (4.14)

where τ_t = carrier transit time.

Carrier lifetime and transit time are not the only contributors to finite bandwidth in the *p-n* junction. The separation of charge in the depletion region that was illustrated in Fig. 4.7(c) is equivalent to a parallel plate capacitor across the reverse-biased junction. The area of the capacitor "plates" corresponds to the area of the depletion region, while the separation between the plates is essentially the depletion region length l_d. The equivalent junction capacitance is then

$$C_j = \frac{\varepsilon_o \varepsilon_r A}{l_d},$$ (4.15)

where

$\varepsilon_o = 8.85 \times 10^{-12}$ farads/meter = permittivity in vacuum,

ε_r = relative permittivity of the semiconductor,

A = area of the depletion region,

l_d = depletion region length.

For example, if the photodiode was fabricated using GaAs with $\varepsilon_r = 13.1$, an area of 1000 square-microns, and a 1 micron long depletion region, it would have a junction capacitance of 160 fF (0.160 pF). Since this is a capacitor in parallel with the junction, it shunts current away from the external circuit and is particularly noticeable at high frequencies. Fortunately, the photodiode is essentially a current source, not a voltage source, and the effects of the capacitance can be mitigated by operating the diode with as low a value load resistor as possible, ideally a short-circuit. The minimum resistance is ultimately determined by the internal series resistance of the *p-n* junction. An internal series resistance arises because of the small but finite contact and bulk series resistance in the semiconductor material. The frequency at which there is a $\sqrt{2}$ decrease in photocurrent due to the *RC* time-constant is

$$\omega_{RC} = \frac{1}{R_s C_j},$$ (4.16)

where

R_s = junction series resistance, C_j = junction capacitance.

Note that the junction capacitance will decrease as the depletion region is made longer, and the associated bandwidth will increase. This is the opposite of what occurred with transit time. In the transit time case, the longer the depletion region, the longer the transit-time and the lower the bandwidth. There is consequently a trade-off between transit-time and junction capacitance. Since the photodiode's quantum efficiency was also improved by lengthening the depletion region, there is a similar trade-off between quantum efficiency and transit-time. Fortunately, a well-designed photodiode can have both a high quantum efficiency and a wide-bandwidth.

A photodiode is not only a photodetector, it is also a diode. Diodes can be operated in an unbiased mode or they can be operated under either a forward or a reverse bias. These three different bias schemes correspond to the three modes in which a photodiode can be operated when connected to an external circuit. Figure 4.10 illustrates the three modes. They are commonly referred to as open-circuit "photovoltaic" operation, short-circuit "photoconductive" operation, and reverse-biased "photoconductive" operation.

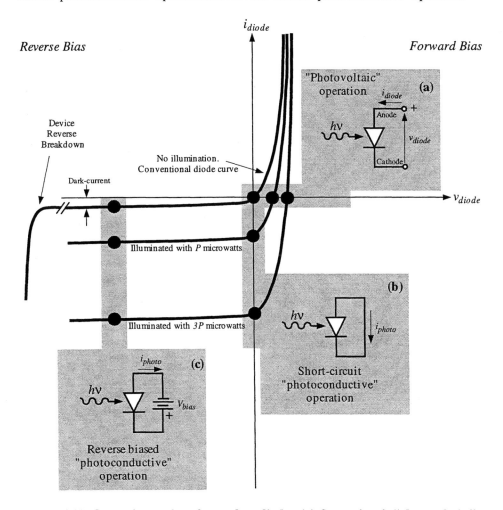

Figure 4.10 Operating regions for a photodiode. (a) Open-circuit "photovoltaic." (b) Short-circuit "photoconductive." (c) Reverse-biased "photoconductive."

Regardless of the mode of operation, the current and voltage observed at the terminals of a photodiode are due to the combination of a light induced photocurrent and the conventional DC current-voltage, or I-V, characteristic for a semiconductor diode. The combination can be written as

$$i_{diode} = i_s \left[\exp\left(\frac{q v_{diode}}{kT} \right) - 1 \right] + i_{leakage} - i_{photo}, \qquad (4.17)$$

where

i_s = reverse saturation current,

$i_{leakage}$ = diode leakage current,

v_{diode} = voltage across the photodiode,

i_{photo} = photocurrent = $\dfrac{\eta q}{h v} P_{rcvd}$,

q = charge of an electron,

k = Boltzmann's constant,

T = absolute temperature in degrees kelvin,

and we have defined a positive i_{diode} to be a current flowing *into* the anode of the photodiode, as shown in Fig. 4.10(a).

The dark-current i_{dark} that flows whether the diode is illuminated or not is given by

$$i_{dark} = i_s \left[\exp\left(\frac{q v_{diode}}{kT} \right) - 1 \right] + i_{leakage}. \qquad (4.18)$$

The reverse saturation current i_s corresponds to the reverse current that flows because of thermally excited carriers and is present in some form in all diodes. The leakage current $i_{leakage}$ has components due to generation-recombination in the depletion region and tunneling between the conduction and valence bands [7]. Dark-current generally increases with applied voltage and temperature. In many materials intended for use at wavelengths of several microns, i_{dark} can be significant and must be reduced by physically cooling the device. Some long wavelength detectors, particularly those used for imaging applications, are cooled to 77°K to help minimize dark-current. In photodiodes used in communication applications at wavelengths under 1550 nm, room-temperature dark-currents less than a few tens of nA are achievable.

The diode characteristic curves drawn in Fig. 4.10 are for three different values of illumination. When there is no illumination a conventional diode curve passing through $v_{diode} = 0$, $i_{diode} = 0$ is obtained. When the photodiode is illuminated with a received optical signal with an average power of P microwatts, the diode curve is offset downward by an amount equal to the resulting average DC photocurrent. With an average received power of $3P$ microwatts, the curve is further offset to correspond to the increased photocurrent.

The open-circuit photovoltaic mode illustrated in Fig. 4.10(a) is commonly used in *p-n* junction solar cells [11]. The received optical signal generates electron-hole pairs in the depletion region. These free carriers separate and recombine with opposite charge carriers, causing an increase in the total number of charges in the junction and a corresponding increase in the diode voltage v_{diode}. Rearranging Eq. 4.17 to solve for the diode voltage with the constraint that the diode current must be zero yields

$$v_{diode} = \frac{kT}{q} \ln\left[\frac{i_{photo} + i_s}{i_s}\right]. \tag{4.19}$$

Since an increase in diode voltage is the primary response of a photovoltaic photodiode when it is illuminated, the responsivity of a photovoltaic-mode photodiode is typically measured in volts per watt instead of the commonly used amps per watt. For the three illuminations illustrated in Fig. 4.10, three operating points are shown. They occur along the horizontal v_{diode} axis and correspond to the three values of voltage obtained across the photodiode for the respective cases of no illumination, P microwatts of illumination, and $3P$ microwatts of illumination. When the solar cell is loaded by a resistance and power is extracted from the received optical signal, a photocurrent also flows and the operating points move up into the forward biased region.

The short-circuit photoconductive mode is illustrated in Fig. 4.10(b). In this mode the photodiode is operated as a current source where the short-circuit current corresponds to the photocurrent and $v_{diode} = 0$. For the three illuminations illustrated in Fig. 4.10 three operating points are again shown. Now they occur along the vertical i_{diode} axis and correspond to the three values of current obtained through the photodiode for the respective cases of no illumination, P microwatts of illumination, and $3P$ microwatts of illumination.

The bandwidths obtained in the open-circuit photovoltaic and short-circuit photoconductive modes are relatively low because the electric field across the depletion region is small and an additional diffusion capacitance occurs, further reducing the achievable bandwidth. The field can be dramatically increased and diffusion capacitance effects avoided by operating the photodiode with a reverse bias, as illustrated in Fig. 4.10(c). This is termed the reverse-biased photoconductive mode. In this mode the photodiode is operated as a current source, the diode current corresponds to the photocurrent, and $v_{diode} = -V_{bias}$. Three operating points, corresponding to the three levels of received optical signal power, are again shown. They now occur along a vertical line parallel to the i_{diode} axis.

Figure 4.10(c) corresponds to the most popular operating region for a photodiode. The reverse bias creates a strong electric field across the depletion region, which has three effects. 1) The high field strength rapidly sweeps photogenerated carriers out of the depletion region, leading to a short transit-time. 2) The field tends to extend the depletion region in which photons can be absorbed, leading to a higher quantum efficiency. 3) The longer depletion region decreases the junction capacitance.

A photodiode is commonly referred to as having a photoconductive mode since in the no illumination case it appears to have a high resistance and the resistance appears to decrease as the power in the received optical signal increases, because photocurrent begins flowing. There is a strong possibility of confusion when describing photodiodes in this way since the term "photoconductive" has now been used to describe the operation of a photodiode in addition to the operation of the "true" photoconductor described in Section 4.2. This terminology, although widely used, is not strictly correct since it is not v_d/i_d that determines the photodiode resistance. Instead, it is the differential resistance, defined as $R_{dif} = \Delta v_d / \Delta i_d$, that is the true definition of a photodiode's resistance. A reverse-biased photodiode's differential resistance is very large and is essentially unaffected by the optical signal strength. The differential resistance for a photoconductor, $R_{dif} = \Delta v_r / \Delta i_r$, does change in direct proportion to the optical signal strength. This difference between the differential resistance of a photodiode operating in a reverse-biased "photoconductive" mode and a true photoconductor is illustrated in Fig. 4.11.

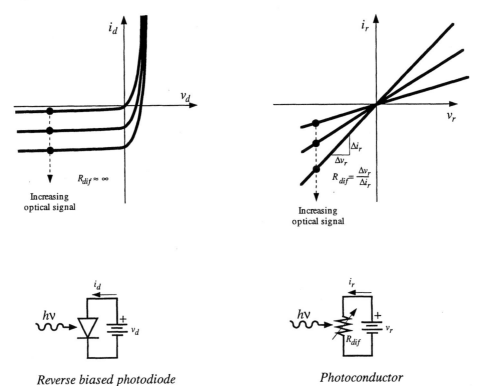

Figure 4.11 I-V characteristics of a photodiode and a photoconductor.

There are other ways to become confused when discussing photodiode operation. Some authors use the term "photovoltaic" to describe any photodiode under any bias condition, reserving the term "photoconductive" only for photoconductors. Others use the terms "photodiode," "photovoltaic," and "photoconductive" to describe different photodiode bias conditions. In this text, the term *photoconductive detector* refers to a semiconductor material used as a photodetector. The term *photoconductive mode* refers to a semiconductor photodiode operated so that we observe the current through the photodiode. The term *photovoltaic mode* refers to a semiconductor photodiode operated so that we observe the voltage across the photodiode.

4.4 Metal-Semiconductor Photodiodes

A variation of the semiconductor *p-n* photodiode is the metal-semiconductor or Schottky-barrier photodiode shown in Fig. 4.12(a). In this type of photodiode a thin semitransparent metal film is used to form one-half of the *p-n* junction. An electric field, known as the barrier-field, is developed between the metal and the semiconductor. This field is used to separate the photogenerated charges just as the space-charge region in a semiconductor *p-n* diode did.

Figure 4.12(b) illustrates the energy-band diagram for a metal-semiconductor photodiode. A photon with energy greater than the barrier energy E_b will be absorbed, causing the formation of electron-hole pairs. The electron-hole pairs are then separated by the barrier field. They ultimately move to the metal contact and into the semiconductor, forming a photocurrent.

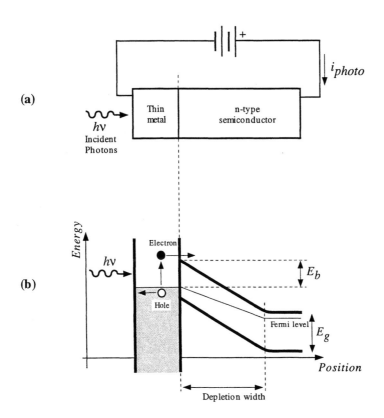

Figure 4.12 Metal-semiconductor photodiode. a) Structure of the photodiode b) Energy-band diagram.

Metal-semiconductor photodiodes can have extremely fast response times. There are examples of Schottky-barrier photodiodes with bandwidths in excess of 100 GHz [12]. Another form of the metal-semiconductor photodiode is the metal-semiconductor-metal or MSM photodiode illustrated in Fig. 4.13. This structure is physically similar to the interdigitated photoconductor that was illustrated in Fig. 4.5(b) except that the metal-semiconductor and semiconductor-metal junctions are fabricated as Schottky barriers instead of ohmic contacts. Being a planar structure, the MSM photodiode lends itself to monolithic integration and can be fabricated using processing steps nearly identical to those required for making field-effect transistors. There have been demonstrations of ultra-wideband MSM photodiodes [13] and complete wide bandwidth monolithic receiver front-ends have been reported [14, 15].

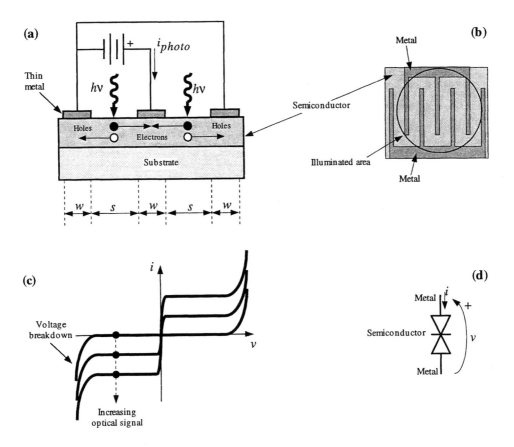

Figure 4.13 Metal-semiconductor-metal photodiode. (a) Cross-section structure of the photodiode. (b) Top view. (c) I-V curves. (d) Schematic symbol.

Figure 4.13(a) illustrates a cross-sectional view of an MSM photodiode. A top view showing the electrode structure is shown in Fig. 4.13(b). An MSM photodiode consists of interdigitated electrodes on a semiconductor substrate. In Figure 4.13(a), the electrode fingers are w wide and separated by a distance s.

Absorbed photons generate electron-hole pairs in the semiconductor. The holes drift with the applied electric field to the negative contacts while electrons drift to the positive contacts. The quantum efficiency of the MSM photodiode is dependent on the shadowing caused by the metal electrodes and is given by

$$\eta = (1-R)\frac{s}{s+w}\left(1-e^{-\alpha l_a}\right), \tag{4.20}$$

where

l_a = length of MSM absorbing region,

R = reflectivity of the semiconductor,

α = absorption coefficient of the semiconductor.

High quantum efficiency requires a high ratio between the area of the semiconductor and the area covered by metal, or $s \gg w$. It is also possible to increase QE by using transparent metal contacts or by back illuminating the device through the substrate [16].

A good approximation to the MSM structure is an infinite series of alternating, infinitely thin parallel electrodes. The capacitance associated with this type of structure has been determined but requires the evaluation of elliptic integrals [17]. For many practical cases the capacitance can be approximated by [18]

$$C \cong (N-1)L\varepsilon_o(1+\varepsilon_r)\frac{\pi}{2}\ln\left(2\sqrt{\frac{1+k'}{1-k'}}\right), \tag{4.21}$$

where

N = number of fingers in the electrodes,

L = length of a single finger,

$k' = 1-k^2$, $\quad k = \tan^2\left(\frac{\pi}{4}\frac{s}{s+w}\right)$.

Typical MSM structures are able to achieve capacitance values of a hundred fF or less (i.e. < 0.1pF) [18, 19].

The transit-time limited bandwidth of an MSM photodiode is complex to evaluate because the variation in the electric field with depth causes carriers produced deep in the semiconductor to have longer transit-times than those produced near the surface [19]. In general transit-times of 10-20 ps and bandwidths on the order of 10-20 GHz are relatively easy to obtain [15, 18, 19].

Figure 4.13(c) illustrates the I-V characteristics for an MSM photodiode for three levels of illumination. The photodiode is essentially a pair of Schottky diodes connected back-to-back. The schematic symbol for an MSM photodiode is illustrated in Fig. 4.13(d). One diode is forward biased while the other is reverse biased. The diode can generally be used with either a positive or a negative bias [20], the only difference being that the polarity of the signal will be reversed. At low bias voltages there is a pronounced "knee" in the curve. This occurs as one of the Schottky diodes becomes forward biased. Typically, voltages of 5 to 10 volts are applied across the MSM photodiode to ensure that one diode is fully forward biased while the other is fully depleted. At normal operating voltages the I-V characteristic is reasonably flat and the photocurrent is a function of only the power in the received optical signal. A gain, similar to a photoconductive gain, is sometimes observed in MSM photodiodes. This is generally considered undesirable since it introduces additional noise and may constrain the achievable device bandwidth [15, 19].

4.5 The *p-i-n* Photodiode

Another variation of the semiconductor *p-n* photodiode is the *p-i-n* photodiode shown in Fig. 4.14. A *p-i-n* photodiode is a *p-n* photodiode with a layer of intrinsic (or lightly doped) semiconductor sandwiched between the *p* and *n* layers. The photodiode structure and an associated bias circuit are illustrated in Fig. 4.14(a).

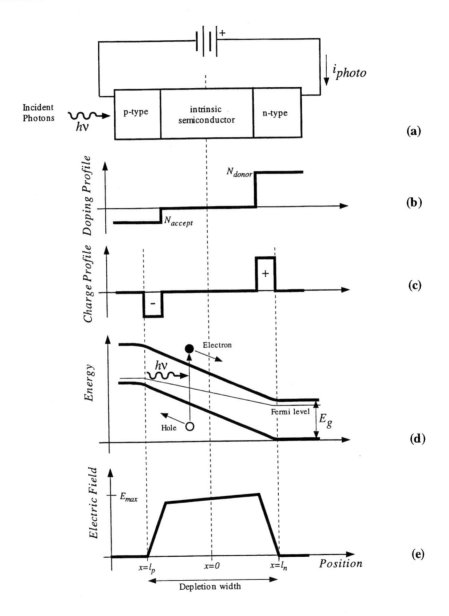

Figure 4.14 Example *p-i-n* photodiode. (a) Diode structure. (b) Doping profile. (c) Charge profile. (d) Energy-band diagram. (e) Electric field profile.

The photodiode consists of three sections of semiconductor, with the doping profile shown in Fig. 4.14(b). The left side is a p-type semiconductor containing a surplus of holes. It has been doped with acceptor atoms to a density of N_{accept}. The right side is an n-type semiconductor with a surplus of electrons. It has been doped with donor atoms to a density of N_{donor}. The middle region is either an intrinsic semiconductor or a semiconductor doped just lightly enough so that its characteristics are well controlled.

The buildup of charge is illustrated in Fig. 4.14(c). The p region develops a net negative charge while the n region develops a net positive charge. The intrinsic region has a small number of free-carriers and is easily depleted of any charge. The point at which the electric field completely depletes the i region is known as the punch-through voltage. The energy diagram for the p-i-n photodiode is shown in Fig. 4.14(d), while the electric field profile corresponding to the space-charge region is shown in Fig. 4.14(e).

Note that the depletion region in a p-i-n photodiode is almost entirely contained in the intrinsic region and that it is longer than the depletion region that was obtained in the p-n photodiode. This increase in the length of the depletion region will tend to increase the photodetector's quantum efficiency. The increase in length also corresponds to a longer distance between the "plates" of the junction capacitance and consequently, the junction capacitance is reduced in a p-i-n photodiode compared to that obtained in a p-n photodiode.

The depletion region occurs between l_p and l_n and has a total length $l_d = l_p + l_n$. Incident photons of sufficient energy cause the formation of electron-hole pairs within the depletion region. The holes and electrons are then transported by the electric field, forming a photocurrent i_{photo} flowing in the external circuitry. The separation of charge in a p-i-n photodiode occurs at the ends of the intrinsic region and so the electric field is nearly constant throughout the depletion region for applied voltages greater than the voltage known as the punch-through voltage. This keeps the photogenerated electrons and holes moving at a high velocity and helps reduce the overall transit-time of the photodiode. Even though the depletion region in a p-i-n photodiode is longer than that obtained in a p-n photodiode, the carriers will be traveling at or near their saturation velocity virtually the entire time they are in the depletion region. This is an improvement over the p-n photodiode case, where the electric field, and consequently the carrier velocity, peaked at the p-n interface and then rapidly diminished. This generally makes the transit-time of the p-i-n photodiode shorter than that obtained in a p-n photodiode even though the depletion region is longer than in the p-n photodiode case.

Since the absorption length is primarily defined by the intrinsic region, the semiconductor layer that the light first penetrates need only be thick enough to provide a suitable ohmic contact to the external circuit. This significantly reduces the contribution of any carrier diffusion effects, and in a well constructed p-i-n structure nearly all the photons that generate electron-hole pairs will be absorbed in the intrinsic region, where the carriers are swept out at a high-speed by the applied electric field.

The performance of a p-i-n photodiode is governed by essentially the same equations that were developed for the photoconductor and the p-n photodiode. The photodetector quantum efficiency will still be determined by the reflection of the semiconductor surface, the intrinsic semiconductor's absorption length, and the length of the depletion region. There will be a transit-time associated with carriers traversing the depletion region and the junction capacitance will still be governed by the relative permittivity of the semiconductor material and the area and length of the depletion region.

Since transit-time, capacitance, and quantum efficiency are interdependent, there will be trade-offs involved in the design of a wide bandwidth *p-i-n* photodiode. As the depletion region length is increased, the quantum efficiency increases, the junction capacitance decreases, which increases bandwidth, and the transit-time increases, which decreases bandwidth. An optimization procedure must be used to simultaneously obtain both a wide bandwidth and a high quantum efficiency.

The two principal bandwidth limits are the bandwidth associated with the carrier transit-time and the bandwidth associated with the RC time-constant. We can obtain an estimate of the overall bandwidth for the photodiode, B, by root-sum-squaring the two individual bandwidths as

$$B = \frac{1}{\sqrt{\left(\frac{1}{f_{RC}}\right)^2 + \left(\frac{1}{f_t}\right)^2}} \quad \text{(Hz)}, \tag{4.22}$$

where

$$f_{RC} = \frac{\omega_{RC}}{2\pi} \quad \text{and} \quad f_t = \frac{\omega_t}{2\pi}.$$

If we model the *p-i-n* photodiode using the simple cylindrical volume structure illustrated in Fig. 4.15(a), we can obtain expressions for the bandwidth and quantum efficiency as a function of the device's photosensitive area and depletion length.

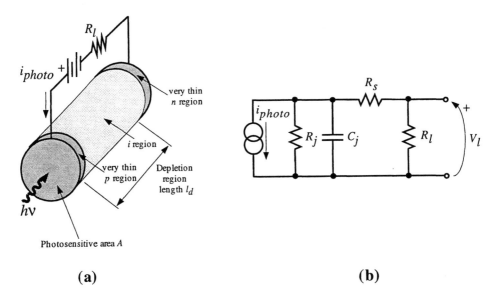

(a)   **(b)**

Figure 4.15 Simple *p-i-n* model. (a) Physical structure. (b) Equivalent circuit.

This simplified model assumes that the *p* and *n* regions are thin and optically transparent. All photon absorption occurs in the *i* region and any photons not absorbed exit through the thin *n* region at the rear of the photodiode and are lost. There are also photodiode designs in which the n region is actually made highly reflecting so that the optical signal makes two trips through the absorbing region [7]. This tends to increase the quantum efficiency at the expense of a decrease in optical bandwidth and an increase in temperature dependence.

Carriers that are created in the i region are transported by the applied electric field at the saturation velocity v_s. As soon as they reach the n and p contact regions they contribute to the observed photocurrent. The equivalent electrical circuit for our simple *p-i-n* photodiode model is shown in Fig. 4.15(b). The generation of photocurrent in a depletion region is equivalent to a current source. There is a parallel junction resistance R_j that is equivalent to the differential resistance of the diode. For a high-quality diode operating in a reverse-bias mode, this resistance is typically greater than 10^6 Ω and can usually be ignored. The junction capacitance C_j arises from the separation of charge in the depletion region and is given by Eq. 4.15. The series resistance R_s is due to bulk and contact resistance and is usually only a few ohms. The load resistor R_l is included to model the loading effects of any following electronic circuit and to provide a convenient way of monitoring the photocurrent through the voltage v_l.

We can use Eq. 4.5 to determine the quantum efficiency and Eq. 4.15 and Eq. 4.16 to determine the bandwidth associated with the RC time-constant. The exact relationship between the transit-time in a *p-i-n* photodiode and the corresponding bandwidth is subject to many subtleties. The semiconductor doping levels, the exact locations in the depletion region at which electron-holes are generated, and the differences in the velocities of the holes and the electrons all influence the actual bandwidth that can be achieved [16]. Equation 4.14 represents a good approximation for *p-i-n* photodiodes where the depletion region is relatively thin, the light enters through the p region first, and the absorption decays away exponentially [7]. Substituting Eqs. 4.5, 4.14, 4.15, and 4.16, into Eq. 4.22 yields an expression for the overall photodiode bandwidth:

$$B = \frac{1}{\sqrt{\left(2\pi R_s \varepsilon_o \varepsilon_r \frac{A}{l_d}\right)^2 + \left(\frac{l}{0.44 v_s}\right)^2}} \quad \text{(Hz)}. \tag{4.23}$$

Let us assume we are utilizing a *p-i-n* photodiode with a photosensitive area A of 100 microns^2. The semiconductor used to fabricate the *p-i-n* has an absorption length α of 1 micron for the wavelength of light we are interested in receiving. The carrier saturation velocity v_s is 10^7 cm/s and the relative permittivity ε_r is 12. The photodiode series resistance R_s is 5 Ω and the photodiode is placed in a circuit with an equivalent load resistance R_l of 20 Ω. The total resistance that will contribute to the RC time-constant is 20 Ω in series with 5 Ω or 25 Ω.

We will also make the assumption that a high quality antireflection coating has been applied to the semiconductor so that the reflection loss at the surface is insignificant. The quantum efficiency is then given by Eq. 4.5 with $x = l_d$ and $R = 0$, or

$$\eta = (1 - e^{-\alpha l_d}). \tag{4.24}$$

Figure 4.16 illustrates the bandwidth of Eq. 4.23 and quantum efficiency of Eq. 4.24 for a range of depletion region lengths. For short lengths, $l \ll l_d$, the quantum efficiency is low, the junction capacitance is large, and the RC time-constant determines the bandwidth. As the length is increased, the quantum efficiency improves, the junction capacitance decreases, and bandwidth improves.

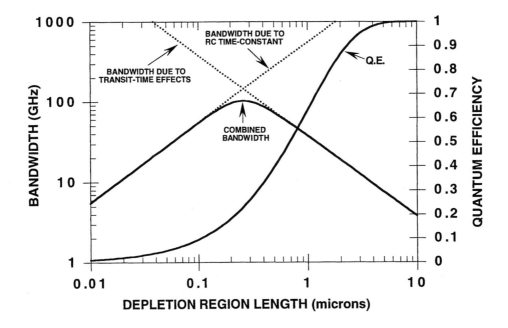

Figure 4.16 **Bandwidth and QE as a function of absorption length for a *p-i-n* photodiode with area = 100 microns^2, $v_s = 10^7$ cm/s, $R_s = 5\Omega$, $R_l = 20\Omega$, $\alpha = 1.0$ micron, $e_r = 12$.**

Eventually the bandwidth reaches a maximum value when the effects of RC time-constant and transit-time are approximately equal. For longer lengths, where $l \gg l_d$, the bandwidth decreases because of transit-time effects. Unlike bandwidth, the photodetection quantum efficiency continuously increases with increasing depletion region length.

It is possible to tailor both the photodetector area and depletion region length in order to obtain the desired performance. If we solve Eq. 4.23 for the area as a function of depletion region length we obtain

$$ A = \frac{l_d}{2\pi R_l \varepsilon_o \varepsilon_r} \sqrt{\frac{1}{B^2} - \left(\frac{l_d}{0.44 v_s}\right)^2}. \qquad (4.25) $$

Rearranging Eq. 4.24 to obtain the value of depletion region length needed to obtain a desired quantum efficiency yields the depletion region length

$$l_d = \frac{-1}{\alpha}\ln(1 - \eta).$$
(4.26)

We can determine the maximum area allowed given a desired bandwidth B and quantum efficiency η by first using Eq. 4.26 to obtain the required depletion region length and then using that length in Eq. 4.25 to solve for the area. Solving Eqs. 4.25 and 4.26 for a range of quantum efficiencies and bandwidths generates a series of constant bandwidth contours, as shown in Fig. 4.17.

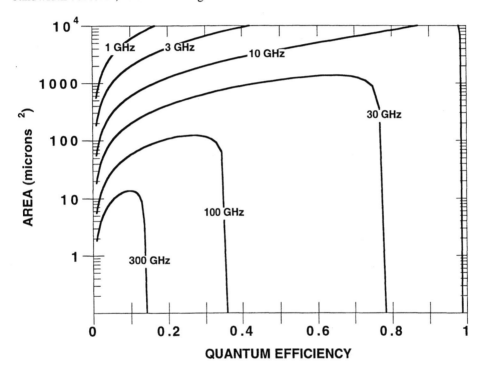

Figure 4.17 Bandwidth contours for a *p-i-n* photodiode with $v_s = 10^7\,$**cm/s,**
$R_s = 5\Omega, R_l = 20\Omega,$ $\alpha = 1.0$ **micron,** $e_r = 12.$

For the photodiode parameters used, a 10 GHz bandwidth and a greater than 0.95 quantum efficiency can be obtained for a wide range of detector areas. Note that although bandwidths of several hundred GHz are possible, a price is paid in terms of substantially reduced quantum efficiency. A 100 GHz bandwidth cannot be obtained unless we are willing to accept less than a 0.35 quantum efficiency.

The values plotted in Fig. 4.17 are for a 25 Ω total load. Receivers are frequently designed to use other impedances. The y-axis can be scaled by $25/R_{actual}$ to obtain the maximum area for other loads. Thus, if the photodiode was operated with a 50 Ω total load, the maximum area would be half that predicted in Fig. 4.17.

4.6 The Avalanche Photodiode

Another type of photodiode is known as the avalanche photodiode, or APD. Unlike conventional *p-n* or *p-i-n* photodiodes that generate a single electron-hole pair in response to the absorption of a photon, an APD can generate many electron-hole pairs from a single absorbed photon. This internal "gain" makes APDs attractive for use in constructing sensitive receivers. The structure of a typical APD is shown in Fig. 4.18(a).

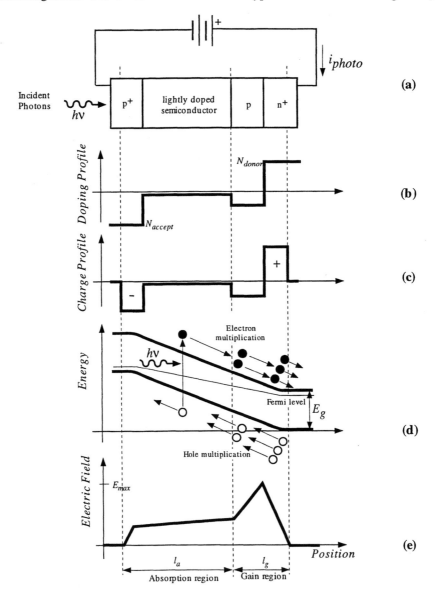

Figure 4.18 Example of an avalanche photodiode (APD). (a) Diode structure. (b) Doping profile. (c) Charge profile. (d) Energy-band diagram. (e) Electric field profile.

The APD consists of four sections of semiconductor, with the doping profile shown in Fig. 4.18(b). In our example the leftmost section is a *p*-type semiconductor that has been doped with acceptor atoms to a density of N_{accept}. The rightmost section is an *n*-type semiconductor that has been doped with donor atoms to a density of N_{donor}. The middle region is a lightly doped semiconductor in combination with a second *p*-type region. The charge density is illustrated in Fig. 4.18(c). The *p* regions develops a net negative charge while the *n* region develops a net positive charge. The energy diagram for the APD is shown in Fig. 4.18(d), while the electric field profile corresponding to the space-charge region is shown in Fig. 4.18(e).

When a *p-n* junction is subjected to a high reverse bias there are two principal breakdown mechanisms that can occur. In the first mechanism, the atoms are directly ionized by the applied field. This is known as *zener breakdown* and is commonly used to make voltage-regulating diodes. The second mechanism is called *avalanche breakdown* and is caused by high velocity carriers causing impact ionizations within the semiconductor that generate additional carriers. This is the technique used in APDs.

In the absorption region shown in Fig. 4.18(e), photons are absorbed in the lightly doped semiconductor material, causing electron-hole pairs to form. The carriers are rapidly transported by the applied electric field to the edges of the absorption region. Carriers that reach the gain region are further accelerated by an even higher electric field. They can achieve a sufficient velocity to cause their kinetic-energy to exceed the bandgap energy of the material and the carriers can actually "kick" additional electrons out of the valence band and up into the conduction band, causing the formation of additional electron-hole pairs. This is the so-called impact ionization process. These additional carriers can then undergo additional impact ionizations, causing an avalanche effect.

Just as in the *p-i-n* photodiode case we would like to keep the absorption region relatively long to maximize the quantum efficiency. Conversely, we would like to maximize the uniformity in the high field multiplication region to minimize the probability of any localized uncontrolled avalanche breakdown effects. This indicates that the multiplication region should be relatively short. These two conflicting requirements can be met in a separate absorption and multiplication (SAM) APD, in which the doping profile is varied in the device to form both a well defined absorption region and a well defined gain region. A variation on the SAM APD is the reach-through APD or RAPD [21]. This type of APD was used as the basis for the example APD of Fig. 4.18.

In effect, an APD is similar to a photomultiplier in that a single absorbed photon causes a "shower" of electrical carriers. The gain or multiplication factor of an APD is given by [7]

$$M = \frac{i_{photo}}{i_{primary}}, \qquad (4.27)$$

where

i_{photo} = observable photocurrent at APD terminals,

$i_{primary}$ = internal photocurrent before multiplication.

The primary photocurrent is obtained using the same approach as was used for *p-n* and *p-i-n* photodiodes and is given by

$$i_{primary} = \frac{\eta q}{h\nu} P_{rcvd},$$

(4.28)

where

$\eta = (1 - R)(1 - e^{-\alpha l_a})$ = photodetector quantum efficiency,
l_a = length of absorption region,
P_{rcvd} = received optical signal power.

Although multiplication factors of a few tens to a few hundreds are typically used, it is possible for the multiplication factor to be as high as 10^3 or 10^4 in specialized applications [21, 22]. Figure 4.19 illustrates carrier multiplication in the avalanche region of an APD.

Let us define ionization coefficients for electrons and holes as α_e and α_h, respectively. These are equivalent to the α ionization coefficient for electrons and the β ionization coefficient for holes that is used by some authors. The ionization coefficients correspond to the probability that the carrier will cause an ionization within a unit length. Thus, the probability that an electron will have an ionizing collision within the length l is given by $\alpha_e l$. The ionization coefficients increase with applied electric field and decrease with increasing device temperature. The increase with field is due to additional carrier velocity, while the decrease with temperature is due to an increase in non-ionizing collisions with thermally excited atoms. These additional collisions reduce the carrier's velocity and lower the probability that it will gain sufficient energy to cause impact ionization. For a given temperature, the ionization coefficients are exponentially dependent on the electric field and have a functional form of

$$\alpha_x = a \exp\left(-\left[\frac{b}{E}\right]^c\right) \quad (\text{cm}^{-1}),$$

(4.29)

where

a, b, c = experimentally determined constants,
  E = magnitude of electric field.

In general, the expression for holes will have different constants than the expression for electrons. For example, in GaAs the ionization coefficients are given by [23]

$$\alpha_e = 1.9 \times 10^5 \exp\left(-\left[\frac{5.8 \times 10^5}{E}\right]^{1.83}\right) \quad (\text{cm}^{-1}),$$

$$\alpha_h = 2.2 \times 10^5 \exp\left(-\left[\frac{6.6 \times 10^5}{E}\right]^{1.75}\right) \quad (\text{cm}^{-1}).$$

(4.30)

Measurements on InP have also been obtained [24].

An important parameter for describing APD performance is the ionization ratio

$$k = \frac{\alpha_h}{\alpha_e}.$$

(4.31)

When holes do not contribute to impact ionization, $k \ll 1$, and the case illustrated in Fig. 4.19(a) is obtained. A photogenerated electron that enters the gain region is rapidly accelerated by the applied field. The high kinetic-energy electron initiates a series of impact ionizations that eventually results in seven electrons being generated from the single photoelectron.

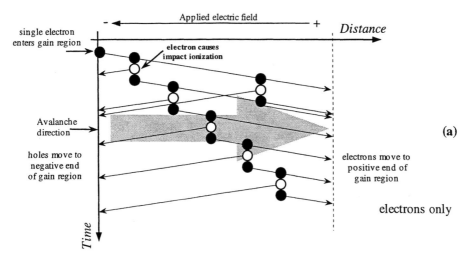

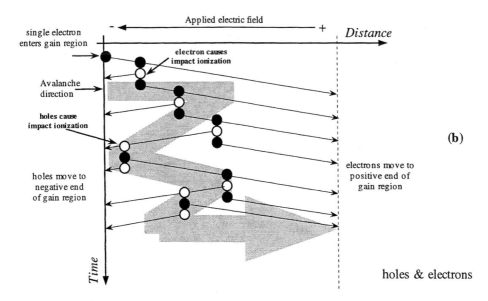

Figure 4.19 Avalanche gain via impact ionization. (a) Only electrons contribute to the impact ionization process. (b) Both holes and electrons contribute.

In Fig. 4.19(a) the avalanche proceeds in a well defined manner from left to right. There is a clearly defined beginning when the photoelectron enters the gain region, and the avalanche process ends when the additional carriers arrive at the ends of the gain region. This description would also be true for the case where only holes are contributing to impact ionization and $k >> 1$, except that things would then proceed from right to left in our figure. In order to simplify APD analysis, it is common practice for APDs in which $k > 1$ to define an effective value of k as $k_{eff} = \alpha_h/\alpha_e = 1/k$. Thus $k_{eff} = k$ in APDs in which $k < 1$. This allows k_{eff} to always be less than one and it is understood that the avalanche is initiated by whichever carrier is more ionizing.

Figure 4.19(b) illustrates the case where $k \approx 1$ and both holes and electrons contribute to impact ionization. An electron again initiates a series of impact ionizations. Sometimes they are caused by high kinetic-energy electrons and sometimes they are caused by high kinetic-energy holes. Electrons moving to the right can cause holes to form that move to the left that cause electrons to form that move to the right that cause holes to form that move to the left, etc. This is essentially a positive feedback mechanism.

The avalanche process now proceeds in a somewhat uncontrolled fashion, bouncing back and forth across the gain region depending on whether holes or electrons are causing the ionizations. There is still a clearly defined beginning when the photoelectron enters the gain region, but there is no longer a clearly defined end point unless one waits until all possible impact ionizations occur. This uncertainty in the multiplication process and the extra time involved in completely clearing all of the ionized carriers reduces APD performance and consequently the $k \approx 1$ case is undesirable. The case of $k = 1$ is not necessarily the worst case operation of an APD. The actual characteristics depend strongly on the exact location at which the carrier enters the gain region and the spatial variation of α_e and α_h within the gain region.

Fortunately, in some semiconductor materials k is inherently different from unity. For example, in Si, electrons are much more ionizing than holes and $\alpha_e >> \alpha_h$. Values of k between 0.003 and 0.01 can be obtained and Si can be used to fabricate extremely high-quality APDs for use at short wavelengths [21]. For materials based on InP and Ge, holes are somewhat more ionizing than electrons and $\alpha_h > \alpha_e$. Unfortunately, in InP as well as Ge and GaAs, k becomes very close to unity for high electric fields, and high quality APDs are difficult to fabricate.

In practice, the APD structure is optimized for whichever carrier is the more ionizing. In materials in which $\alpha_e > \alpha_h$ we would want the photons to be absorbed near the p side of the junction and would want the semiconductor to have a value of k as low as possible. In materials in which $\alpha_h > \alpha_e$ we would want the photons to be absorbed near the n side of the junction and the semiconductor should then have as large a value of k as possible.

Even in the ideal case of $k = 0$ or ∞ the exact number of impact ionizations that a carrier will experience has an amount of randomness associated with it. This randomness arises because the total number of impact ionizations depends on where the carrier was generated, the type of carrier (hole or electron), the local magnitude of the electric field, the local doping density of the semiconductor, and the path the carrier takes through the semiconductor. Consequently, the avalanche multiplication process is inherently statistical and there are unavoidable fluctuations in the exact number of carriers that reach the circuit for each photon that is absorbed. This is equivalent to a noise source within the APD and is known as multiplication noise or APD excess noise [7, 21]. Just as we did with the G-R noise that occurs in photoconductors, we will defer until Chapter 5 our discussion of the effects that multiplication noise have on receiver performance.

The expressions for hole multiplication and electron multiplication in a gain region of length l_g are [7, 21, 25]

$$M_e = \frac{1}{1 - \int_0^{l_g} \alpha_e \exp\left(-\int_0^x (\alpha_e - \alpha_h)dx'\right)dx},$$

$$M_h = \frac{\exp\left(-\int_0^{l_g} (\alpha_e - \alpha_h)dx\right)}{1 - \int_0^{l_g} \alpha_e \exp\left(-\int_0^x (\alpha_e - \alpha_h)dx'\right)dx}.$$

(4.32)

Exact solutions to Eq. 4.32 are difficult to obtain because both α_e and α_h in a practical device are functions of the position within the gain region. Least-squares fits to experimental data are frequently used. In certain cases closed form solutions are available. One is for the case where $\alpha_e \neq \alpha_h$, only electrons are being injected into the gain region, the electric field in the gain region is uniform, and the number of ionizing collision per primary carrier is large. For this case, the multiplication factor for electrons is given by

$$M_e = \frac{1-k}{\exp(-(1-k)\alpha_e l_g) - k},$$

(4.33)

where l_g = length of the gain region. For the case of $k = 1$, Eq. 4.33 reduces to

$$M_e = \frac{1}{1 - \alpha_e l_g}.$$

(4.34)

An instability will occur for $l_g = 1/\alpha_e$ and this condition must be avoided.

The multiplication in an APD does not occur instantaneously. There is an inherent time delay because the avalanche requires a certain amount of time to build-up. The ultimate bandwidth of an APD is limited by avalanche multiplication build-up time τ_m in addition to the *RC* time-constant and transit time effects associated with all photodiodes. Again, the exact solution is difficult to obtain unless simplifying assumptions are made [26].

For the case where $k \ll 1$ and electrons are the carriers of interest, the transit-time consists of three components, the time required for an electron to transit the absorption region τ_a, the time an electron requires to transit the gain region, and the time the holes generated during the avalanche multiplication require to transit both the gain and absorption region τ_h. In most APDs the gain region is much shorter than the absorption region and its contribution to the transit-time can be ignored. The remaining two transit-times are

$$\tau_a = \frac{l_a}{v_e}, \quad \tau_h = \frac{l_a}{v_h}, \tag{4.35}$$

where

v_e = mean velocity of the electrons,

v_h = mean velocity of the holes.

The gain build-up time is approximated by [7, 21, 26]

$$\tau_m \cong \frac{M_e k l_g}{v_e}. \tag{4.36}$$

The magnitude of the frequency response of the multiplication gain in an APD is then given by [26, 27]

$$M(\omega) = \frac{M_o}{\sqrt{1 + \left(\dfrac{\omega}{\omega_c}\right)^2}}, \tag{4.37}$$

where

ω_c = multiplication cutoff frequency = $\dfrac{1}{\tau_{eff}}$,

$\tau_{eff} = \tau_a + \tau_h + \tau_m$,

ω = electrical frequency,

M_o = DC gain.

Equation 4.37 is reasonably accurate for the case where the DC gain exceeds the value of k_{eff}. If the DC gain is less than k_{eff}, the multiplication gain is approximately independent of frequency. The *RC* time-constant and carrier transit-time then determine the ultimate bandwidth achievable. Since the gain build-up time increases as the gain increases, the APD cut-off frequency ω_c decreases with increasing gain. Thus a gain-bandwidth product is evident in APDs. Typical values for the achievable gain-bandwidth in an APD are 100 to 200 GHz. The effects of the roll-off in APD gain at high frequencies can be reduced but not eliminated through electrical equalization techniques [28].

It is much harder to fabricate a high performance APD for use at long-wavelengths such as 1310 nm or 1550 nm than it is to fabricate one for use at short-wavelengths such as 850 nm. Much of the difficulty is because of the need to use relatively narrow band-gap materials such as Ge, InGaAs, or InGaAsP to absorb the relatively low-energy long-wavelength optical signal. This allows the narrow bandgap semiconductor to undergo tunneling effects at relatively low levels of applied field and causes a substantial increase in dark-current [29]. This problem is substantially reduced in the SAM APD structure since a narrow bandgap material such as InGaAs can be used in the absorption region and a wider bandgap material such as InP can be used in the gain region [30]. An example of a SAM APD structure suitable for use at long wavelengths is illustrated in Fig. 4.20

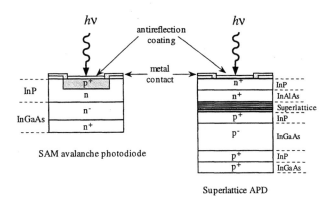

Figure 4.20 Long wavelength APD structures.

Unfortunately, the difference in bandgap also leads to an accumulation of carriers at the InGaAs/InP interface, which tends to slow the device, and the bandwidth may be reduced to only a few hundred MHz [31]. This problem can be reduced by retaining the SAM structure and adding a graded bandgap profile using a superlattice that is located between the absorption and the gain regions. This helps match the bandgap of InGaAs to the bandgap of InP. This type of APD is often termed a SAGM-APD, where the G stands for grading [32]. Another approach that has yielded good results is the superlattice-APD, or SL-APD, in which the superlattice is used as part of the gain region [33]. Superlattices such as shown in Fig. 4.20 are able to "artificially" increase the ratio between ionization coefficients, resulting in smaller values for k_{eff}. Another variation, known as the "staircase" APD, has the potential to behave much like a solid-state photomultiplier, with extremely good performance even at long wavelengths [34, 35].

4.7 Small Signal Equivalent Circuit Model for a Photodetector

In practice photodiodes do not have the simple physical structure that we used in Figs. 4.7(a), 4.12(a), 4.14(a), and 4.18(a). They are typically fabricated using the structures illustrated in Fig. 4.21. Note that antireflection coatings are almost universally used to obtain high values of quantum efficiency.

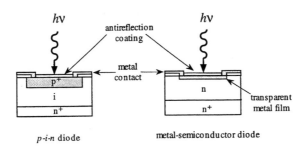

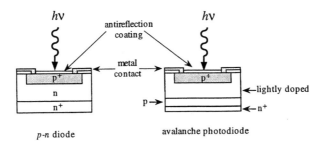

Figure 4.21 Examples of photodiodes. (Based on Fig. 13 in [7])

The photodetectors illustrated in Fig. 4.21 are called vertically illuminated since the optical signal travels in a direction that is perpendicular to the semiconductor junctions. There are also photodetectors in which the optical signal travels parallel to the junction. These are known either as waveguide photodetectors [36] or traveling-wave photodetectors [37]. They have the potential for high bandwidths and are suitable for some configurations of monolithically integrated receivers.

The need to make physical contact to the photodetector often governs the ultimately achievable bandwidth. An unpackaged wideband photodiode chip (or die) may have a junction capacitance on the order of 50-500 fF. Packaging can significantly increase the overall capacitance of the device and can also introduce other stray effects such as a series inductance due to bond-wires. In addition, a low-noise bias supply that is well bypassed to a local RF ground should be used to minimize the pickup of any stray electrical signals that may corrupt the photodetector's output.

Figure 4.22 illustrates examples of several mounting techniques that have been used with photodiodes. In Fig. 4.22(a) the photodetector is operated in a back illuminated mode. Back illuminated requires that the photodetector have a transparent substrate. The device can be mounted directly to the wall of the receiver housing. The optical fiber is held in place in a small hole in the housing by am ultra-low expansion epoxy.

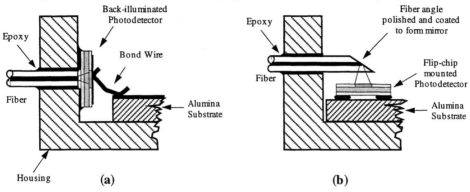

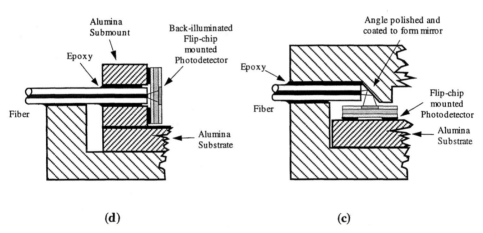

Figure 4.22 Examples of photodetector mounting techniques. (a) Back-side illumination. (b) Polished fiber as a mirror. (c) Reflected illumination. (d) Submount approach.

In Fig. 4.22(b) the end of the fiber is polished so that the light reflects off the polished end, passes through the cladding, and then illuminates the photodetector. The photodetector is typically made in a flip-chip mount with both contacts on one surface so that a bond-wire, which might interfere with the fiber placement, is not required. Figure 4.22(c) illustrates a variation on the reflection technique of Fig. 4.22(b). Here the reflection is accomplished using a reflective surface within the receiver housing. Precise mechanical alignment of the fiber or the reflective surface and the photodetector is required in these reflection schemes to guarantee that the photodetector is correctly aligned.

Figure 4.22(d) illustrates an approach in which a submount is used to hold the photodetector and the fiber. The detector is flip-chip mounted and back-illuminated and the electrical contacts run along the submount before connecting to the rest of the receiver circuitry. The fiber runs though a hole in the submount. The fiber is epoxied to the submount and the fiber-submount assembly is then mounted to the receiver substrate.

Regardless of the mounting technique used, some amount of stray inductance and capacitance is introduced. This results in a practical photodetector having the simplified equivalent circuit shown in Fig. 4.23.

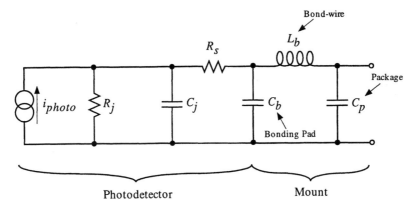

Figure 4.23 Simple small-signal equivalent circuit of a photodetector.

We complete the equivalent circuit of Fig. 4.23 by adding the components necessary to accommodate current multiplication and by including dark-current generators, as shown in Fig. 4.24.

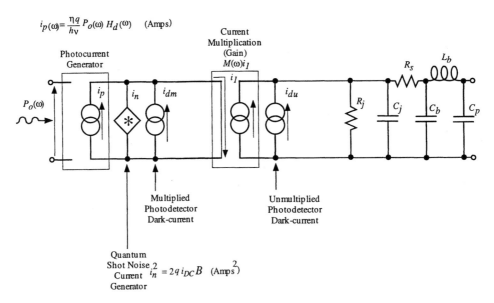

Figure 4.24 Complete small-signal equivalent circuit of a photodetector.

The dark-current observed will generally consist of two components. The first component is i_{du} and corresponds to the portion of dark-current that is not multiplied by the photodetector gain. The second component, i_{dm}, corresponds to the portion of the dark-current that is amplified by the gain. In APDs the i_{du} term comes from surface leakage currents, while the i_{dm} term comes from bulk leakage. In *p-i-n*'s there is no multiplied dark current.

A simple expression for the photocurrent can be obtained by combining the equations that we have used to describe the operation of the various photodetectors. For most practical communications applications, the magnitude of the photocurrent as a function of electrical frequency is well-approximated by

$$i_{photo}(\omega) \cong RP_{rcvd}G\frac{1}{\sqrt{1+\left(\dfrac{\omega}{\omega_c}\right)^2}} + i_{dark} + noise,$$

(4.38)

where

ω = electrical freqeuncy,

ω_c = bandwidth of the photodetector,

P_{rcvd} = received optical signal power,

$R = \dfrac{\eta q}{h\nu}$ = responsivity of the semiconductor material,

G = gain,

= 1 for a *p-n* and *p-i-n* photodiode,

$\gg$ 1 for an APD and a photoconductor,

i_{dark} = dark current.

In Eq. 4.38, the photodetector bandwidth ω_c is determined by the combination of transit-time effects, RC time-constant, and any gain-bandwidth product that may be evident.

Exact values for the electrical components in the small signal model are often very difficult to obtain analytically. An approach that often reveals the principal contributors to the equivalent circuit is the construction of a small-signal impedance model of the mounted photodetector. This can be done by measuring the microwave scattering parameter S_{11} [38]. This is equivalent to measuring the magnitude and phase of the microwave reflection coefficient for the device. An example of a measured S_{11} for a Si *p-i-n* photodetector that is mounted on a simple silica substrate is shown in Fig. 4.25. The device is an RCA (now EG&G) model C30971E. The device was operated with sufficient reverse-bias to guarantee that the *i* region was fully depleted (i.e. above the punch-through voltage.)

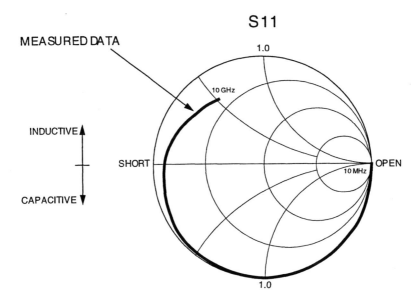

Figure 4.25 S$_{11}$ measurement of a photodetector.

The measured data indicate that at low frequencies the photodetector is high impedance. It appears to be a capacitor for frequencies up to several GHz. At about 7.5 GHz a resonance occurs and at higher frequencies the photodetector appears inductive. Either values for the output impedance Z_{out} or the values for S_{11} can be used to derive an equivalent circuit model. Impedance is frequently the more intuitive measurement and will be the technique used here. The measured values of S_{11} are converted into output impedance values via

$$Z_{out} = Z_o \frac{1 + S_{11}}{1 - S_{11}}, \tag{4.39}$$

where Z_o = characteristic impedance of the measurement system (usually 50 Ω).

Using the optimization and data fitting abilities of any of a variety of commercially available microwave design program, the equivalent circuit shown in Fig. 4.26(a) can be obtained. The impedance characteristic is dominated by the junction capacitance C_j, the series resistance R_s, and the bond-wire inductance L_b. The good agreement between the measured impedance values and the values calculated using the equivalent circuit model are shown in Fig. 4.26(b).

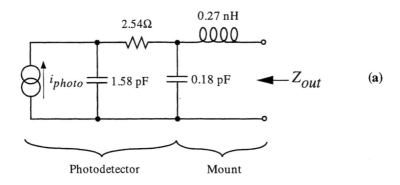

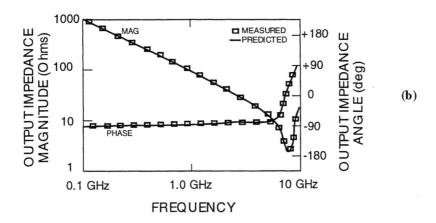

**Figure 4.26 Small signal impedance model of a photodiode. (a) Equivalent circuit.
(b) Measured and computed output impedance.**

The values found in Fig. 4.26 are representative for a moderate bandwidth *p-i-n* photodiode. The Si *p-i-n* device we used and its mounting structure are comparatively highly capacitive. The C30971E is designed for operation below 1 GHz. Wider bandwidth photodetectors would naturally have smaller values of junction capacitance.

Table 4.2 presents a summary of the characteristics of some popular photodetectors that were discussed in this chapter. The values given are for devices that would typically be used for communications applications and do not necessarily represent fundamental limits to performance for a material or device type. In many cases it is possible to improve the performance in one specific area, such as bandwidth, if performance in another area, such as quantum efficiency, is sacrificed.

Table 4.2 Summary of Photodetector Characteristics

	Si PIN	*Si APD*	*Ge APD*	*InGaAs PIN*	*InGaAs APD*
Useful λ Region (nm)	400 - 1150 nm	400 - 1150 nm	800 - 1750 nm	900 - 1700 nm	900 - 1700 nm
QE	60 - 90%	70 - 80%	50 - 80%	70 - 90%	60 - 90%
Gain	1	50 - 300	10 - 100	1	10 - 40
Carrier Ionization Ratio (k_{eff})	-	0.01 - 0.10	0.6 - 0.9	-	0.2 - 0.5
Unmultiplied Dark-current (nA)	1 nA	0.1 nA	1 - 50 nA	1 - 10 nA	0.5 - 5 nA
Multiplied Dark-current (nA)	-	0.1 - 1.0 nA	5 - 100 nA	-	0.5 - 5 nA
Detector Capacitance (pF)	1 - 5 pF	1 - 5 pF	1 - 5 pF	0.2 - 2 pF	0.2 - 2 pF
Response Time (ns)	0.3 - 3 ns	0.5 - 5 ns	0.3 - 3 ns	0.05 - 1 ns	0.1 - 1 ns

4.8 References

1. R.W. Engstrom, "Multiplier Phototube Characteristics: Applications to Low Light Levels," Journal of the Optical Society of America, 1947, vol. 37, pp. 420-424.

2. M.B. Fisher and R.T. McKenzie, "A Traveling-wave Photomultiplier," IEEE Journal of Quantum Electronics, 1966, vol. QE-2, pp. 322-327.

3. J.R. Lesh. "Power Efficient Communications for Space Applications," in *International Telemetering Conference*. 1982, International Foundation for Telemetering, Vol. XVIII.

4. E. Dereniak and D. Crowe, *Chap. 4 - Photoconductors* in *Optical Radiation Detectors,* 1984, John Wiley & Sons: New York.

5. E. Dereniak and D. Crowe, *Chap. 3 - Photovoltaic Detection Theory* in *Optical Radiation Detectors,* 1984, John Wiley & Sons: New York, Chapters 3,4, & 5.

6. R.J. Keyes, *Chapter 2* in *Optical and Infrared Detectors,* 1977, Springer-Verlag: New York.

7. S.M. Sze, *Chap. 13 - Photodetectors* in *Physics of Semiconductor Devices,* 1981, John Wiley & Sons: New York.

8. H. Melchoir, *et al.*, "Photodetectors for Optical Communication Systems," Proceedings of the IEEE, 1970, vol. 58, no. 10, pp. 1466-1490.

9. S.R. Forrest, "The Sensitivity of Photoconductor Receivers for Long-Wavelength Optical Communications," Journal of Lightwave Technology, 1985, vol. LT-3, no. 2, pp. 347-360.

10. R.H. Kingston and D. Spears, "Anomalous noise behavior in wide-bandwidth photodiodes in heterodyne and background-limited operation," Applied Physics Letters, 1979, vol. 39, no. 9.

11. S.M. Sze, *Chap. 14 - Solar Cells* in *Physics of Semiconductor Devices,* 1981, John Wiley & Sons: New York.

12. S.Y. Wang and D.M. Bloom, "100 GHz Bandwidth Planar GaAs Schottky Photodiode," Electronics Letters, 1983, vol. 19, pp. 554-555.

13. B.J. Van Zeghbroeck, "105-GHz Bandwidth Metal-Semiconductor-Metal Photodiode," IEEE Electron Device Letters, 1988, vol. 9, no. 19, pp. 527-529.

14. J.S. Wang, *et al.*, "11 GHz Bandwidth Optical Integrated Receivers Using GaAs MESFET and MSM Technology," IEEE Photonics Technology Letters, 1993, vol. 5, no. 3, pp. 316-318.

15. E.H. Bottcher, "Ultrafast Semi-insulating InP:Fe-InGaAs:Fe-InP:Fe MSM Photodetectors: Modeling and Performance," IEEE Journal of Quantum Electronics, 1992, vol. 28, no. 10, pp. 2343-2357.

16. J.H. Kim, *et al.*, "High-Performance Back-Illuminated InGaAs/InAlAs MSM Photodetector with a Record Responsivity of 0.96 A/W," IEEE Photonics Technology Letters, 1992, vol. 4, no. 11, pp. 1241-1244.

17. Y. Lim and R. Moore, "Properties of Alternatively Charged Coplanar Parallel Strips by Conformal Mapping," IEEE Transactions on Electron Devices, 1968, vol. ED-15, no. 3, pp. 173-179.

18. M. Ito and O. Wada, "Low Dark-current GaAs Metal-Semiconductor-Metal (MSM) Photodiodes Using WSi_x Contacts," IEEE Journal of Quantum Electronics, 1986, vol. QE-22, no. 7, pp. 1073-1077.

19. J.B.D. Soole and H. Schumacher, "InGaAs Metal-Semiconductor-Metal Photodetectors for long Wavelength Optical Communications," IEEE Journal of Quantum Electronics, 1991, vol. 27, no. 3, pp. 737-752.

20. S.M. Sze, "Current Transport in Metal-Semiconductor-Metal Structures," Solid State Electronics, 1971, vol. 14, no. 12, pp. 1209-1218.

21. P.P. Webb, *et al.*, "Properties of Avalanche Photodiodes," RCA Review, 1974, vol. 35, pp. 234-278.

22. X. Sun and F. Davidson, "Photon Counting with Avalanche Photodiodes," Journal of Lightwave Technology, 1992, vol. 10, no. 8, pp. 1023-1032.

23. G.E. Bulman, *et al.*, "Experimental Determination of Impact Ionization Coefficients in (100) GaAs," IEEE Electron Device Letters, 1983, vol. EDL-4, pp. 181-182.

24. N. Tabatabaie, *et al.*, "Impact Ionization Coefficients in (111) InP," Applied Physics Letters, 1985, vol. 46, pp. 182-184.

25. G.E. Stillman and C.M. Wolfe, *Avalanche Photodiodes*, in *Semiconductors and Semimetals,*R.K. Williamson and A.C. Beer eds. Vol. 12, 1966, Academic Press: New York, pp. 290-295.

26. R. Emmons, "Avalanche Photodiode Frequency Response," Journal of Applied Physics, 1967, vol. 38, no. 9, pp. 3705-3714.

27. S.R. Forrest, "Gain-Bandwidth Limited Response in Long-Wavelength Avalanche Photodiodes," Journal of Lightwave Technology, 1984, vol. LT-2, no. 1, pp. 34-39.

28. B.L. Kasper and J.C. Campbell, "Multigigabit-per-second Avalanche Photodiode Lightwave Receivers," Journal of Lightwave Technology, 1987, vol. LT-5, pp. 1351-1364.

29. S.R. Forrest, *et al.*, "Evidence for Tunneling in Reverse-biased III-V Photodetector Diodes," Applied Physics Letters, 1980, vol. 36, pp. 580-582.

30. K. Nishida, *et al.*, "InGaAsP Heterojunction Avalanche Photodiode with High Avalanche Gain," Applied Physics Letters, 1979, vol. 35, pp. 250-252.

31. S.R. Forrest, *et al.*, "Optical Response Time of InGaAs/InP Avalanche Photodiodes," Applied Physics Letters, 1983, vol. 41, pp. 95-98.

32. J.C. Campbell, *et al.*, "High Performance Avalanche Photodiode with Separate Absorption, Grading, and Multiplication Regions," Electronics Letters, 1983, vol. 19, pp. 818-820.

33. I. Watanabe, *et al.*, "High-speed and Low Dark-Current Flip-Chip InAlAs/InAlGaAs Quaternary Well Superlattice APD's with 120 GHz Gain-Bandwidth Product," IEEE Photonics Technology Letters, 1993, vol. 5, no. 6, pp. 675-677.

34. K. Matsuo, *et al.*, "Noise Properties and Time Response of the Staircase Avalanche Photodiode," IEEE Transactions on Electron Devices, 1985, vol. ED-32, pp. 2615-2623.

35. F. Capasso, "Staircase Solid-State Photomultipliers and Avalanche Photodiodes with Enhanced Ionization Rates Ratio," IEEE Transactions on Electron Devices, 1983, vol. ED-30, no. 4, pp. 381-390.

36. A. Alping, "Waveguide PIN Photodetectors: Theoretical Analysis and Design Criteria," IEE Proceedings Part J, 1989, vol. 136, no. 3, pp. 177-182.

37. K.S. Giboney, *et al.*, "Traveling-Wave Photodetectors," IEEE Photonics Technology Letters, 1992, vol. 4, no. 12, pp. 1363-1365.

38. G.F. Vendelin, *Chap. 1 - The S-Parameters* in *Design of Amplifiers and Oscillators by the S-Parameter Method*, 1982, John Wiley & Son: New York.

Chapter 5
RECEIVER NOISE MODELING

An understanding of the origins of noise is required for a receiver's performance to be accurately characterized. The amount of noise present in a receiver will be the primary factor that determines the receiver's sensitivity. In this chapter, we will first review the definitions and analysis techniques needed to understand the effects of noise on a receiver's performance. The noise sources that are commonly found in an optical receiver are then discussed, including noises that are of optical as well as electrical origin. Our goal is to develop equivalent circuit models that will accurately describe the noise performance of an optical receiver. Once we have established an acceptable model, we will derive expressions for the noise-density and total noise present in the receiver.

5.1 Fundamentals of Noise Analysis

Noise, defined in the broadest practical terms, is any signal present in the receiver other than the desired signal. Another widely used definition is that noise is any *unwanted* disturbance that masks, corrupts, reduces the information content of, or interferes with the desired signal. We note an important exception to these definitions. Distortion effects that arise within the receiver circuitry are not included in our definition of noise. Receiver distortion is a circuit design problem, not a noise problem.

The sources of noise present in a receiver circuit can be broadly divided into two classes. The first are intrinsic noise sources arising from fundamental physical effects in the optoelectronic and electronic devices used to construct the receiver. The second are coupled noise sources arising from interactions between the receiver circuitry and the surrounding environment. Examples of intrinsic noise sources are the thermal-noise found in resistors, electronic shot-noise and thermal-noise in transistors, and the quantum shot-noise inherent in photodetection. These noise sources are found in all optical receivers. Coupled noise originates in solar, cosmic, or atmospheric disturbances, nearby electrical transmission lines, power supplies, fast switching logic circuitry, etc. Coupled noise that degrades receiver performance is often termed interference or circuit cross-talk. The amount of coupled noise present in a receiver is dependent on the exact physical location and orientation of the circuitry with respect to the surrounding electromagnetic fields. In a well designed receiver with adequate shielding, coupled noise can usually be made to be negligible when compared to intrinsic noise [1].

There is also a type of noise-like disturbance, called microphonics, that can arise from mechanical vibrations causing small changes in receiver components that have a critical mechanical alignment. Most optical receivers are subject to some amount of mechanical vibrations, whether from power supply cooling fans, nearby road-traffic, or motion of a vehicle containing the receiver. Microphonics can occur in an optical receiver where there is a tuned circuit whose resonant frequency is sensitive to a mechanical spacing. As the receiver vibrates, the resonant frequency varies and the receiver response becomes time varying. Microphonics can also occur if there are substantial co-aligned reflections from both the photodetector surface and from the incoming fiber. This allows a resonant cavity similar to a Fabry-Perot interferometer to form [2]. The transmission of this cavity will be sensitive to variations in mechanical spacing as small as a fraction of a micron and can, in certain circumstances, introduce additional noise into the receiver [3]. Vibrations of the receiver package will cause the transmission of the cavity to change, inducing a time varying optical signal power at the photodetector surface. Microphonics, when present, can be reduced by careful electro-mechanical design so that they are negligible when compared to intrinsic noise.

Noise in a receiver can also be described as being either additive or signal dependent. An additive noise is a source of noise that is observed whether there is a signal present at the receiver or not. A signal dependent noise is one that is observed only when there is a signal present at the receiver. Signal dependent noises are also frequently termed multiplicative noise since the noise is effectively multiplied by the signal power. Thus, when the signal is zero, so is the noise. The amount of a signal dependent noise that is observed is also frequently proportional to the signal power.

Noise, being a random process, cannot be described as an explicit function of time. Regardless of the number of times we observe the noise, there is no way to determine exactly what value a noise process will have at any future time. Noise in the time-domain can be characterized in probabilistic terms such as a mean and a variance [4]. Another widely used characteristic is the root-mean-square (RMS) value. For an observed noise process $n(t)$ the RMS value is defined as

$$n_{rms} = \sqrt{\overline{n^2(t)}} = \sqrt{\frac{1}{T} \int_0^T n^2(t)dt}, \qquad (5.1)$$

where the horizontal bar indicates a time average.

Of particular interest is the noise *power,* which is proportional to the square of the RMS value of the noise. Whenever two noise sources are added, the instantaneous sum is the sum of the individual instantaneous values. The total average noise power N resulting from the summation of two noise processes $n_1(t)$ and $n_2(t)$ is given by

$$N = \overline{n_{total}^2(t)} = \overline{[n_1(t) + n_2(t)]^2} = \overline{n_1^2(t)} + 2\gamma\overline{n_1(t)n_2(t)} + \overline{n_2^2(t)}$$

$$\qquad (5.2)$$

$$= \overline{n_1^2(t)} + 2\gamma\sqrt{\overline{n_1^2(t)}}\sqrt{\overline{n_2^2(t)}} + \overline{n_2^2(t)} \qquad \text{(watts).}$$

The term containing the cross product of the two noises contains the coefficient γ that is used as a measure of the correlation between the noise processes. For uncorrelated noise processes, $\gamma =$ zero and the total noise power will be proportional to the sum of the squared values of the individual noise processes:

$$N = n_{1_{rms}}^2 + n_{2_{rms}}^2 \qquad \text{(watts).} \qquad (5.3)$$

Noise can also be characterized in the frequency-domain. Frequency-domain techniques are in fact preferred for noise analysis. The most important frequency-domain characteristic of a noise process is its power spectral-density (PSD). Power spectral-density is defined as the Fourier transform of the time-domain autocorrelation function [5]. The impact of this relationship is that power spectral-density corresponds to the time-averaged noise power that is present in a 1 Hz bandwidth around the measurement frequency. A PSD will therefore have the units of watts-per-Hz. Equation 5.2, expressed in terms of spectral-densities, is

$$N(\omega) = |n_1(\omega)|^2 + 2\gamma(\omega)\sqrt{|n_1(\omega)|^2}\sqrt{|n_2(\omega)|^2} + |n_2(\omega)|^2 \quad (\frac{\text{watts}}{\text{Hz}}). \tag{5.4}$$

We will use the notation $n^2(\omega)$ to represent a PSD even though PSD is actually a magnitude squared as shown in Eq. 5.4. For simplicity in writing equations (and by common convention) we will forego the use of vertical bars to indicate magnitude when discussing noise-power spectral-density unless they are required for clarity.

A noise source's PSD will generally have some frequency dependence. Some commonly observed frequency dependencies have specific names. If the power spectral-density is constant with frequency, that is, an equal amount of noise power is present in every 1 Hz bandwidth, the noise is referred to as "white-noise." "Pink-noise" refers to noise whose power spectral density varies inversely with frequency as $1/f$. This characteristic results in noise with an equal amount of noise power in each decade of frequency. "Red-noise" varies as $1/f^2$, and the term "blue-noise" has been used to describe a noise power spectrum that increases with frequency. If a noise source has a white PSD, with a power spectral-density of n_{input}, the total noise present in a B Hz bandwidth is simply $n_{input}B$. When white noise of power spectral-density n_{input} watts per Hz passes through a network described by a power transfer function given by $G(f)$, the network will shape the noise power spectral-density so that it is no longer white. The total noise power N present at the output of the network will be given by

$$N = n_{input} \times \frac{1}{G_o} \int_0^\infty G(f)df = n_{input} \times NEB, \tag{5.5}$$

where G_0 is the peak power gain, $G(f)$ is the power gain as a function of frequency, and *NEB* is the noise equivalent bandwidth.

Noise equivalent bandwidth has the units of Hz. If voltage-noise density or current-noise density values were being used, the values would be multiplied by the square-root of the noise-equivalent-bandwidth to arrive at the total current-noise I_n or total voltage-noise V_n present in the circuit.

Since the power gain in a network is proportional to the network transfer function squared, the noise equivalent bandwidth can also be written in the form

$$NEB = \frac{1}{H_0^2} \int_0^\infty |H(f)|^2 \, df \quad \text{(Hz)}, \tag{5.6}$$

where H_0 is the peak magnitude of the transfer function $H(f)$.

We note that the magnitude-squared of a transfer function is easily obtained with the use of the identity

$$|H(\omega)|^2 = H(\omega) \times H^*(\omega) = H(j\omega) \times H(-j\omega), \tag{5.7}$$

where $\omega = 2\pi f$ and * indicates the complex conjugate.

The noise equivalent bandwidth of a network is conceptually equivalent to the bandwidth that an ideal rectangular filter would have if it were to produce the same amount of RMS noise power at its output as the network in question. The NEB of a network is generally *not* equal to the 3-dB bandwidth. The 3-dB bandwidth refers to the point(s) in the frequency-domain at which the power gain transfer function falls to half its maximum value. Noise equivalent bandwidth is the *area* under the power gain transfer function normalized to the peak-power gain, as illustrated in Fig 5.1.

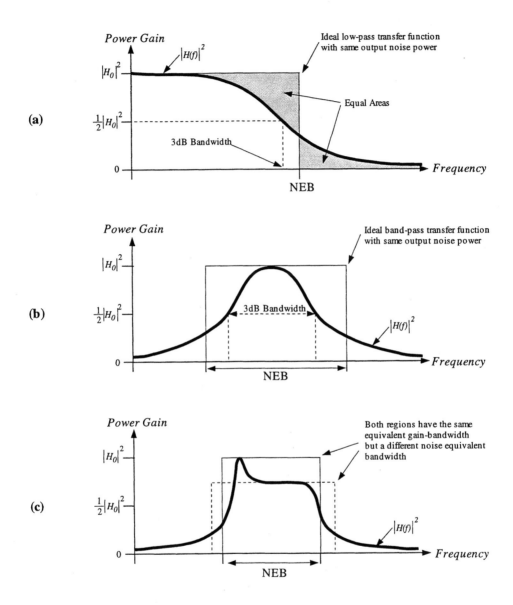

**Figure 5.1   Noise equivalent bandwidth. (a) Simple low-pass transfer function.
(b) Symmetrical band-pass transfer function. (c) Asymmetrical bandpass transfer function.**

Figure 5.1(a) illustrates the NEB for a low-pass transfer function. Note that the NEB
is larger than the 3 dB (i.e. half-power) bandwidth. For a simple single pole roll-off low-
pass filter network with a 6 dB per octave roll-off, the equivalent noise bandwidth is
nominally 1.57 times the 3-dB bandwidth. The order of the network would need to be
larger that six, with a greater than 36 dB per octave roll-off, for the difference between

the NEB and 3-dB bandwidths to be less than 10%.

Figure 5.1(b) illustrates the NEB for a symmetrical band-pass transfer function. The noise bandwidth is again larger than the 3 dB bandwidth. Additional care must be exercised whenever the transfer function is asymmetrical. As shown in Fig. 5.1(c), two different noise-equivalent-bandwidths can be obtained, depending on the value used for peak power gain. If the true peak is used, an NEB corresponding to the solid rectangle is obtained. If the peak corresponding to the flat portion of the passband is used, a different, somewhat larger NEB corresponding to the dotted rectangle is obtained. Since both the solid and the dotted regions will have the same amount of noise power at their outputs, either NEB is considered correct. Either can be used as long as throughout the receiver design, we are consistent in the way the peak value of a transfer function is determined.

5.2 Noise Sources in an Optical Receiver

Figure 5.2(a) illustrates a simple model for an optical receiver. The received signal and any optical background that may be present is photodetected and then amplified in a linear signal path. The demodulator observes the amplified waveforms and recovers the transmitted information. Figure 5.2(b) shows the six major contributors to the noise present in the receiver. For simplicity, we will model all noise sources as additive even though some noise sources such as photodetector excess-noise or optical excess-noise are in fact usually signal dependent.

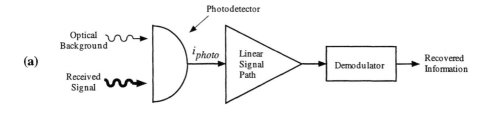

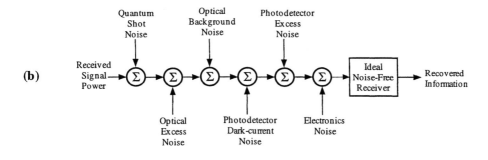

**Figure 5.2   Noise sources in an optical receiver. (a) Simple receiver model.
(b) Noise sources in the receiver.**

The six major noise sources are 1) quantum shot-noise, 2) optical excess noise, 3) optical background noise, 4) photodetector dark-current noise, 5) photodetector excess noise, and 6) electronics noise. The implications of each noise source will be addressed in the following sections. We will start with the noise source that we have, in some sense, the most control over, the noise in the receiver electronics.

5.3 Electronics Noise

Receiver electronics noise consists of three primary components, thermal-noise, electronic shot-noise, and 1/f noise. Thermal-noise is the noise most often associated with receivers. In fact, some authors refer to *all* of a receiver's electronics noise as thermal-noise even though many sources of electronics noise are not directly associated with temperature.

In addition to these three principal noise components, electronic devices can also exhibit generation-recombination noise due to the random trapping of charge carries in semiconductors, popcorn or burst noise due to semiconductor defects, contact noise at metal semiconductor interfaces, avalanche noise in devices such as zener diodes, and hot-carrier noise from hot electrons in high fields [6-8].

5.3.1 Thermal-noise

Thermal-noise (also called Johnson noise after its discoverer [9]) is a result of thermally induced random fluctuations in the charge carriers in a resistance and is similar to the Brownian motion of particles. Carriers are in random motion in all resistances at a temperature higher than absolute zero. The amount of motion is a direct function of the absolute temperature of the resistance. Taken as a whole, the resistor will exhibit charge neutrality; however, on a microscopic level, the agitated free carriers give rise to locally fluctuating charge gradients, which produce a wideband random voltage fluctuation appearing in series with the resistance. The power spectral-density for thermal-noise is essentially that of white-noise for frequencies up to near-infrared frequencies, and since thermal-noise inherently results from the accumulated effect of a large number of individual charge motions, it exhibits Gaussian statistics. Nyquist showed that the open-circuit RMS voltage produced by a resistance R is given by [10]

$$V_n = \sqrt{4kTBR} \quad \text{(volts rms)}, \tag{5.8}$$

where k is Boltzmann's constant (1.38×10^{-23} joules/°K), T is absolute temperature in °K, and B is our observation bandwidth in Hz.

Thermal-noise is present in all electronic elements containing a resistance that dissipates energy but is not present in purely dynamic resistances that arise in active devices. Thermal-noise can be modeled by replacing the noisy resistor with the combination of a noise-free resistor either in series with a voltage-noise generator or in parallel with a current noise generator.

In both the voltage-noise and current-noise models, the noise source is a zero mean Gaussian noise generator with a white PSD. In the voltage-noise model, the spectral-density *increases* with increasing resistance. The opposite occurs in the current-noise model. Current noise spectral-density *decreases* with increasing resistance.

Resistors also have an excess noise term associated with them that is a measure of the uniformity of their composition and their terminal contact characteristics [8, 11]. Carbon resistors are particularly well-known for exhibiting excess noise and are often avoided in favor of metal film resistors in critical low-noise applications. Capacitors and inductors do not generate thermal-noise unless they have significant resistive components or are suffering from dielectric breakdown.

In many circuit design problems we are interested in transferring power from a source to a load. Figure 5.3 illustrates the case of power transfer from a noisy resistor R to a load resistor R_l.

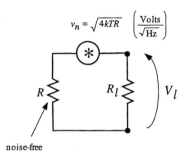

$$v_n = \sqrt{4kTR} \quad \left(\frac{\text{Volts}}{\sqrt{\text{Hz}}} \right)$$

noise-free

Figure 5.3   Maximum power transfer from a noisy resistor.

The thermal-noise power dissipated in the load resistor is given by

$$n_t^2(f) = \frac{V_l^2}{R_l} = \frac{\left(\dfrac{R_l}{R + R_l} \sqrt{4kTR} \right)^2}{R_l} \quad (\frac{\text{watts}}{\text{Hz}}). \tag{5.9}$$

The maximum power transfer occurs for a matched load when $R_l = R$. Under this constraint the thermal-noise power in the load resistor is given by

$$n_t^2(f) = \frac{\left(\dfrac{R}{R + R} \sqrt{4kTR} \right)^2}{R} = \frac{\left(\dfrac{1}{2} \sqrt{4kTR} \right)^2}{R} = \frac{kTR}{R} = kT \quad (\frac{\text{watts}}{\text{Hz}}). \tag{5.10}$$

For a 290°K noise temperature (approximating room temperature), the resulting noise power spectral-density is the frequently quoted -174 dBm/Hz that is used extensively in microwave systems.

Nyquist's results as used in obtaining Eqs. 5.8 through 5.10 are actually a simplification of a more general result. The exact quantum mechanical expression for the available thermal-noise power spectral-density is given by [12, 13]

$$n_t^2(f) = \frac{h\nu}{\exp\left(\dfrac{h\nu}{kT} \right) - 1} \quad (\frac{\text{watts}}{\text{Hz}}), \tag{5.11}$$

where h is Planck's constant (6.625×10^{-34} J - s) and ν is the frequency in Hz.

At room temperature and for frequencies below approximately 5×10^{12} Hz, the photon energy $h\nu$ is much less than the thermal energy kT, and Eq. 5.11 reduces to Eq. 5.10. At higher frequencies the thermal-noise power spectral-density falls off exponentially.

From Eq. 5.11 we see that as frequency increases into the optical region, thermal-noise ceases to be a concern. We have also seen that quantum shot-noise forms the fundamental noise limit at optical frequencies. It is possible to combine the thermal-noise and quantum shot-noise contributions to establish the fundamental limitations on the noise density in an ideal receiver. Oliver has shown that the fundamental limit to the noise density present at the input to a receiver using an ideal linear amplifier is given by [12]

$$n^2(f) = \frac{h\nu}{\exp\left(\dfrac{h\nu}{kT}\right) - 1} + h\nu \quad (\frac{\text{watts}}{\text{Hz}}). \tag{5.12}$$

Equation 5.12 is plotted in Fig. 5.4 for the two cases of a receiver operating at room temperature (290 degrees Kelvin) and another operating at 77 degrees Kelvin.

An important conclusion can be drawn from Fig. 5.4. An optical receiver is inherently "noisier" than a microwave receiver. Consequently, in terms of its ability to detect the smallest possible signal, a microwave receiver will generally be more sensitive than an optical receiver. This difference occurs simply because $h\nu >> kT$ in most optical communication applications.

Let us compare the smallest signal that a microwave receiver operating at 10 GHz can detect to the smallest signal that an optical receiver operating at 300 THz (1.0 micron) can detect. The microwave receiver is operating at 290°K and to keep the comparison fair, we will use a 1 GHz noise bandwidth for each receiver.

Since we are well into the region where thermal-noise is dominant, the total noise present in the microwave receiver is given by

$$N = kTB = (1.38 \times 10^{-23})(290)(1 \times 10^9) = 4 \times 10^{-12} \text{ watts} \tag{5.13}$$

while the noise in the optical receiver is given by

$$N = \int_{\nu_0 - B/2}^{\nu_0 + B/2} h\nu d\nu = h\nu_o B \tag{5.14}$$

$$= (6.625 \times 10^{-34})(3 \times 10^{14})(1 \times 10^9) = 2 \times 10^{-10} \text{ watts.}$$

Assuming that a receiver can detect the presence of a signal when the signal power is equal to the total noise power, the microwave receiver can detect a signal as small as 4 pW. The optical receiver requires 50 times more power, or 200 pW.

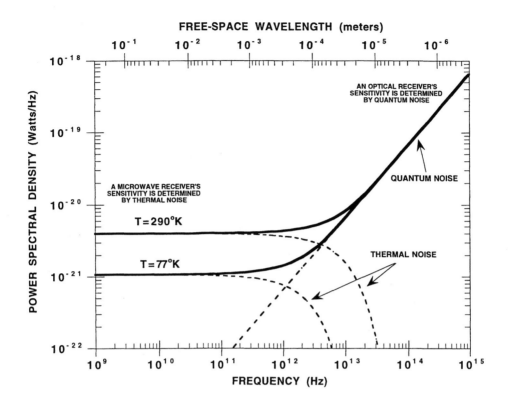

Figure 5.4 Fundamental limit of receiver noise density.

5.3.2 Electronic Shot-Noise

The flow of an electrical current is not smooth and continuous. It is composed of individual carriers each carrying q coulombs of charge through the circuit. The sum of all the individual impulses forms the average current flow. When charge carriers cross a potential charge barrier, the statistics that describe a charge's crossing determine the noise characteristics associated with the current flow. If the charges cross the barrier in a periodic, predictable fashion, the current flow is uniform and noise is not generated. If the number of carriers that cross the barrier is random and independent of the number of carriers preceding or following, the current is characterized by a Poisson distribution [15]. This is the same distribution that we used to describe photon counting in Chapter 3.

The noise associated with the passage of carriers across a potential barrier is known as shot-noise. In his original work with current-noise in potential barriers, Schottky used an analogy between current flow and the patter of small shot falling into a container and his description was adopted as the name for the noise [16]. Potential barriers occur in *p-n* junctions in semiconductor diodes and junction transistors. In a receiver, we would expect there to be electronic shot-noise associated with each diode or junction transistor used in the receiver's circuitry.

There will also be shot-noise associated with any current flow through a receiver's photodiode. This means that any photocurrent in the photodiode will have electronic shot-noise associated with it. This leads us to an important observation. Based strictly on the characteristics of the noise that is observed, it is generally impossible to distinguish between the quantum shot-noise arising from photons being detected with a photodiode and the electronic shot-noise arising from the photocurrent flowing through the *p-n* junction in the photodiode. Quantum shot-noise in the photodiode occurs because photon detection is random and each photon is detected independently of the others. Electronic shot-noise in the *p-n* junction occurs because carrier flow through the junction is random and carriers cross independently of each other.

When the number of events that occur per unit observation time is large, the Poisson distribution may be replaced by a Gaussian distribution [17]. Approximately 6×10^{12} carriers cross a *p-n* junction every second for each milliamp of DC current the junction is carrying. This is such a large number of events per observation time that in most practical cases encountered in electronic devices, electronic shot-noise is accurately described by a Gaussian distribution with a white PSD.

Because electronic shot-noise is associated with current flow, it is naturally modeled as a current-noise source in parallel with the device in which the electronic shot-noise occurs. The total electronic shot-noise associated with a current I_{DC} flowing through a potential barrier is given by [16]

$$I_{shot} = \sqrt{2qI_{DC}B} \quad \text{(amps)}, \tag{5.15}$$

where q = electronic charge, 1.602×10^{-19} coulombs, and B = observation bandwidth.

We emphasize that Eq. 5.15 is accurate when there is no interaction between carriers. In certain space-charge limited devices (such as vacuum diodes) there is correlation between the charge carriers and it is possible to obtain partial shot-noise [18].

The normalized power spectral-density of electronic shot-noise is given by

$$i_n^2(f) = 2qI_{DC} \quad (\frac{\text{amps}^2}{\text{Hz}}). \tag{5.16}$$

Note that this is exactly the equation used in Chapter 3 to describe the PSD of quantum shot-noise associated with photodetection.

As an example of electronic shot-noise, consider the semiconductor diode illustrated in Fig. 5.5(a). The diode has a DC current of I_d and a DC bias voltage of V_d.

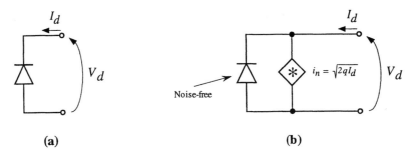

(a)   (b)

**Figure 5.5   Shot and thermal-noises in a diode. (a) Diode schematic symbol.
(b) Noise equivalent model.**

Figure 5.5(b) illustrates the noise equivalent model. A current-noise source that accounts for the shot-noise in the diode has been placed in parallel with the now noise-free diode [19].

5.3.3 1/f Noise

1/f noise has been observed in many physical systems. Low frequency fluctuations in the resistance of a semiconductor, long-term variations in a geographic region's average seasonal temperature, changes in the rate of traffic flow, and certain types of biologic phenomena have all been described using 1/f noise [20-22]. When used to describe vacuum tubes it has been called "flicker" noise. In resistors it is called "excess" noise. In switches and relays it has been called "contact" noise. The power spectral-density of 1/f noise varies inversely with frequency as

$$n(f) = \frac{a}{f^b} \quad (\frac{\text{watts}}{\text{Hz}}), \tag{5.17}$$

where a is a constant that sets the absolute level and b is an exponent that is usually close to 1.0.

In many semiconductor devices, the absolute amount of 1/f noise is related to the amount of current flowing in the device, and Eq. 5.17 is modified as

$$n(f) = I_{DC}^d \times \frac{a f_c}{f^b} \quad (\frac{\text{watts}}{\text{Hz}}), \tag{5.18}$$

where the constant d is included to set the relationship between the DC current and the level of the 1/f noise and f_c is the 1/f noise's corner frequency.

For classic 1/f noise $b = 1$ and Equation 5.18 indicates that the noise power decreases linearly with frequency. The total noise power present in the frequency range f_1 to f_2 for a classic 1/f noise process is given by

$$N = \int_{f_1}^{f_2} \frac{a}{f} df = a \ln \frac{f_2}{f_1} \quad (\text{watts}). \tag{5.19}$$

Note that for any decade in frequency, that is, whenever $f_2 = 10 \times f_1$, the noise power will be identical and equal to $2.3a$. Thus there is the same amount of 1/f noise power in the 1 Hz to 10 Hz frequency band as there is in the 1 MHz to 10 MHz frequency band. Equal power in each decade is a fundamental characteristic of 1/f noise.

Even though 1/f noise appears in virtually all electronic devices, the exact origins of 1/f noise are not well understood. Unlike thermal-noise and shot-noise, which are fundamental in nature, 1/f noise in electronic devices is usually thought to be caused by imperfections in the semiconductor materials [22].

The presence of a 1/f noise component in what would otherwise be a noise source with a constant spectral-density can be accounted for by including an additional noise source in parallel with the original noise source, as shown in Fig. 5.6. This is illustrated in Fig. 5.6(a) where the constant spectral-density current-noise source has been augmented with a 1/f spectral-density current-noise source. The two noise sources are uncorrelated and the total noise power spectral-density is simply the sum of the individual noise power spectral densities. Thus, an equation of the form

$$n_t(f) = i_n^2 + i_n^2 \frac{f_c}{f} \quad (\frac{\text{amps}^2}{\text{Hz}}).,$$ (5.20)

will account for the effects of the 1/f noise.

The resulting power spectral-density is illustrated in Fig. 5.6(b). For frequencies below f_c the noise PSD has a distinct 1/f characteristic while for frequencies above f_c, the noise has a white PSD. A 1/f noise process is not always as obvious as shown in Fig. 5.6(a). In cases where the underlying noise is proportional to f, a 1/f contribution would cause the spectral density to be constant for frequencies less than f_c. Similarly, for an underlying noise process that was originally proportional to f^2, a 1/f contribution would cause a change in slope from f^2 to f for frequencies below f_c.

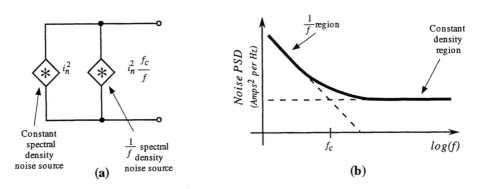

Figure 5.6   Including 1/f noise in a shot-noise generator. (a) Noise sources. (b) Corresponding noise spectral-density.

1/f noise becomes a concern in optical receiver design when the receiver is required to have a low frequency cut-off that is less than a few tens of MHz [23]. The amount of 1/f noise present will then depend on the choice of transistors used in the first stages of the receiver. Silicon bipolar transistors generally exhibit small amounts of 1/f noise. The corner frequency at which a Si bipolar transistor's 1/f noise becomes noticeable is usually in the tens of kHz region. GaAs field-effect transistors (FETs) can have significantly larger amounts of 1/f noise [23-26]. In some GaAs devices, 1/f noise corner frequencies can be as high as 100 MHz.

5.3.4   Noise Model of an Electronic Amplifier

Amplification of low-level signals is a critical function of any receiver. Because of thermal, shot, and 1/f noises in electronic components, the output of any realizable electronic amplifier will be corrupted by noise. The overall noise performance of an amplifier is intimately related to the noise characteristics of the individual devices and components that form the amplifier circuit.

The concept of noise figure is a well established and widely used indicator of the noise performance of an amplifier [27]. It is particularly popular for describing the noise characteristics of microwave amplifiers, where the source and load impedances are usually well defined resistive terminations. An amplifier's noise factor F is defined as the ratio of the amplifier's input and output signal-to-noise ratios, or

$$F = \frac{SNR_{input}}{SNR_{output}}. \tag{5.21}$$

An amplifier's noise figure NF is simply the noise factor expressed in terms of dB, or

$$NF = 10\log_{10}[F] = 10\log_{10}\left[\frac{SNR_{input}}{SNR_{output}}\right] \quad \text{(dB)}. \tag{5.22}$$

An equivalent expression for an amplifier's noise figure is the ratio between the amplifier's output noise power normalized by the amplifier power gain and the amount of noise power originally at the input to the amplifier. This relationship can be seen by expanding Eq. 5.22 as

$$NF = 10\log_{10}\left[\frac{\dfrac{S_{in}}{N_{in}}}{\dfrac{S_{out}}{N_{out}}}\right] = 10\log_{10}\left[\frac{S_{in}}{S_{out}} \times \frac{N_{out}}{N_{in}}\right] = 10\log_{10}\left[\frac{1}{G} \times \frac{N_{out}}{N_{in}}\right], \tag{5.23}$$

where G is the power gain of the amplifier.

By definition, an amplifier's noise figure is measured using a resistive source impedance, and the only noise present at the amplifier's input is thermal-noise originating in the resistive source. The source is further specified to be at a temperature of $290^{\circ}K$ [28]. The most common technique for measuring noise figure uses matched terminations. If a 50 Ω input impedance amplifier is being measured, a 50 Ω source impedance is used. A 3.0 dB noise figure would indicate that the amplifier contributed the same amount of noise as there was thermal-noise already present at the amplifier input. The output SNR would therefore by reduced by 3.0 dB.

The use of noise factor is particularly helpful in understanding the overall noise performance of a system, such as a receiver, that consists of a number of individual subsystems. Figure 5.7 illustrates the noise factor associated with an amplifier that consists of three separate stages. There is a resistive source impedance R with an associated thermal-noise. The first stage of the amplifier has a noise factor F_1 and power gain G_1. The second has a noise factor F_2 and power gain G_2 while the third has a noise factor F_3 and power gain G_3. In his original work on noise figure, Friis showed that the overall noise factor for this combination is given by [27]

$$F = F_1 + \frac{F_2 - 1}{G_1} + \frac{F_3 - 1}{G_1 G_2}. \tag{5.24}$$

The conclusion that we can draw from Eq. 5.24 is that if the gain of the first stage is large, the noise performance of a cascade of subsystems will be dominated by the noise of the first-stage. In an optical receiver, the combination of the photodetector, any associated photodetector biasing circuitry, and the first stage(s) of the amplifier immediately following the photodetector are known as the receiver front-end. It is this receiver front-end that will generally be the principal factor in determining the overall noise performance and sensitivity of the optical receiver.

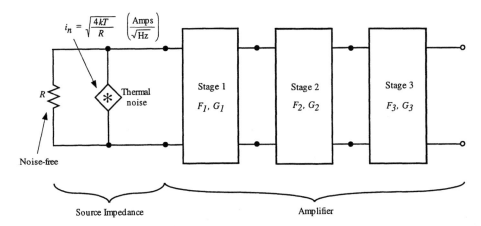

Figure 5.7 Noise figure of a cascade of subsystems.

In spite of its popularity, noise figure has several drawbacks when used to describe the noise performance of amplifiers intended for use in optical receivers. Noise figure cannot describe the noise performance of an amplifier for a variety of source impedances. Noise figure is defined using a resistive source. It can provide misleading results when used with high impedance current sources or low impedance voltage sources and neglects capacitive or inductive effects in sources. The small-signal equivalent circuit for a photodiode that we used in Chapter 4 indicated that a photodiode has much different impedance characteristics than a 50 Ω resistor. A photodiode generally appears to be a capacitive current source to the first stage amplifier in the receiver. Being dominated by capacitance, the magnitude of the source impedance is not constant but varies continuously with frequency. This presents quite a challenge to any analysis based on noise figures determined using a constant value resistive termination.

A technique that overcomes the drawbacks of noise figure is to model an amplifier using equivalent noise sources, as illustrated in Fig. 5.8. The noisy amplifier shown in Fig. 5.8(a) is replaced by an independent voltage-noise generator, an independent current-noise generator, and a noise-free amplifier, as shown in Fig. 5.8(b). The voltage-noise has a noise spectral-density of $v_n(\omega)$ volts per-root-Hz, while the current-noise source has a noise-spectral density of $i'_n(\omega)$ amps per-root-Hz.

In addition to the two noise generators, a complex correlation coefficient that accounts for the real and imaginary parts of the cross-correlation between the sources is required. This modeling technique has been shown to completely characterize the noise performance of an electronic amplifier [29, 30]. Note that both types of noise sources are required for the amplifier noise to be accurately represented for a wide range of source impedances. For example, if only the current-noise source was used and there was a short circuit across the amplifier's input, the noise equivalent model would predict zero noise at the output, an unrealistic result. Similarly, if only the voltage-noise source was used and the amplifier input was open-circuited, the model would again predict noise-free operation.

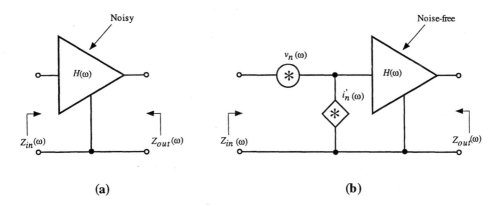

Figure 5.8 **Noise model of an electronic amplifier. (a) Noisy amplifier.**
(b) Voltage-noise and current-noise equivalent model .

The correlation between the voltage-noise source and the current-noise source can be accounted for by splitting the current-noise source into two parts as [29-31]

$$i_n''(\omega) = i_n(\omega) + i_{corr}(\omega). \tag{5.25}$$

The first term, $i_n(\omega)$, is uncorrelated with the voltage-noise while the second term, $i_{corr}(\omega)$, is correlated with the voltage source. The amount of correlation can be expressed by a complex correlation admittance $Y_c(\omega)$ as

$$i_{corr}(\omega) = v_n(\omega) Y_c(\omega). \tag{5.26}$$

This correlation admittance can be related to the conventional complex correlation coefficient $\gamma(\omega)$ used in Eq. 5.4 by [30]

$$Y_c(\omega) = \gamma(\omega) \sqrt{\frac{\overline{v_n^2(\omega)}}{\overline{i_n^2(\omega)}}} = \frac{\overline{i_n(\omega)v_n^*(\omega)}}{\overline{v_n^2(\omega)}}. \tag{5.27}$$

The use of the $v_n - i_n$ model is illustrated in Fig. 5.9. Figure 5.9(a) illustrates a receiver consisting of a signal source such as a photodetector and an amplifier. For the time being we will ignore any noise sources that are internal to the source and we will concentrate solely on amplifier noise. The signal-source is modeled as a current-source $i_s(\omega)$ in parallel with a source-admittance $Y_s(\omega)$. The amplifier has an input admittance $Y_{in}(\omega)$ and a voltage transfer function $H(\omega)$. The noise performance of the amplifier is represented by the voltage-noise spectral-density $v_n(\omega)$, the current-noise spectral-density $i_n(\omega)$, and the correlated current-noise spectral-density $i_{corr}(\omega)$.

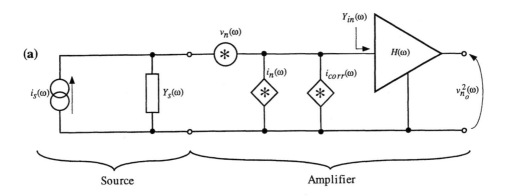

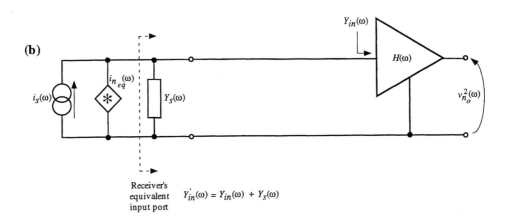

Figure 5.9   An amplifier and a signal source. (a) Noise model. (b) Noise model using an equivalent input current noise.

We are interested in determining the overall noise performance of the circuit. In particular we would like to obtain an equivalent input current-noise $i_{n_{eq}}(\omega)$ as shown in Fig. 5.9(b). This single equivalent source can completely characterize the noise performance of the source and amplifier combination. Equivalent input noise is a powerful circuit analysis tool. It allows a direct comparison between the signal and the noise and can be used to directly compare the performance of various source-amplifier combinations. It allows amplifiers to be directly compared regardless of each amplifier's absolute gain, impedance, or transfer function shape. Using an equivalent input current noise (as opposed to an equivalent input voltage-noise) is particularly advantageous for optical receivers because it allows a direct comparison between the photocurrent and the noise currents.

Our approach to analyzing noise in linear circuits follows the generally accepted noise analysis principles discussed by Motchenbacher [32]. We first identify the circuit's output node and determine the total noise power at that point. The total noise power resulting from a combination of independent noise sources is determined by using the principle of superposition in linear circuits [33]. When determining the contribution for a specific noise source, all other independent voltage-noise sources are replaced by short-circuits and all other independent current-noise sources are replaced by open circuits. The total noise power is then the sum of the noise powers obtained with each independent source. Once the total output noise power is determined, we will refer the noise back to the input of the circuit to obtain an equivalent input noise density. This is done by dividing by the appropriate power transfer function that relates the selected output node to the input port.

For the circuit in Fig. 5.9(a), we need to obtain an expression for the total noise at the output of the amplifier as a function of frequency. We start with the noise at the output due to the current-noise. Superposition allows us to replace the voltage-noise source by a short-circuit and the correlated current-noise source by an open-circuit. A conventional circuit analysis then shows that the noise power at the amplifier output is given by

$$v_{n_{o-1}}^2(\omega) = i_n^2(\omega) \frac{|Z_s(\omega)|^2 |Z_{in}(\omega)|^2}{|Z_s(\omega) + Z_{in}(\omega)|^2} |H(\omega)|^2. \tag{5.28}$$

For the voltage-noise source, the current-noise source is replaced by an open-circuit. The correlated current-noise must be retained since it is a dependent source of the voltage-noise. The output noise for the voltage-noise and the portion of the current-noise that is correlated with the voltage-noise is given by

$$v_{n_{o-2}}^2(\omega) = v_n^2(\omega) \frac{|Z_{in}(\omega)|^2 [1 + Z_s(\omega)Y_c(\omega)] \times [1 + Z_s^*(\omega)Y_c^*(\omega)]}{|Z_s(\omega) + Z_{in}(\omega)|^2} |H(\omega)|^2. \tag{5.29}$$

The transfer function that relates the output voltage $v_{n_o}(\omega)$ to the input signal current $i_s(\omega)$ is a given by

$$Z_t(\omega) = \frac{v_{n_o}(\omega)}{i_s(\omega)} = \frac{Z_s(\omega)Z_{in}(\omega)}{Z_s(\omega) + Z_{in}(\omega)} H(\omega) \quad (\Omega). \tag{5.30}$$

Note that Eq. 5.30 has the dimensions of an impedance. Since it relates the output voltage to the input current, it is known as the *transimpedance* of the receiver.

The total output noise is simply the sum of Eq. 5.28 and Eq. 5.29. The equivalent input current-noise is obtained by dividing the total output noise by the magnitude-squared of the receiver transimpedance as

$$i_{n_{eq}}^2(\omega) = \frac{v_{n_{o-1}}^2(\omega) + v_{n_{o-2}}^2(\omega)}{|Z_t(\omega)|^2}. \tag{5.31}$$

By substituting Eqs. 5.28-5.30 into 5.31, multiplying the resulting expression out, and collecting like terms, the equivalent input current-noise due to electronics noise is

$$i_{elec}^2(\omega) = i_{n_{eq}}^2(\omega) = i_n^2(\omega) + v_n^2(\omega)|Y_s(\omega) + Y_c(\omega)|^2. \tag{5.32}$$

The fact that the multiplier for the voltage-noise term contains the complex source admittance gives rise to a circuit design procedure known as noise-matching or noise-tuning. Conceivably, if it were possible to use a matching network to make $Y_s(\omega) + Y_c(\omega)$ nearly zero over most of the receiver bandwidth while maintaining adequate amounts of signal photocurrent flowing into the amplifier, the receiver's total input noise would be reduced, possibly by a substantial amount. This is one of the motivations for the use of resonant circuits in low-noise receivers [34]. Their specific use in optical receivers will be further described in Chapter 6.

If we rearrange Eq. 5.25 and Eq. 5.26 as

$$i_n(\omega) = i_n'(\omega) - i_{corr}(\omega),$$

$$Y_c(\omega) = \frac{i_n'(\omega) - i_n(\omega)}{v_n(\omega)},$$

(5.33)

and substitute the results into Eq. 5.32, we obtain an alternative expression for the equivalent input current-noise due to electronics noise as

$$i_{elec}^2(\omega) = i_{n_{eq}}^2(\omega) = i_n'^2(\omega) + v_n^2(\omega)|Y_s(\omega)|^2 + 2\operatorname{Re}\left[Y_s(\omega)Y_c^*(\omega)\right]\left|v_n^2(\omega)\right|^2.$$

(5.34)

In obtaining Eq. 5.34 we have made use of the following identities that govern the products of the correlated and uncorrelated noise sources:

$$i_n(\omega) \times i_{corr}(\omega) \equiv 0,$$
$$i_n(\omega) \times v_n^*(\omega) \equiv 0,$$
$$i_n'^*(\omega)v_n(\omega) \equiv Y_c^*(\omega)|v_n(\omega)|^2.$$

(5.35)

There are two important conclusions that we want to draw from Eq. 5.32 and Eq. 5.34. The first comes from the fact that the input impedance of the amplifier does not appear in either equation. This means that the input impedance of the first stage amplifier does not influence the noise performance of the receiver. A common misconception is that low input-impedance amplifiers are inherently "noisier" than high input-impedance amplifiers. In fact, the input impedance does not directly influence the noise performance [34]. The type and number of components used to construct the various amplifiers makes the difference. This fact also leads us to a simpler way of obtaining the equivalent input current-noise when $v_n(\omega)$, and $i_n(\omega)$, are known. The equivalent input current-noise $i_{n_{eq}}(\omega)$ can be obtained by determining the short-circuit current appearing at the input of the noiseless amplifier. This is the same procedure one would use when obtaining a Norton equivalent circuit for the source and is often a simpler approach than computing both the total output voltage-noise and the transimpedance.

The second conclusion comes from the presence of the source admittance in the expression for the total-noise. Since a photodetector is essentially a capacitive current-source, its corresponding source admittance is approximately $Y_s(\omega) = j\omega C_d$. Substituting this into Eq. 5.32 and ignoring correlation for the time being by setting $Y_c(\omega) = 0$, we obtain

$$i_{elec}^2(\omega) = i_n^2(\omega) + v_n^2(\omega)\omega^2 C_d^2.$$

(5.36)

Note that the influence of the voltage-noise increases with frequency. The detector capacitance acts as a high-pass filter to the voltage-noise source of the amplifier. At low frequencies the impedance of the capacitor is large and the contribution from the voltage-noise to the overall current flowing in the circuit is small. At high frequencies the capacitor's impedance decreases, causing the amount of circulating current due to the voltage-noise source to increase. It is not just the photodiode capacitance that determines the contribution from the amplifier's voltage-noise. Any capacitance present across the input node that allows the voltage-noise to generate a circulating current will play a role. Consequently, any stray capacitance must also be minimized if low-noise operation is to be achieved.

5.4 Photodetector Dark-Current Noise

A photodetector's dark-current was introduced in Chapter 4. Dark-current flows whether the photodetector is illuminated or not and results from the presence of current leakage paths in the photodetector.

In photoconductors, the leakage path comes from the residual conductance that is always present in the semiconductor material. In photodiodes, the leakage currents come from surface recombination and bulk leakage. Leakage occurs because of the small reverse current that flows because of thermally excited carriers, generation-recombination in the depletion region, and tunneling between the conduction and valence bands. Dark-current generally increases with both applied photodiode bias voltage and with photodiode temperature. Conversely, it can be reduced by cooling the photodetector.

Since the dark-current in a photodiode flows through a *p-n* junction just as the signal photocurrent does, the dark-current will also have a shot-noise term associated with it. The shot-noise due to dark-current can be accounted for by a current-noise source in parallel with the dark-current generator, as shown in Fig. 5.10

For photodiodes such as the APD, where the dark current consists of both a multiplied dark-current i_{dm} and an unmultiplied dark current i_{du}, there will be a shot-noise generator associated with both dark current terms.

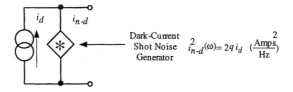

$$i_{n\text{-}d}^2(\omega) = 2q\, i_d \quad (\frac{\text{Amps}^2}{\text{Hz}})$$

Figure 5.10 Photodetector dark-current noise model.

In photoconductors, there is no potential barrier that the dark current must cross, so these types of photodetectors are not modeled using a dark-current shot-noise generator. The noise associated with the dark-current in a photoconductor is instead governed by the generation-recombination noise as described in Section 5.6.

5.5 Photodetector Thermal-Noise

Since thermal-noise is present in all electronic elements that contain dissipative resistance, there will be thermal noise present in the resistances in a photodetector. These resistances were first illustrated in Figure 4.15. There is a parallel resistance R_p and a series resistance R_s. The thermal noise associated with these resistances can be accounted for with a current-noise generator in parallel with R_p and a voltage-noise generator in series with R_s, as illustrated in Fig. 5.11.

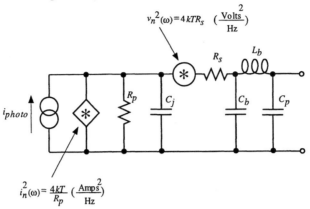

Figure 5.11 Thermal-noise sources in a photodetector.

The impact of photodetector thermal noise will be influenced by how the photodetector is terminated. In most cases, the effect of photodetector thermal noise is best accounted for by incorporating the noise sources into the calculations for the equivalent input current noise due to the electronics, i_{elec}.

In a photodiode, the parallel resistance is the junction leakage resistance and is typically so large that it and any associated current-noise can be safely ignored. The series resistance is the contact resistance of the semiconductor and series resistance in any bond-wires and mounting circuitry. It is typically less than 10 Ω. In a photoconductor, the parallel resistance represents the combination of the residual conductance and the photo-induced change in the semiconductor's conductance and can be an important contributor to the overall noise performance. The separate series resistance from mounting and contact resistance is usually ignored since, to first order, any series resistance will appear simply as a decrease in the residual conductance. The total conductance of a photoconductor depends on the amount of illumination. The amount of thermal noise present is therefore also dependent on illumination.

5.6   Photoconductor Generation-Recombination Noise

In addition to the thermal noise produced in a photoconductor there is also an associated current noise. This noise can be thought of as random fluctuations in the mean photoconductance. The fluctuations arise from the statistical nature of carrier generation recombination and trapping in semiconductors and insulators [18]. Generation-recombination noise is frequently confused for electronic shot-noise because the amount of G-R noise observed is a function of the total current flow. In fact, there is no electronic shot-noise in a photoconductor and G-R noise is somewhat more similar to thermal-noise than it is to shot-noise [18, 35]. In spite of any similarity, it is a distinctly different mechanism. In G-R noise, the conductance of the material fluctuates as the number of carriers randomly changes. In thermal noise, the instantaneous velocity of the carriers is changing, but the number of carriers is relatively constant.

For a photoconductor operating with a constant bias voltage and an average DC current (photocurrent plus dark-current) of i_{DC}, the random fluctuations in conductance due to generation, recombination, and trapping of carriers will induce random fluctuations on the DC photocurrent. This can be modeled as a parallel current-noise generator with a value of

$$i_{GR}^2(\omega) = 4qi_{DC}\left(\frac{\tau_{carrier}}{\tau_{transit}}\right)\frac{1}{1+\left(\dfrac{\omega}{\omega_c}\right)^2} \cong 4q|G(\omega)|^2 RP_{rcvd} \quad \left(\frac{\text{amps}^2}{\text{Hz}}\right), \tag{5.37}$$

where

ω_c = cutoff frequency = $\dfrac{1}{\tau_{carrier}}$ = the bandwidth of the photoconductor,

$i_{DC} = \bar{i}_p + i_d$,

$\bar{i}_p$ = average photocurrent = $\bar{P}_{rcvd}RG$,

P_{rcvd} = average received optical power,

R = responsivity = $\dfrac{\eta q}{hv}$,

G = photoconductor gain = $\left(\dfrac{\tau_{carrier}}{\tau_{transit}}\right)$.

The spectrum of the noise that is found in a photoconductor is illustrated in Fig. 5.12. At low frequencies there will be a 1/f noise component. The midband region is dominated by G-R noise. Above the photoconductor cutoff frequency that is determined by the carrier lifetime, the thermal noise in the photoconductor is the dominant noise source. The highest signal-to-noise ratios are usually obtained in the region dominated by G-R noise. This is where the photoconductive gain will be amplifying the signal and quantum shot-noise and other receiver noise terms will be less significant.

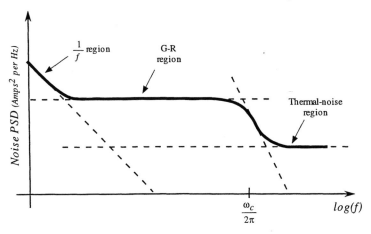

Figure 5.12 Spectrum of noise sources found in a photoconductor.

If we combine the current noise from generation-recombination with the thermal noise from the photoconductor, the equivalent circuit of Fig. 5.13 is obtained [35-38].

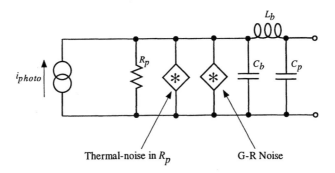

Figure 5.13 Equivalent circuit of photoconductor with G-R and thermal-noise.

It is important to note that a photoconductor does not exhibit electronic shot-noise as do photodiodes because there is no potential barrier that charges must cross. Shot-noise is proportional to $2qI_{DC}$, while G-R noise is proportional to $4qI_{DC}$. Some authors refer to G-R noise in photoconductors as "shot-noise" or "twice shot-noise" to be consistent with the need for quantum shot-noise to be included in our model for photodetection. Regardless of the terminology used, the quantum shot-noise associated with the signal can be thought of as being embedded within the G-R noise in a photoconductor.

5.7 APD Excess Noise

We saw in Chapter 4 that the avalanche multiplication process in an APD can result in the generation of a multitude of photocarriers for each photon that is absorbed. When a constant photocurrent is amplified by a current multiplication factor M, the DC signal power is equal to the square of the mean (i.e. time average) of the DC photocurrent times the square of the expected value (i.e. statistical average) of the multiplication as

$$P_s = \bar{i}_{DC}^2 E\{M\}^2 \quad \text{(amps}^2\text{)}. \tag{5.38}$$

The noise power is given by the variance in the photocurrent due to quantum shot noise times the expected value of the square of the multiplication:

$$P_n = 2q\bar{i}_{DC}E\{M^2\}B \quad \text{(amps}^2\text{)}. \tag{5.39}$$

The SNR after multiplication is given by

$$SNR = \frac{P_s}{P_n} = \frac{\bar{i}_{DC}^2 E\{M\}^2}{2q\bar{i}_{DC}E\{M^2\}B} = \frac{i_{DC}}{2q\bar{i}_{DC}B} \times \frac{E\{M\}^2}{E\{M^2\}}. \tag{5.40}$$

For a deterministic noise-free multiplication process, the square of the expected value is equal to the expected value of the square and the SNR after multiplication is identical to the SNR before multiplication. The signal power and the noise power are both simply amplified by M^2.

The avalanche multiplication process in an APD is statistical in nature. The number of electrical carriers resulting from the absorption of a single photon is dependent upon where in the absorption region the photon was absorbed, the type of carrier (hole or electron), the local magnitude of the electric field, the local doping density of the semiconductor, and the path the carriers travel through the semiconductor. This results in the mean of the square being larger than the square of the mean, and the SNR after multiplication is lower than it was before multiplication.

We can account for the reduction in SNR due to the multiplication process by modifying Eq. 5.39 to be

$$P_n = 2q\bar{i}_{DC}M^2 F(M)B \quad (\text{amps}^2) \tag{5.41}$$

where M is understood to be the expected value of the multiplication gain and $F(M)$ is an excess noise factor defined as the ratio of $E\{M^2\}$ to $E\{M\}^2$ [36-40].

The value of the excess noise factor depends on the APD semiconductor material, whether holes and/or electrons are the carriers undergoing impact ionization, and the electric field profile in the APD. Excess noise is particularly sensitive to the ratio of the ionization coefficients for electrons and holes. It is smallest in devices where only one type of carrier undergoes impact ionizations. It is largest in devices in which both holes and electrons produce ionizing collisions [39, 40]. The excess noise factor can be expressed in terms of the ionization coefficient k_{eff} and the mean multiplication as [39]

$$F(M) = k_{eff}M + \left(1 - k_{eff}\right)\left(2 - \frac{1}{M}\right). \tag{5.42}$$

Equation 5.42 is sometimes loosely approximated by

$$F(M) \cong M^x \quad (0.1 < x < 1.0). \tag{5.43}$$

Figure 5.14 illustrates the excess noise factor of Eq. 5.43 for a range of multiplication factors and effective ionization ratios.

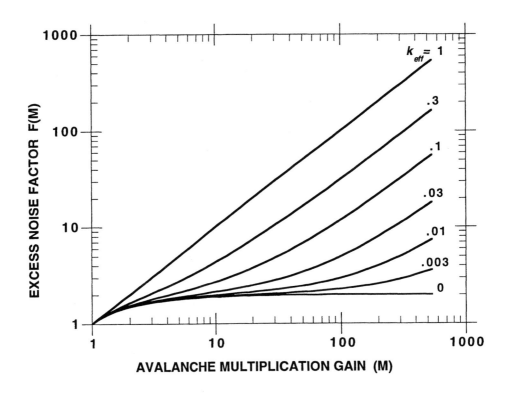

Figure 5.14 APD noise.

Modifying our APD equivalent circuit as illustrated in Fig 5.15(a) allows us to account for the noise sources within the APD. An APD's dark current consists of both a multiplied dark-current i_{dm} and an unmultiplied dark current i_{du} and there is a shot-noise generator associated with each dark current term. The shot noise associated with the multiplied dark-current is increased by the excess-noise factor $F(M)$. The signal photocurrent i_s is multiplied by the average multiplication factor M, while the shot-noise power associated with the signal is multiplied by the square of the multiplication gain and the excess-noise factor.

There is a subtle assumption in our use of the same excess noise factor for both the signal quantum shot-noise and the multiplied dark-current electronic shot-noise. Signal carriers are typically generated near the edge of the depletion region. Bulk dark currents are generated throughout the entire device volume. Thus carriers associated with bulk dark current may see a different effective multiplication than do signal related carriers. In many cases the difference is not substantial, but when the dark-current electronic shot-noise is comparable to the amount of signal quantum shot-noise, improved accuracy can be obtained if different values of $F(M)$ are used for each noise source [41].

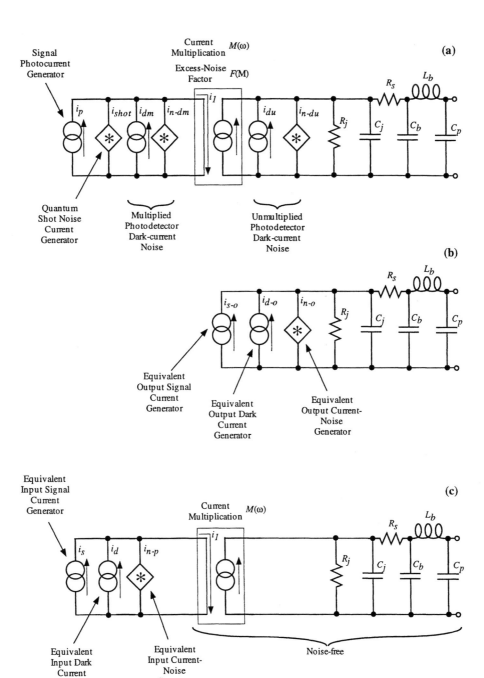

Figure 5.15 Equivalent circuit of APD. (a) Complete model. (b) Simplified output model with equivalent current sources. (c) Simplified input model with equivalent current sources.

We can simplify the equivalent-circuit for the APD by defining equivalent current generators as illustrated in Fig 5.15(b). They roughly correspond to the short circuit currents that would be measured at the output of the APD. In fact they represent currents that have been referred to the output of the multiplication region but before the photodetector small-signal model. The equivalent output signal current $i_{s_{out}}(\omega)$, the equivalent output DC dark current generator $i_{d_{out}}$, and the power spectral-density of the equivalent output current-noise generator $i_{n_{out}}(\omega)$ are given by

$$
\begin{aligned}
i_{s_{out}}(\omega) &= M(\omega)\frac{\eta q}{hv}P_{rcvd}(\omega), \\
i_{d_{out}} &= i_{du} + i_{dm}M(0), \\
i_{n_{out}}^2(\omega) &= 2qi_{du} + 2qi_{dm}M^2(\omega)F\big(M(\omega)\big) + i_{shot}^2M^2(\omega)F\big(M(\omega)\big),
\end{aligned}
\tag{5.44}
$$

where i_{shot} corresponds to the quantum shot-noise associated with the primary (unmultiplied) photocurrent generated by the received signal.

Alternatively, we could define equivalent input current generators as shown in Fig. 5.15(c). In this case, the equivalent input signal current $i_s(\omega)$, the equivalent input DC dark current generator i_d and the power spectral-density of the equivalent input photocurrent current-noise generator $i_{n_{eq}}(\omega)$ are given by

$$
\begin{aligned}
i_s(\omega) &= \frac{\eta q}{hv}P_{rcvd}(\omega), \\
i_d &= \frac{i_{du}}{M(0)} + i_{dm}, \\
i_{n_{eq}}^2(\omega) &= \frac{2qi_{du}}{M^2(\omega)} + \Big[2qi_{dm} + i_{shot}^2\Big]F\big(M(\omega)\big).
\end{aligned}
\tag{5.45}
$$

These are termed input currents since they correspond to the equivalent internal APD currents that occur at the input to the multiplication region. They are not measurable currents, but they are calculable once the avalanche multiplication is known. The benefit to the use of input current generators is that the effects of the avalanche multiplication transfer function and APD excess noise factor can be immediately accounted for in the input current-noise calculation. It is possible to do accurate receiver noise calculations with either model as long as we keep track of the location that we are calling the input.

The excess noise factor by itself is not sufficient to fully characterize the effects of the excess noise in an APD. The statistics of the multiplication must also be known. In Chapter 7 we will complete our analysis of excess-noise in an APD by examining the probability density function's influence on digital receiver design.

5.8 Optical Excess-Noise

Optical excess-noise can be broadly defined as any noise that appears along with the received signal, other than quantum shot-noise. It is particularly important for the receiver designer to be aware of the sources of these noises because their impact is generally first observed at the receiver, and they are frequently mistaken for other receiver effects. The exact amount of optical excess-noise that will be present in a receiver is difficult to predict because of the variety of possible sources and the dependence of the excess-noise sources on the system layout and installation [42]. Regardless of the source of the noise, we can account for optical excess noise as an additional current-noise source, as illustrated in Fig. 5.16.

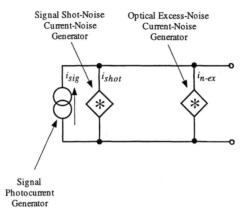

Figure 5.16 Optical excess-noise current-noise generator.

Optical excess-noise can arise from a variety of sources, including imperfections in the laser, cross-coupling between laser frequency-noise and amplitude-noise due to fiber nonlinearity, mixing between waveguide modes in a multimode fiber or polarization-modes in single-mode fiber, frequency-modulation to amplitude-modulation conversions in cavities formed by small reflections, or optical feedback that destabilizes a laser's output. The most common types of optical excess-noise can be grouped into three major categories, laser intensity-noise, modal noise, and mode-partition-noise.

5.8.1 Laser Intensity-Noise

The output amplitude of a laser is not perfectly constant. It is actually constantly fluctuating around an average value [43-47]. The fluctuations are due to temperature and acoustic disturbances as well as fundamental quantum-mechanical considerations. These fluctuations can be accounted for by modeling the output power from a laser as the sum of a constant term and a time-varying term as

$$P(t) = P_o + \delta P(t) \quad \text{(watts)}. \tag{5.46}$$

The long-term time-average of the time-varying term is zero. The average value is then simply P_o.

If the optical signal described by Eq. 5.46 is detected it will generate a photocurrent given by

$$i_p(t) = \frac{\eta q}{h\nu} P(t) = \frac{\eta q}{h\nu} P_o + \frac{\eta q}{h\nu} \delta P(t) = i_{DC} + i_{AC}(t). \tag{5.47}$$

The photocurrent now consists of a DC term and an AC term. The fluctuations in photocurrent represented by the AC photocurrent term amount to a noise source that can exceed the quantum shot-noise due to the average DC photocurrent. As with any noise source, excess-noise will have an average noise power and a power spectral density.

The amount of noise present in an optical source is described by a parameter known as the relative-intensity-noise, or RIN [44]. RIN is defined as the mean-square noise power divided by the mean power squared. In the frequency domain, the expression for RIN is given by

$$RIN(\omega) = \frac{\overline{|\delta P(\omega)|^2}}{\overline{P}^2}. \tag{5.48}$$

In terms of the measured photocurrent, RIN can be expressed as

$$RIN(\omega) = \frac{i_n^2(\omega)}{i_p^2}, \tag{5.49}$$

where $i_n^2(\omega)$ is the measured current-noise power spectral density in amps^2/Hz that is associated with the average DC photocurrent i_p. Since RIN is the ratio of a noise-power spectral-density to a power, it has the units of inverse frequency. RIN is usually expressed in dB per Hz as

$$RIN(\omega) = 10\log\left(\frac{i_n^2(\omega)}{i_p^2}\right) = \quad (\frac{dB}{Hz}). \tag{5.50}$$

Since there is an inherent quantum shot-noise associated with the generation of photocurrent, the quantum shot-noise will be the lower limit for RIN. The lower limit, or quantum shot-noise limited RIN, in terms of observed photocurrent is given by

$$RIN_{QL} = \frac{i_n^2(\omega)}{i_p^2} = \frac{2qi_p}{i_p^2} = \frac{2q}{i_p} = \frac{3.2\times10^{-19}}{i_p} \quad (\frac{1}{Hz}). \tag{5.51}$$

At a photocurrent of 0.1 mA, the quantum-limited RIN is -145 dB/Hz. In terms of optical power in milliwatts and optical wavelength in microns, RIN is given by

$$RIN_{QL} = \frac{2h\nu}{P_{opt}} = \frac{3.96\times10^{-16}}{\lambda(\text{microns})\,P_{opt}(\text{mW})} \quad (\frac{1}{Hz}). \tag{5.52}$$

At a wavelength of 1.55 microns and a laser output power of 1.0 mW, the quantum limited RIN is -156 dB/Hz. The amount of RIN present in a laser is dependent upon semiconductor material properties and device structure and can vary depending on the technique used to modulate the laser [45, 46]. Realistic semiconductor lasers operating well above threshold have RIN values between -150 and -130 dB/Hz [44]. Solid-state lasers such as the Nd:YAG have very low RIN levels, approaching that of the quantum-noise limited RIN.

The impact of RIN on a receiver can be accounted for by including an additional current-noise generator in parallel with the quantum shot-noise current-noise generator. An associated DC photocurrent generator is not needed to model RIN. The value of the RIN current-noise generator is given by

$$i_{RIN}^2(\omega) = RIN(\omega)i_p^2 - 2qi_p \quad (\frac{\text{amps}^2}{\text{Hz}}). \tag{5.53}$$

In this definition we subtract-off the quantum shot-noise term from the RIN. Thus, if we had a quantum shot-noise limited laser source, i_{RIN}^2 would equal zero. In this definition we do not allow values of i_{RIN}^2 to be less than zero. Another definition of RIN that is also in use does not remove the shot-noise component.

The primary impact of RIN is to increase the amount of noise present for a given received optical signal level. RIN is a signal-dependent noise and its effects increase with increasing signal power. Consequently, it directly degrades the receiver's SNR. Analog systems generally require higher received signal powers and higher SNRs to provide an acceptable quality-of-service than do digital systems. Analog systems are therefore more sensitive to RIN effects than are digital systems and their ultimate performance can in fact be limited by the RIN of the transmitter laser [48]. Digital systems can often tolerate RIN levels as high as -120 to -100 dB/Hz.

5.8.2 Modal-Noise

Another signal-dependent noise is known as modal noise. There is a possibility of modal-noise whenever there is more than one propagating mode present in the communication system [49]. When a monochromatic source, such as a laser, is used as the transmitter in a multimode fiber system, microscopic variations in the fiber cause the laser power to be distributed among the various waveguide modes that the fiber allows to propagate. Each mode will travel at a slightly different velocity, and the transmitter's energy is consequently spread across both frequency and time. The distribution of power among the modes is altered when the signal passes through discontinuities such as splices, couplers, and connectors. As the various modes propagate they interfere with each other. At the receiver, this is observed as a speckle pattern [3, 50] on the surface of the photodetector. This relationship to laser speckle has caused modal-noise to also be termed speckle-noise.

The structure of the speckle pattern depends on the phase and amplitude relationships between the various modes and is subject to subtle variations in source frequency, the temperature and vibration of the fiber, and any associated discontinuities in the fiber. Changes in modal power distribution at the discontinuities in the system introduce an additional intensity-noise into the received signal that is similar to RIN. Since most of the disturbances are environmental in nature they will be at comparatively low-frequencies. The magnitude of the fluctuations can be large, however, causing substantial degradation in performance.

The amount of modal-noise present is highly dependent on the exact details of the link design [51]. A good technique to combat modal noise in multimode fiber systems is to introduce rapidly fluctuating modal-noise into the system [52]. This is essentially a form of time-averaging. The amplitude of the noise within the receiver bandwidth is reduced by spreading the noise over a wider frequency spectrum. This is typically accomplished by using broadband sources such as low-coherence lasers, superluminescent diodes, or LEDs [53].

Modal-noise can also occur in single-mode fibers. Here the effect is generally due to mixing between the two polarization modes [54], operating the fiber in a regime where there are actually two modes propagating [55], or multiple reflections in the system causing interferometers to form [56].

5.8.3 Mode-Partition-Noise

A third signal-dependent noise is known as mode-partition-noise [43-47, 57-61]. Whereas modal noise referred to the various propagating modes in a fiber, mode-partition-noise refers to effects caused by the side-modes present in the transmitter laser. Mode-partition-noise is related to laser relative intensity-noise and in some sense they are different manifestations of the same effect, the constant fluctuations that are present in a laser's output power. In discussing intensity-noise, we were concerned with the amount of fluctuation present in the entire output spectrum of the lasers. With mode-partition-noise we are concerned with how the fluctuating power is distributed (the partitioning) among the components in the laser's output spectrum.

The output of a laser is not a single frequency; it is instead composed of a main lasing mode and a number of relatively low-amplitude side-modes. The power fluctuations in the overall output are comparatively low but the power fluctuations in a single mode can be quite large. As these modes propagate through a system, any chromatic dispersion that will separate these modes in time, or any filtering that will change the amplitude and phase relationships among of the modes, will introduce an additional noise source in the receiver when the signal is ultimately photodetected.

In digital systems, substantial mode-partition-noise can limit the achievable SNR, causing the formation of a bit-error-rate floor. In analog systems, mode-partition-noise can appear as an increase in the apparent RIN of the source. The increase in RIN can be particularly dramatic if the laser's main mode is filtered out by an optical filter or grating [43, 47].

5.9 Optical Background Noise

The noise in an optical receiver can also be influenced by additional non-signal related sources of optical energy that fall on the photodetector. Possible sources of an optical background include amplified spontaneous emission from optical amplifiers and crosstalk from other transmitters. It also possible to model a laser with an imperfect on-off ratio as the combination of an ideal laser and a constant background. Because there is an additional source of illumination, the equivalent circuit for optical background contains both a DC photocurrent generator and a current-noise generator, as shown in Fig. 5.17.

In general, unless the background itself contains optical excess-noise or is modulated, the noise associated with the background is usually the quantum shot-noise associated with i_b. In certain circumstances it may also be necessary to account for the specific counting probabilities associated with incoherent background light such as the Bose-Einstein or Laguerre distributions we first encountered in Chapter 3.

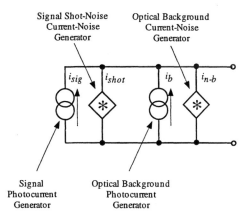

Figure 5.17 Equivalent circuit for a signal and an optical background

In free-space systems, there are easily observed sources of background noise. Broadband illumination from the sky, sun, moon, stars, or planets that are within the receiver's field-of-view can also introduce additional noise. Celestial background sources exhibit components due to both blackbody radiation and reflected light [62]. The relative contributions from each component are frequently time varying and worst-case conditions are usually assumed to bound the potential background noise degradation.

The amount of background radiation collected by a free-space optical receiver is dependent on the receiver's field-of-view as well as its optical bandwidth. To control the amount of background collected, the use of a narrow field-of-view in combination with a narrow optical band-pass filter is desirable. The receiver field-of-view cannot be arbitrarily reduced since the spatial tracking system must still be able to support tracking with the increasingly narrow beamwidths. Similarly, narrow optical filters typically have relatively high insertion losses that will attenuate both the desired signal as well as the undesired background, and they impose absolute frequency requirements on the transmitter laser frequency. As a compromise, interference filters with a few tens of Angstroms of bandwidth and insertion losses of 1-2 dB are often used.

An example of the time varying nature of an optical background is the surface of the Earth as seen by a receiver at geosynchronous Earth-orbit that is communicating with a satellite in low-Earth-orbit. With a few microradians receiver field-of-view, only a few square kilometers of surface area contribute to the optical background at any time, but the background is continuously changing as the GEO tracks the LEO satellite across the Earth. The background radiation will be dependent on the local cloud cover present, the position of the Sun, and the reflectivity of surface features such as snow and ice.

The worst case occurs for one satellite looking at another satellite that may have the Sun in the background. A single spatial-mode receiver operating near 1.0 microns that is looking at the Sun will have a background photon arrival rate of approximately 10^9 photons per second per Angstrom of optical bandwidth [63]. This is a very intense background and can cause the communication link to fail if not accounted for. A more common case is a GEO satellite looking down at LEO that has the Earth in the background. In this case the single spatial-mode receiver will experience a background count-rate of approximately 10^4 photons per second per Angstrom of optical bandwidth.

5.10   Noise Equivalent Circuit for an Optical Receiver

We can summarize the implications of the previous sections with the functional block diagram for the front-end of an optical receiver that is illustrated in Fig. 5.18. Each block in the receiver modifies the signal and introduces unwanted noise. The photon-to-photoelectron converter converts all incoming optical signals into an electrical current.

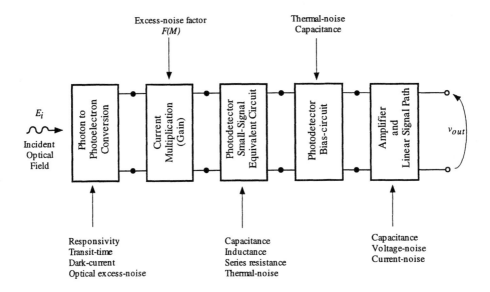

Figure 5.18 Functional block-diagram for an optical receiver front-end.

The photocurrent will contain unwanted optical-excess noise and background noise effects in addition to the desired signal and quantum shot-noise. The photodetector may have internal multiplication and introduce excess-noise. Once the photocurrents are generated, the small-signal equivalent circuit adds elements such as junction capacitance, bond-wire inductance, and series resistance with thermal-noise. The photodetector may have a bias circuit that introduces additional capacitance, resistance, and thermal noise. The resulting signals are then amplified by an electronic amplifier that introduces additional voltage-noise and current-noise.

By substituting the various equivalent circuits that we have developed in the previous sections for the functional blocks of Fig. 5.18, we can obtain the equivalent noise model illustrated in Fig. 5.19(a).

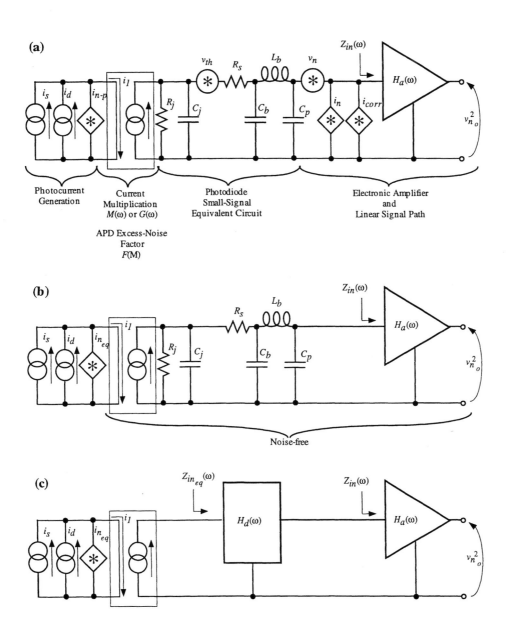

**Figure 5.19 Noise model for an optical receiver front-end. (a) Complete model.
(b) Equivalent input current-noise model. (c) Transfer-function model.**

The complete model can be simplified by using the equivalent input current-noise model of Fig. 5.19(b). The current-noise source $i_{n_{eq}}$ will contain contributions from signal shot-noise, optical excess-noise, optical background-noise, photodetector excess-noise, photodetector thermal-noise, and electronics-noise. All of the remaining elements in the receiver are noise-free. In particular the APD excess noise is accounted for in the equivalent input current-noise generator and the multiplication process in Figs. 5.19(b) and (c) are noise free.

For numerical analysis purposes the complete model can be further simplified by converting it into a transfer function model, as shown in Fig. 5.19(c). The avalanche multiplication or photoconductor gain contributes one transfer function, the small-signal model for the photodetector contributes another, and the receiver front-end (or linear-signal path) contributes a third. There is now an effective input impedance $Z_{in_{eq}}(\omega)$ that accounts for the input impedance of the amplifier as well as the loading effects due to the photodetector equivalent circuit.

The expressions for the current sources in the models illustrated in Fig. 5.19(b) and 5.19(c) depend on the type of photodetector being used in the receiver. When the photodetector is an APD there is internal gain. The expressions for the APD's signal current $i_s(\omega)$, the receiver's equivalent input current-noise spectral density $i_{n_{eq}}(\omega)$, and the effective dark current i_{dk} are then

$$i_s(\omega) = \frac{\eta q}{h\nu} P_{rcvd}(\omega),$$

$$i_{n_{eq}}^2(\omega) = \frac{i_{elec}^2(\omega) + 2qi_{du}}{|M(\omega)|^2} + \left[2qi_{dm} + i_{shot}^2 + i_{excess}^2 + i_{nb}^2\right]F(M(\omega)),$$

(5.54)

$$i_{dk} = \frac{i_{du}}{M} + i_{dm} + i_b,$$

where

$M(\omega)$ = avalanche multiplication transfer-function,
$i_{elec}(\omega)$ = equivalent input current-noise from the receiver electronics including photodiode thermal noise,
i_{du} = unmultiplied dark-current,
i_{dm} = multiplied dark-current,
i_{shot} = quantum shot-noise associated with the signal photocurrent,
i_{excess} = optical excess-noise,
i_b = DC photocurrent due to optical background,
i_{nb} = optical background noise.

We note a very important benefit of a photodetector with internal gain. The impacts of the electronics noise and unmultiplied dark-current noise are reduced by the gain in the photodetector. Since gains of 10 - 100 are possible, substantial reductions in the effects of receiver electronics noise are possible. This is a particularly significant when high sensitivity is desired.

When the photodetector is a *p-n*, *p-i-n*, or MSM photodiode, there is no internal gain and the expressions for noise and dark current change as

$$i_{n_{eq}}^2(\omega) = i_{elec}^2(\omega) + 2qi_{dk} + i_{shot}^2 + i_{excess}^2 + i_{nb}^2,$$

$$i_{dk} = i_{du} + i_b,$$

(5.55)

where i_{dk} = photodiode dark-current.

In the case of a photoconductor, there is also an internal gain mechanism but it is different from that in an APD, and the shot-noise must be replaced by the G-R noise. The expression for noise then become

$$i_{n_{eq}}^2(\omega) = \frac{i_{elec}^2(\omega)}{|G(\omega)|^2} + \frac{4kT}{R_d} + 4qi_{DC} + i_{excess}^2 + i_{nb}^2, \tag{5.56}$$

where

$G(\omega)$ = photoconductive gain transfer - function,
$i_{elec}(\omega)$ = equivalent input current - noise from the receiver electronics excluding photodetector thermal noise,
R_d = resistance of photoconductor,
$i_{DC} = i_s + i_b + Gi_d$,
i_s = DC photocurrent due to signal = $\frac{\eta q}{h\nu}\bar{P}_{rcvd}$,
i_b = DC photocurrent due to background,
i_d = dark current measured across photoconductor material,
i_{excess} = optical excess - noise,
i_{nb} = optical background noise other than G - R noise resulting from i_b.

We have now obtained expressions for $i_{n_{eq}}(\omega)$ in which the sources of noise associated with photodetection are explicitly identified and the sources of noise associated with the receiver electronics are grouped into $i_{elec}(\omega)$. When analyzing the equivalent circuit illustrated in Fig. 5.19(a) to obtain an expression for $i_{elec}(\omega)$ in terms of the photodetector small-signal equivalent circuit, the photodetector thermal-noise v_{th}, and the equivalent input noise sources for the amplifier $v_n(\omega)$, $i_n(\omega)$, and $i_{corr}(\omega)$, we follow the standard noise analysis procedures we first used with Fig. 5.9. We first compute the contribution to the output noise voltage from each noise source and combine the results. This gives the total output voltage noise density. If this output noise density is divided by the square of the magnitude of the equivalent transimpedance that relates the output voltage to the equivalent input photocurrent, the equivalent input current-noise source $i_{elec}(\omega)$ that represents all of the electronic noise sources in the receiver is obtained.

The general expression for $i_{n_{eq}}(\omega)$ expressed in terms of each individual noise source and the photodetector small-signal model is generally quite complicated. However, once the complete expression for $i_{n_{eq}}(\omega)$ is evaluated using the actual photodiode and amplifier parameters and we convert from radian frequency ω to frequency f in Hz, the equivalent input current-noise can be expressed in a power series of the form

$$i_{n_{eq}}^2(f) = i_{shot}^2 + \sum_{j=-N}^{M} a_j f^j, \tag{5.57}$$

where i_{shot} is the spectral density of the unmultiplied quantum shot-noise associated with the signal and the coefficients a_j may, depending on the details of the receiver and system implementation, be proportional to received signal power, optical background, APD gain, temperature, etc. In Eq. 5.57 we have implicitly assumed that the spectral density of the quantum shot-noise from the signal is neither frequency nor time dependent in the bandwidth and observation-time of interest. Should this not be accurate, we would simply use the expression for the frequency-dependent and/or time-dependent spectral density instead.

In many practical cases, only three or four terms are needed for an accurate representation and Eq. 5.57 can be limited to [64]

$$i_{n_{eq}}^2(f) = i_{shot}^2 + a_0 + a_1 f + a_2 f^2. \tag{5.58}$$

We note we have excluded explicit 1/f noise in this expression. This is generally a valid assumption unless a FET is used at low-frequencies [23, 25, 65].

5.11   Degradation from the Quantum Shot-Noise Limited Noise-Density

Using Eq. 5.58, we can define a figure of merit for the optical receiver that can be used to describe the noise performance of coherent, subcarrier, and broadband analog systems where we can assume that the amount of signal related quantum shot-noise is essentially constant. Let us rewrite Eq. 5.58 as

$$i_{n_{eq}}^2(f) = i_{shot}^2 + i_{rcvr}^2(f), \tag{5.59}$$

where i_{shot}^2 is the unmultiplied quantum shot-noise associated with the signal and $i_{rcvr}^2(f)$ is the equivalent input current-noise due to all of the other noise sources present in the receiver.

Now let us define a parameter $\kappa(f)$ that is indicative of the difference between the receiver's total equivalent input current noise-density and the noise-density due to the quantum shot-noise of the signal as

$$\kappa(f) = \frac{i_{n_{eq}}^2(f)}{i_{shot}^2} = \frac{i_{shot}^2 + i_{rcvr}^2(f)}{i_{shot}^2}. \tag{5.60}$$

The form of Eq. 5.60 is similar to that of a microwave amplifier's noise-figure, but we emphasize that it is not a noise-figure in the traditional sense. The receiver degradation can be expressed in dB as

$$\kappa_{dB}(f) = 10\log_{10}\left(\kappa(f)\right) = 10\log_{10}\left(\frac{i_{shot}^2 + i_{rcvr}^2(f)}{i_{shot}^2}\right) \quad \text{(dB)}. \tag{5.61}$$

Equation 5.61 compares the receiver's actual noise density to the noise density due solely to the quantum shot-noise of the signal. A receiver with $\kappa = 0$ dB would be operating exactly at the quantum limit. In practice quantum shot-noise limited performance can be difficult to obtain. A ratio of about 10:1 is required between shot-noise and receiver-noise if the receiver is to operate within 1 dB of the quantum shot-noise limit. Note that it is also possible to define the receiver degradation in terms of the *total* equivalent input current- noise instead of the noise density as we have done here.

5.12   Total Equivalent Noise Power in the Receiver

In many cases we are interested in the *total* equivalent input current-noise power. Total noise power directly impacts receiver performance since it, in combination with signal power, determines the receiver's signal-to-noise ratio. The total noise at the output of the receiver in Fig. 5.19 is given by

$$V_{n_o}^2 = \int_0^\infty v_{n_o}^2(f)\,df = \int_0^\infty i_{n_{eq}}^2(f)\left|Z_t(f)\right|^2 df \quad \text{(volts}^2\text{)}, \qquad (5.62)$$

where

$Z_t(f) = Z_{in_{eq}}(f)H_d(f)H_a(f)$ is the overall transimpedance of the receiver,

$Z_{in_{eq}}(f)$ = equivalent input impedance of receiver,

$H_d(f)$ = transfer function of photodetector small-signal model,

$H_a(f)$ = transfer function of receiver linear signal-path.

It is convenient to express the transimpedance as the product of the maximum value of transimpedance times an amplitude normalized frequency dependent part whose peak value is equal to unity. Assigning R_t as the maximum value of transimpedance and $H_t(\omega)$ as the amplitude normalized frequency dependent portion allows the transimpedance to be written as

$$Z_t(f) = R_t H_t(f). \qquad (5.63)$$

Substituting Eq. 5.63 into Eq. 5.62 yields

$$V_{n_o}^2 = R_t^2 \int_0^\infty i_{n_{eq}}^2(f)\left|H_t(f)\right|^2 df \quad \text{(volts}^2\text{)}. \qquad (5.64)$$

The equivalent input current noise is then simply

$$I_{n_{eq}}^2 = \frac{V_{n_o}^2}{R_t^2} = \int_0^\infty i_{n_{eq}}^2(f)\left|H_t(f)\right|^2 df \quad \text{(amps}^2\text{)}. \qquad (5.65)$$

If we use our truncated power series expression for the equivalent input current-noise spectral density, Eq. 5.65 can be written as

$$I_{n_{eq}}^2 = i_{shot}^2 \int_0^\infty \left|H_t(f)\right|^2 df + a_0 \int_0^\infty \left|H_t(f)\right|^2 df$$
$$+ a_1 \int_0^\infty \left|H_t(f)\right|^2 f\,df + a_2 \int_0^\infty \left|H_t(f)\right|^2 f^2 df. \qquad (5.66)$$

Recalling our definition of noise-equivalent-bandwidth that was given by Eq. 5.6, we see that the integrals in Eq. 5.66 correspond to a series of NEBs. The first integral is the NEB of the quantum shot-noise. The remaining integrals are the NEB of the frequency independent receiver noise, the NEB of the receiver noise that increases with f, and the NEB of the receiver noise that increases with f^2. The techniques we use to evaluate Eq. 5.66 will vary depending on the type of receiver being analyzed and whether the shot-noise is constant or a function of time.

5.12.1   Broad-Band Systems

In broad-band analog or coherent detection systems, the signal-shot noise can usually be considered to be a constant. In broadcast systems the quantum shot-noise is determined by the average received signal, while in coherent systems it is set by the comparatively strong local oscillator laser.

In these broad-band systems, the most desirable receiver transimpedance is often one that is maximally flat over as broad a bandwidth as practical. Let us model the frequency dependent portion of the transimpedance to be an ideal low-pass filter as

$$H_t(f) = \begin{cases} 1 & \text{for} \quad 0 < f < B, \\ 0 & \text{for} \quad f > B. \end{cases} \tag{5.67}$$

Using this model, the signal bandwidth and the noise-equivalent bandwidth will both be equal to B. The total equivalent input current-noise is

$$I_{n_{eq}}^2 = i_{shot}^2 B + a_0 B + \frac{1}{2} a_1 B^2 + \frac{1}{3} a_2 B^3. \tag{5.68}$$

The relatively rapid increase of noise with an increase in receiver bandwidth due to the terms proportional to the square and cube of the receiver bandwidth are the limiting factors in achieving low-noise performance. We will see in Chapter 6 that much of the effort involved in the design of a low noise optical receiver centers on making the coefficients a_1 and a_2 as small as possible. Note that in a realistic system we cannot expect the infinitely sharp roll-off we have assumed here. In a realistic system with a realizable transfer-function, we would use the equivalent noise bandwidth for B.

5.12.2   Narrow-Band Systems

In narrow-band systems, we can also usually assume that the shot noise is constant. The most desirable receiver transimpedance is usually one that is maximally flat over a narrow bandwidth centered around a frequency f_o. Let us model the frequency dependent portion of the transimpedance to be an ideal band-pass filter as

$$H_t(f) = \begin{cases} 0 & \text{for} \quad f < f_o - \dfrac{B}{2} \\[2mm] 1 & \text{for} \quad f_o - \dfrac{B}{2} < f < f_o + \dfrac{B}{2}, \\[2mm] 0 & \text{for} \quad f > f_o + \dfrac{B}{2}. \end{cases} \tag{5.69}$$

Using this model, the signal bandwidth and the noise-equivalent bandwidth will again both be equal to B. This total equivalent input current-noise can then be written as

$$I_{n_{eq}}^2 = i_{shot}^2 B + a_0 B + a_1 B f_o + a_2 \left(f_o^2 B + \frac{B^3}{12} \right). \tag{5.70}$$

Just as in the broadband case, the terms proportional to the square and cube of the receiver bandwidth are the limiting factors in achieving low-noise performance, although now they both modify a single coefficient a_2. Thus if it were possible to make a_2 nearly zero within our bandwidth we would obtain a lower noise receiver. This is another indication of the potential benefits of the technique known as noise-tuning that we first alluded to in Section 5.3.4.

5.12.3   Base-Band Digital Systems

With base-band digital detection the quantum shot-noise varies directly with the signal being received. This is simply a confirmation of the fact that the quantum shot-noise of a conditional Poisson process is nonstationary, a fact we first alluded to in Chapter 3. In analyzing a base-band digital receiver we must therefore account for the symbol to symbol (i. e. pulse-to-pulse) variation in the quantum shot-noise level. This will include the quantum shot-noise contribution from the symbol that is currently being received as well as the quantum shot-noise from any other nearby symbols that may overlap in time with the current symbol.

By combining our statistical model for photodetection from Chapter 3 with the model for a photodiode, we can express the signal and noise in the photocurrent appearing at the output of the photodiode as

$$i_s(t) = E\{i(t)\} = \frac{\eta q}{h v} M \int\limits_{-\infty}^{+\infty} P_{rcvd}(\tau) h'_d(t - \tau) d\tau \quad \text{(amps)},$$

$$i_n^2(t) = \text{var}\{i(t)\} = q \frac{\eta q}{h v} M^2 F(M) \int\limits_{-\infty}^{+\infty} P_{rcvd}(\tau) h'^2_d(t - \tau) d\tau \quad \text{(amps}^2), \tag{5.71}$$

where $h'_d(t)$ = normalized impulse response of the photodetector, $1 = \int\limits_{-\infty}^{\infty} h'_d(t) dt.$

Note that the format of Eq. 5.71 is equivalent to a convolution [66] of the received power waveform with the photodetector impulse response.

Let us investigate an uncoded binary system in which the transmitter sends one of two possible symbols every T seconds. Since no coding is used, the data-rate R is equal to the symbol rate, which is equal to $1/T$. Whenever the data bit being sent is a *zero*, the symbol waveform $P_0(t)$ is received. When the data bit is a one, the symbol waveform $P_1(t)$ is received. Let the only observable difference between the two symbols be that the power in $P_1(t)$ is higher than the power in $P_0(t)$. The receiver observes the stream of received symbols with a photodiode and must decide every T seconds if a *zero* or a *one* was transmitted.

Since the symbols are identical in all respects except their power, we can express each symbol as the product of a constant describing the power in the jth symbol and a normalized time function $h_{rcvd}(t)$ that describes the waveform of the symbol as

$$P_j(t) = P_j h_{rcvd}(t), \tag{5.72}$$

where the waveform is normalized to have an area of unity,

$$\frac{1}{T} \int\limits_{-\infty}^{+\infty} h_{rcvd}(\tau) d\tau = 1.$$

The received signal can then be expressed as a series of symbols of the form

$$P_{rcvd}(t) = \sum_{j=-\infty}^{\infty} P_j h_{rcvd}(t - jT), \tag{5.73}$$

where the parameter j denotes the jth symbol that was received and P_j can have either the value P_0 or P_1.

The linear signal path in a digital receiver is usually designed to maximize the signal-to-noise ratio at the input to the demodulator. This allows the demodulator to make the best estimate of the received symbol and minimizes the bit-error-rate. The relationship between the received symbol and the transmitted symbol depends on the characteristics of the channel. For example, if significant dispersion is encountered along the channel, the received symbols will be smeared out in time. The linear signal path in the receiver can then be used not only to amplify and filter but to equalize the channel induced effects to improve demodulator performance [67, 68].

Let the desired symbol waveform at the output of the linear signal path be $h_{out}(t)$. The symbol waveform at the input to the receiver is given by $h_{rcvd}(t)$. As with any linear system, the received and output waveforms and the overall response of the linear signal path $h_z(t)$ are related in the time-domain and frequency-domain by Fourier transform pairs as

$$
\begin{aligned}
h_{out}(t) &= h_{rcvd}(t) * h_z(t) = h_{rcvd}(t) * R_t h_t(t), \\
H_{out}(f) &= H_{rcvd}(f) Z_t(f) = H_{rcvd}(f) R_t H_t(f).
\end{aligned}
\tag{5.74}
$$

Note that the photodetector's impulse response $h_d(t)$ is now contained within the overall receiver's transimpedance impulse response $h_z(t)$. We can express the receiver's transimpedance in terms of the received symbol waveform and the desired symbol waveform as

$$
Z_t(f) = R_t H_t(f) = R_t \frac{H_{out}(f)}{H_{rcvd}(f)},
\tag{5.75}
$$

where
$H_{out}(f) = \mathrm{FT}\{h_{out}(t)\} =$ Fourier transform of output pulse shape,
$H_{rcvd}(f) = \mathrm{FT}\{h_{rcvd}(t)\} =$ Fourier transform of received pulse shape.
The definition of $Z_t(f)$ as expressed by Eq. 5.75 is equivalent to our original definition of $Z_t(f)$ as expressed by Eq. 5.30. The benefit to Eq. 5.75 is that the relationship between the received pulse and the desired pulse is explicitly stated.

The quantum shot-noise at the output of the linear signal path referred back to the output of the APD multiplication region is then

$$
i_{shot}^2(t) = q \frac{\eta q}{h\nu} M^2 F(M) \int_{-\infty}^{+\infty} P_{rcvd}(\tau) h_t^2(t-\tau) d\tau \quad (\mathrm{amps}^2).
\tag{5.76}
$$

The received signal is a series of symbol waveforms. Substituting Eq. 5.73 for the received signal allows the shot-noise to be written as

$$
i_{shot}^2(t) = q \frac{\eta q}{h\nu} M^2 F(M) \sum_{j=-\infty}^{+\infty} P_j \int_{-\infty}^{+\infty} h_{rcvd}(\tau - jT) h_t^2(t-\tau) d\tau.
\tag{5.77}
$$

If we Fourier transform Eq. 5.77 and add in the other receiver noise terms from Eq. 5.66, we can obtain a single-sided frequency domain representation for the receiver's total equivalent current-noise referred to the output of the multiplication region as

$$I^2_{n_{eq}} = 2q\frac{\eta q}{h\nu}F(M)M^2 \sum_{j=-\infty}^{+\infty} P_j \int_0^\infty \left[H_{rcvd}(f)H_t(f)*H_t(f)\right]df$$

$$+a_0\int_0^\infty |H_t(f)|^2 df + a_1\int_0^\infty |H_t(f)|^2 f df + a_2\int_0^\infty |H_t(f)|^2 f^2 df. \tag{5.78}$$

The Fourier transforms in Eqs. 5.75 and 5.78 are clearly dependent upon the duration of the received symbols. We can remove this time dependence by following the approach of Personick [68, 69], in which we define a normalized dimensionless frequency variable y as

$$y = f\tau, \tag{5.79}$$

where τ is the time response of the filter. This will usually be equal to either the symbol duration T or the inverse of the 3 dB frequency, depending on the type of receiver filter employed.

This allows the transforms in Eqs. 5.75 and 5.78 to be written as [70]

$$H'_{rcvd}(f) = \frac{1}{\tau}H_{rcvd}\left(\frac{y}{\tau}\right), \quad H'_{out}(y) = \frac{1}{\tau}H_{out}\left(\frac{y}{\tau}\right), \quad H'_t(y) = \frac{H'_{out}\left(\frac{y}{\tau}\right)}{H'_{rcvd}\left(\frac{y}{\tau}\right)}. \tag{5.80}$$

Substituting the normalized transforms into Eq. 5.78, multiplying out, and collecting terms allows the equivalent current-noise for the jth symbol to be expressed as

$$I^2_{n_{eq}} = 2q\frac{\eta q}{h\nu}F(M)M^2 B\left[P_j I_1\right] + 2q\frac{\eta q}{h\nu}F(M)M^2 B\left[P_1(\Sigma_1 - I_1)\right]$$

$$+a_0 B I_2 + a_1 B^2 I_f + a_2 B^3 I_3, \tag{5.81}$$

where

P_j = received power in the jth symbol, P_1 = received power in the symbol for a *one*,

$$I_1 = \text{Re}\left\{\int_0^{+\infty} H'_{rcvd}(f)H'_t(f)*H'_t(f)\right\}, \quad \text{Re}\{X\} = \text{real part of the complex number } X,$$

$$\Sigma_1 = \frac{1}{2}\sum_{k=-\infty}^{+\infty}\text{Re}\{H'_{rcvd}(k)H'_t(k)*H'_t(k)\},$$

$$I_2 = \int_0^{+\infty}|H'_t(f)|^2 df, \quad I_f = \int_0^{+\infty}|H'_t(f)|^2 f df, \quad I_3 = \int_0^{+\infty}|H'_t(f)|^2 f^2 df,$$

B = receiver bandwdith defined as $= \frac{1}{\tau}$.

The integrals in Eq. 5.81 are normalized equivalent noise bandwidths for the various noise sources and have been termed Personick integrals by some authors. The first term in Eq. 5.81 accounts for the quantum shot-noise associated with the jth received symbol. The second term corresponds to the combined quantum shot-noise from all the other received symbols. The second term is multiplied by the coefficient P_1, which is the power in a symbol for a *one*. This represents the worst case condition, in which all other symbols received $(\Sigma_1 - I_1)$ are *ones*. The third, fourth and fifth terms are the normalized noise bandwidths for the noise sources that are constant with frequency, linear with frequency, and quadratic with frequency, respectively.

In base-band digital systems, the receiver transfer function is usually designed to maximize the signal to noise ratio at the input to the demodulator while minimizing intersymbol interference. In practice, the actual transfer function used depends on the pulse shape and format of the modulation being used and the presence of any channel effects such as dispersion.

The integrals of Eq. 5.81 have been evaluated by Personick for the case where the receiver includes a filter that produces 100% raised-cosine pulses at its output when the received optical pulse-shapes are rectangular, Gaussian, and exponential [68, 69]. We will have more to say about raised-cosine pulses in our next chapter. They are a family of pulse-shapes that are commonly used in digital communication systems because of their good SNR and intersymbol interference properties.

The rectangular input pulse-shape is expected when the receiver and transmitter have short rise and fall times and the channel is distortionless. The Gaussian pulse-shape is frequently found in multimode fiber systems, where there is substantial mode mixing. The exponential-pulse is included to account for systems where there is significant dispersion and the transmitted pulse spreads in time as it propagates through the channel.

In many cases we can invoke fast optoelectronic components, single-mode fiber, and negligible dispersion. This considerably simplifies Eq. 5.81. Without significant dispersion to spread the signal shot-noise across symbol boundaries, the term $(\Sigma_1 - I_1)$ is nearly zero and the second term in Eq. 5.81 can be neglected. The received symbol pulse-shape will then generally be rectangular in time and will be formatted to occupy either an entire symbol-time or a portion of the symbol-time. When the pulse occupies the entire symbol-time the signaling is termed non-return-to-zero (NRZ) signaling. When the pulse occupies only a portion of the symbol-time we have return-to-zero (RZ) signaling. The percentage of the symbol-time that is occupied is known as the duty-cycle of the pulse. The most common type of RZ signaling is one with a 50% duty cycle.

In addition to NRZ and RZ signaling analyzed by Personick, there are many other possible formats and pulse shapes. For example Muoi has evaluated these integrals for the case of a receiver using Manchester coding [71]. We will revisit Manchester coded signals as well as NRZ and RZ signals in the beginning of Chapter 7. Here we summarize the published results. The numerical values for the normalized noise-bandwidth integrals used in Eq. 5.81 for the NRZ, 50% duty-cycle RZ, and Manchester coded symbol waveforms are tabulated in Table 5.1 [64, 68, 69, 71].

These values are applicable when the receiver contains a filter that generates a 100% raised-cosine pulse at the input to the decision circuit. For these cases, the receiver bandwidth B used in Eq. 5.81 is defined to be the inverse of the symbol-time. In many cases the symbol time is the same as bit-time and so B usually corresponds to the bit-rate. This is not always true however. In systems that use *m*-ary signaling, the symbol rate can be different from the bit-rate, and when evaluating the total receiver noise the symbol-time, not the bit-time, should be used.

Note that under this definition, the receiver bandwidth that is to be used in Eq. 5.81 does not refer to the raised-cosine filter's 3 dB bandwidth. In fact, for the NRZ case, the 3-dB bandwidth of the raised-cosine filter is approximately 58% of the symbol-rate. In the 50% duty cycle RZ case, the 3-dB bandwidth is approximately 40% of the symbol-rate [64]. Bandwidth here refers to the base-band width of the spectrum of a 100% raised-cosine pulse waveform. The one-sided spectral width of a 100% raised-cosine pulse is equal to the inverse of the pulse's duration T.

Table 5.1 **Values of normalized noise-bandwidth integrals (also known as Personick integrals) used in Eq. 5.81 or Eq. 5.83 when the receiver implements a 100% raised-cosine filter function. The bandwidth B that is to be used in the equations is the inverse of the symbol-time.**

	I_1	I_2	I_f	I_3
NRZ	0.56	0.56	0.18	0.087
50% Duty-cycle RZ	0.50	0.40	0.098	0.036
Manchester	0.59	0.59	0.51	0.50

Using the values from Table 5.1, the total current-noise of the jth symbol in an optical receiver using symbols that are rectangular NRZ pulses with negligible dispersion and a filter that generates 100% raised cosine waveforms at the input to the decision circuit is

$$I_{n_{eq}}^2 = 2q\frac{\eta q}{h\nu} F(M)M^2 BP_j[0.56]$$

$$+a_0 B[0.56]+a_1 B^2[0.18]+a_2 B^3[0.087] \quad (\text{amps}^2), \tag{5.82}$$

where P_j is the optical power in the jth symbol.

Evaluating the I_1 integral can sometimes prove problematic since it requires the computation of a convolution. A simplified approach has been developed that provides excellent accuracy in many applications [72]. In this technique, Eq. 5.81 which expresses the quantum shot-noise as a function of time during the symbol-time, is replaced by the quantum shot-noise associated with the average photocurrent during the symbol. Since the quantum-shot noise is now assumed to be a constant during our observation-time, its contribution to the equivalent input current-noise is of the same form as the a_o component of the receiver electronics noise. Both are multiplied by the equivalent noise bandwidth integral I_2. The simplified approach allows Eq. 5.81 to be written as

$$I_{n_{eq}}^2 \cong \left[2q \frac{\eta q}{h\nu} P_{rcvd} F(M) M^2 + a_0 \right] B I_2 + a_1 B^2 I_f + a_2 B^3 I_3, \qquad (5.83)$$

where we can include all of the potential noise sources as

$$I_{n\,eq}^2 = I_{shot}^2 + I_{rcvr}^2,$$

$$I_{rcvr}^2 = I_{elec}^2 + I_{excess}^2 + I_{nb}^2,$$

$$I_{excess}^2 = I_{RIN}^2 + I_{modal}^2.$$

In many cases the receiver does not actually implement a 100% raised-cosine filter. It is common for a simple low-pass filter to be used or for the receiver to be constructed to be as wideband as possible even though it is expected to be operated at a data-rate that is less than the receiver bandwidth. For these cases, the normalized noise-bandwidths tabulated in Table 5.2 are useful. The bandwidth definitions for use in Eq. 5.83 are also listed.

Table 5.2   **Values of normalized noise-bandwidth integrals used in Eq. 5.83 when the receiver implements a conventional low-pass filter function. The bandwidth B that is to be used in Eq. 5.81 varies depending on the type of filter. Raised cosine with NRZ signaling is included for comparison.**

	B	I_1 , I_2	I_f	I_3
1st Order Lowpass	f_{3dB}	1.57	∞	∞
2nd Order Butterworth Lowpass	f_{3dB}	1.11	0.79	1.11
3rd Order Butterworth Lowpass	f_{3dB}	1.05	0.61	0.52
Ideal Lowpass	f_c	1.00	0.50	0.33
Sinc	$1/T$	0.50	0.36	∞
NRZ Raised-Cosine	$1/T$	0.56	0.18	0.087

The I_f and I_3 terms in the case of the first order filter and the I_3 term in the case of the sinc filter are essentially infinite. This is a consequence of the slow roll-off of these filters. The noise that is proportional to the square of frequency is increasing faster than the filter is rolling-off and the effective noise bandwidth is therefore infinite. In practice these filters are used in conjunction with a higher-order low-pass filter to limit the high frequency noise content.

The bandwidths used in Eq. 5.83 are the conventional 3 dB filter bandwidths for the Butterworth filters. The bandwidth of the ideal low-pass is simply the filter's cut-off frequency while the bandwidth of the ideal sinc filter is the inverse of the symbol-time.

5.13   Noise Equivalent Power and Detectivity

We conclude this chapter by mentioning some other techniques that have evolved for specifying receiver noise performance. The terms noise-equivalent-power, detectivity, and specific detectivity are figures-of-merit that have historically been used as aids in specifying the performance of an optical receiver. These terms were developed as indicators of the minimum quasi-CW optical signal that could be detected by imaging devices, optical radars, early analog communication systems, and the like. They have little direct significance in digital communication systems unless additional information is known.

Noise-equivalent-power is defined as the optical signal power needed to make the electrical signal-to-noise ratio in an optical receiver equal to unity in a 1 Hz bandwidth. NEP can be dominated by detector dark-currents, signal shot-noise, amplifier noise terms, optical backgrounds, or excess noise present on the received signal. It can also be strongly frequency dependent. The use of NEP is most suited to analog receivers, where SNR is a natural measure of performance. For digital communication systems NEPs must be converted into equivalent noise photon counts or noise current densities. The general form of NEP is given by

$$NEP(\omega) = \frac{i_{n_{eq}}(\omega)}{\dfrac{\eta q}{h\nu}}.$$
(5.84)

As an example, the NEP for a simple *p-i-n* photodetector whose noise performance is dominated by dark-current is given by

$$NEP = \frac{\sqrt{2qi_d}}{\dfrac{\eta q}{h\nu}}.$$
(5.85)

The units of NEP are the somewhat unusual Watts per root Hz. This comes about because NEP is the ratio of current-noise density to responsivity and the units are

$$NEP = \frac{\dfrac{amps}{\sqrt{Hz}}}{\dfrac{amps}{watt}} = \frac{watts}{\sqrt{Hz}}.$$
(5.86)

Detectivity, or D, is the inverse of the NEP and is thought to be a more intuitive figure-of-merit since it increases with improved sensitivity. Specific detectivity or D^* is the value of D normalized to a detector area of one square-centimeter. D^* is sometimes used in background limited systems. The term BLIP is also sometimes used to indicate that a receiver is operating in the background-limited photodetection region. BLIP is a term that is frequently used with sensor systems where identifying a target against a background is a critical function.

5.14   References

1.   H. Ott, *Chap. 6 - Shielding Effectiveness of Metallic Sheets* in *Noise Reduction Techniques in Electronic Systems,* 1976, John Wiley & Sons: New York.

2.   A. Yariv, *Chap. 4 - Optical Resonators* in *Optical Electronics,* 1985, Holt, Rinehart and Winston: New York.

3.   K. Petermann, *Chap. 8 - Noise in Interferometers* in *Laser Diode Modulation and Noise,* 1988, Kluwer Academic Publishers: Boston.

4.   M.J. Buckingham, *Chap. 2 - Mathematical Techniques* in *Noise in Electronic Devices and Systems,* 1983, Halsted Press Division of John Wiley & Sons: New York.

5.   W.B. Davenport and W.L. Root, *Chap. 6 - Spectral Analysis* in *An Introduction to the Theory of Random Signals and Noise,* 1958, McGraw-Hill: New York.

6.   M.J. Buckingham, *Chap. 7 - Burst Noise* in *Noise in Electronic Devices and Systems,* 1983, Halsted Press Division of John Wiley & Sons: New York.

7.   M.J. Buckingham, *Chap. 10 - Hot Electron Devices* in *Noise in Electronic Devices and Systems,* 1983, Halsted Press Division of John Wiley & Sons: New York.

8.   C. D. Motchenbacher and F. C. Fitchen, *Chap. 9 - Noise in Passive Components* in *Low-Noise Electronic Design,* 1973, John Wiley and Sons: New York.

9.   J.B. Johnson, "Thermal Agitation of Electricity in Conductors," Physical Review, July 1928, vol. 32, pp. 97-109.

10.   H. Nyquist, "Thermal Agitation of Electric Charge in Conductors," Physical Review, July 1928, vol. 32, no. 1, pp. 110-113.

11.   H. Ott, *Chap. 5 - Passive Components* in *Noise Reduction Techniques in Electronic Systems,* 1976, John Wiley & Sons: New York.

12.   B.M. Oliver, "Thermal and Quantum Noise," Proceedings of the IEEE, May 1965, vol. 53, pp. 436-454.

13. M.J. Buckingham, *Chap. 11 - Quantum Mechanics and Noise* in *Noise in Electronic Devices and Systems,* 1983, Halsted Press Division of John Wiley & Sons: New York.

14. R.E. Ziemer and W.H. Tranter, *Appendix A - Physical Noise Sources and Noise Calculations in Communication Systems* in *Principles of Communications,* 1976, Houghton Mifflin Company: Boston.

15. W.B. Davenport and W.L. Root, *Chap. 7 - Shot Noise* in *An Introduction to the Theory of Random Signals and Noise,* 1958, McGraw-Hill: New York.

16. W. Schottky, "Uber Spontae Stromschwankungen in Verscheidenen Elektrizitatsleitern," Annals of Physics (Leipzig), 1918, vol. 57, pp. 541-567.

17. W. Davenport and W. Root, *Chap. 8 - The Gaussian Process* in *An Introduction to the Theory of Random Signals and Noise,* 1958, McGraw-Hill: New York.

18. J. Cohen, "Introduction to Noise in Solid State Devices," 1982, NBS Technical Note # 1169, U. S. Department of Commerce/National Bureau of Standards.

19. M.J. Buckingham, *Chap. 4 - Inherent Noise in Junction Diodes and Bipolar Transistors* in *Noise in Electronic Devices and Systems,* 1983, Halsted Press Division of John Wiley & Sons: New York.

20. M.J. Buckingham, *Chap. 6 - 1/f Noise* in *Noise in Electronic Devices and Systems,* 1983, Halsted Press Division of John Wiley & Sons: New York.

21. F.N. Hooge, "1/f Noise," Physica, 1976, vol. 83, pp. 14-23.

22. M. Keshner, "1/f Noise," Proceedings of the IEEE, March 1982, vol. 70, no. 3, pp. 212-218.

23. M. Park, *et al.*, "Analysis of Sensitivity Degradations Caused by the Flicker Noise of GaAs-MESFET's in Fiber-Optic Receivers," Journal of Lightwave Technology, May 1988, vol. 6, no. 5, pp. 660-667.

24. P. Folkes, "Thermal-noise Measurements in GaAs MESFETs," IEEE Transactions on Electron Devices, 1985, vol. 6, no. 12, pp. 620-623.

25. C. Su. "1/f Noise in GaAs MESFETs," in *Proceedings of the International Electron Devices Meeting.* 1983, IEEE.

26. S.M. Sze, *Chap. 6 - JFET and MESFET* in *Physics of Semiconductor Devices,* 1981, John Wiley & Sons: New York.

27. H.F. Friis, "Noise Figures for Radio Receivers," Proceedings of the IRE, 1944, vol. 32, no. 7, pp. 419-422.

28.   H.A. Haus, *et al.*, "IRE Standards on Methods of Measuring Noise in Linear Twoports, 1959," Proceedings of the IRE, January 1960, vol. 48, no. 1, pp. 60-68.

29.   H. Haus, *et al.*, "Representation of Noise in Linear Twoports," Proceedings of the IRE, January 1960, vol. 48, pp. 69-74.

30.   H. Rothe and W. Dahlke, "Theory of Noisy Fourpoles," Proceedings of the IRE, June 1956, vol. 44, no. 6, pp. 811-818.

31.   M.J. Buckingham, *Chap. 3 - Noise in Linear Networks* in *Noise in Electronic Devices and Systems*, 1983, Halsted Press Division of John Wiley & Sons: New York.

32.   C. D. Motchenbacher and F. C. Fitchen, *Chap. 2 - Amplifier Noise* in *Low-Noise Electronic Design*, 1973, John Wiley and Sons: New York.

33.   H.W. Bode, *Chap. 1 - Mesh and Nodal Equations for an Active Circuit* in *Network Analysis and Feedback Amplifier Design*, 1945, Van Nostrand Company: New York.

34.   Y. Netzer, "The Design of Low-Noise Amplifiers," Proceedings of the IEEE, 1981, vol. 69, no. 6, pp. 728-741.

35.   E. Dereniak and D. Crowe, *Chap. 4 - Photoconductors* in *Optical Radiation Detectors*, 1984, John Wiley & Sons: New York.

36.   H. Melchoir, *et al.*, "Photodetectors for Optical Communication Systems," Proceedings of the IEEE, October 1970, vol. 58, no. 10, pp. 1466-1490.

37.   S.M. Sze, *Chap. 13 - Photodetectors* in *Physics of Semiconductor Devices*, 1981, John Wiley & Sons: New York.

38.   M. DiDomenico and O. Svelto, "Solid-State Photodetection: A Comparison Between Photodiodes and Photoconductors," Proceedings of the IEEE, 1964, vol. 52, pp. 136.

39.   R.J. McIntyre, "Multiplication Noise in Uniform Avalanche Diodes," IEEE Transactions on Electron Devices, January 1966, vol. ED-13, no. 1, pp. 164-169.

40.   P.P. Webb, *et al.*, "Properties of Avalanche Photodiodes," RCA Review, June 1974, vol. 35, pp. 234-278.

41.   N.Z. Hakim, *et al.*, "Signal-to-Noise Ratio for Lightwave Systems Using Avalanche Photodiodes," Journal of Lightwave Technology, March 1991, vol. 9, no. 3, pp. 318-320.

42.   M.J. Garvey and S.W. Quinn, "Sources of Intensity Noise in Wide-Band Analog Optical Communication Systems: Their Elimination and Reduction," Journal of Lightwave Technology, September 1986, vol. LT-4, no. 9, pp. 1285-1289.

43.   T.K. Yee, "Analysis of the Intensity Noise of Nearly Single-Longitudinal-Mode Semiconductor Lasers," IEEE Journal of Quantum Electronics, February 1985, vol. QE-22, pp. 275-285.

44.   K. Petermann, *Chap. 7 - Noise Characteristics of Solitary Laser Diodes* in *Laser Diode Modulation and Noise,* 1988, Kluwer Academic Publishers: Boston.

45.   Y. Yamamoto, "AM and FM Quantum Noise in Semiconductor Lasers - Part II: Comparison of Theoretical and Experimental Results for AlGaAs Lasers," IEEE Journal of Quantum Electronics, January 1983, vol. QE-19, no. 1, pp. 47-58.

46.   Y. Yamamoto, "AM and FM Quantum Noise in Semiconductor Lasers - Part I: Theoretical Analysis," IEEE Journal of Quantum Electronics, January 1983, vol. QE-19, no. 1, pp. 34-46.

47.   P.L. Liu, *Chap. 10 - Photon Statistics and Mode Partition Noise of Semiconductor Lasers* in *Coherence, Amplification and Quantum Effects in Semiconductor Lasers,* Y. Yamamoto, Editor. 1991, John Wiley and Sons, Inc.: New York.

48.   W.E. Stephens and T.R. Joseph, "A 1.3 micron Microwave Fiber-Optic Link Using a Direct-Modulated Laser Transmitter," Journal of Lightwave Technology, April 1985, vol. LT-3, no. 2, pp. 308-315.

49.   R.E. Epworth. "The Phenomena of Modal Noise in Analogue and Digital Optical Fiber Systems," in *4th European Conference on Optical Communication (ECOC).* 1978, Geneva, Italy: Instituto Internazionale delle Communicazioni.

50.   D. O'Shea, *et al., Sec. 2.5 - Light Interference: The Laser Speckle Pattern* in *Introduction to Lasers and Their Applications,* 1978, Addison-Wesley: Reading, MA.

51.   T. Kanada, "Evaluation of Modal Noise in Multi-Mode Fiber-Optic Systems," Journal of Lightwave Technology, February 1984, vol. LT-2, no. 1, pp. 11-18.

52.   T.H. Wood and L.A. Ewell, "Increased Received Power and Decreased Modal Noise by Preferential Excitation of Low-Order Modes in Multimode Optical-Fiber Transmission Systems," Journal of Lightwave Technology, April 1986, vol. LT-4, no. 4, pp. 391-395.

53.   R.E. Epworth, "Modal Noise - Causes and Cures," Laser Focus, September 1981, pp. 109-114.

54.   D.N. Payne, *et al.*, "Development of Low- and High-Birefringence Optical Fibers," IEEE Journal of Quantum Electronics, April 1982, vol. QE-18, pp. 477-488.

55.   F.M. Sears, *et al.*, "Probability of Modal Noise in Single-Mode Lightguide Systems," Journal of Lightwave Technology, June 1986, vol. LT-4, pp. 652-655.

56.   J.L. Gimlett, *et al.*, "Degradations in Gb/s DFB Laser Transmission Systems Due to Phase-to-Intensity Noise Conversion by Multiple Reflection Points," Electronics Letters, 1988, vol. 24.

57.   R.M. Gagliardi and S. Karp, *Chap. 2 - Counting Statistics* in *Optical Communications*, 1976, John Wiley & Sons: New York.

58.   S. Karp, *et al.*, *Section 3.5 - Counting Statistics* in *Optical Channels*, 1988, Plenum Press: New York.

59.   K. Ogawa, "Analysis of Mode Partition Noise in Laser Transmission Systems," IEEE Journal of Quantum Electronics, 1982, vol. QE-18, pp. 849-855.

60.   C.H. Henry, *et al.*, "Partition Fluctuations in Nearly Single-Longitudinal-Mode Lasers," Journal of Lightwave Technology, June 1984, vol. LT-2, pp. 209-216.

61.   G.J. Meslener, "Mode-Partition Noise in Microwave Subcarrier Transmission Systems," Journal of Lightwave Technology, January 1994, vol. 12, no. 1, pp. 118-125.

62.   W.K. Pratt, *Chap. 6 - Background Radiation* in *Laser Communication Systems*, 1969, John Wiley & Sons: New York.

63.   M. Ross, *Chap. 7 - Background Energy Considerations* in *Laser Receivers*, 1966, John Wiley & Sons: New York, pp. 195-205.

64.   T.V. Muoi, *Chap. 12 - Receiver Design of Optical-Fiber Systems*, in *Optical Fiber Transmission*, E.E.B. Basch ed. 1987, H. W. Sams & Co. Div. of Macmillan, Inc.: Indianapolis.

65.   K. Ogawa, "Noise Caused by GaAs MESFETs in Optical Receivers," Bell System Technical Journal, July-August 1981, vol. 60, pp. 923-928.

66.   R.N. Bracewell, *Chap. 3 - Convolution* in *The Fourier Transform and Its Applications*, 1978, McGraw-Hill Book Company: New York.

67.   R. Dogliotti, *et al.*, "Baseband Equalization in Fibre Optic Digital Transmission," Optical and Quantum Electronics, 1976, vol. 8, pp. 343-353.

68. S.D. Personick, "Receiver Design for Digital Fiber Optic Communication Systems, I & II," Bell System Technical Journal, July-August 1973, vol. 52, no. 6, pp. 843-886.

69. R.G. Smith and S.D. Personick, *Receiver Design for Optical Fiber Communication Systems*, in *Semiconductor Devices for Optical Communication,* 2nd, H. Kressel ed. Vol. 39, 1982, Springer Verlag: New York, pp. 89-160.

70. R.N. Bracewell, *Chap. 6 - The Basic Theorems* in *The Fourier Transform and Its Applications,* 1978, McGraw-Hill Book Company: New York.

71. T.V. Muoi, "Receiver Design for Digital Fiber Optic Transmission Systems Using Manchester (Biphase) Coding," IEEE Transactions on Communications, May 1983, vol. COM-31, no. 5, pp. 608-619.

72. D.R. Smith and I. Garrett, "A Simplified Approach to Digital Optical Receiver Design," Optical and Quantum Electronics, 1978, vol. 10, pp. 211-221.

Chapter 6

RECEIVER FRONT-END DESIGN

In Chapter 5 we saw that the front-end plays a major role in determining the noise performance of a receiver. In this chapter, we will explore four principal types of front-end designs that are used in optical receivers. We will review the use of bipolar and field-effect transistors in front-end amplifiers and we will examine representative examples of receiver front-ends using *p-i-n* photodiodes and APDs.

6.1 Front-End Architectures

An optical receiver's front-end design can usually be grouped into one of four basic configurations: 1) resistor termination with a low-impedance voltage amplifier, 3) high-impedance amplifier, 4) transimpedance amplifier, and 5) noise-matched or resonant amplifier. Any of the configurations can be built using contemporary electronic devices such as operational amplifiers, bipolar junction transistors, field-effect transistors, or high electron mobility transistors. The receiver performance that is achieved will depend on the devices and design techniques used.

The names typically used to describe these four configurations can be somewhat misleading. Fundamentally, the front-end of an optical receiver responds to an optical signal by generating a photocurrent with a photodetector. The photocurrent is then converted to a voltage. Electronic signal processing stages process the recovered voltage to extract the desired information. The dimensions of the transfer function associated with the front-end will consequently be volts per amp or ohms. Thus the transfer functions of virtually all optical receivers are actually transimpedance in nature.

6.1.1 Low Impedance Voltage Amplifier

A simple optical receiver front-end is illustrated in Fig. 6.1. It consists of a photodetector, a load resistor, and a low input-impedance voltage amplifier.

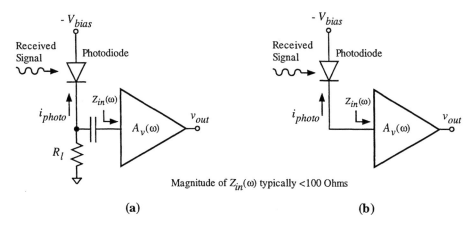

Magnitude of $Z_{in}(\omega)$ typically <100 Ohms

(a)   (b)

Figure 6.1 Voltage amplifier receiver front-end. (a) AC coupled. (b) DC coupled.

The photodiode can be either AC coupled or DC coupled to the amplifier. In the AC coupled case, a separate load resistor is used to derive a voltage proportional to the photocurrent and to provide a path for the DC photocurrent to flow. The low-frequency components of the photocurrent see a load resistor R_l while the high-frequency components see a load resistance that is the parallel combination of R_l and the amplifier input impedance $Z_{in}(\omega)$. There are a wide variety of commercially available high gain wideband amplifiers that are AC coupled. The most common type is the 50 Ω input, 50 Ω output RF/microwave amplifier. Bandwidths of tens of GHz with voltage gains in excess of 20 dB are available. Typical low frequency cut-offs are in the range of a few kHz to a few tens of MHz.

One impact of an AC coupled front-end is that low-frequency components in the photocurrent that may contain useful information can be lost. Loss of low frequency information can often be substantially reduced by using line-coding in a digital system [1-3] or subcarrier modulation in an analog system [4, 5]. Both techniques minimize the amount of information in the low-frequency portions of the photocurrent and allow the use of AC coupling with minimal impact. In some cases, *any* reduction in low frequency information is intolerable and DC coupling is required.

In the DC coupled amplifier, the amplifier itself can provide the DC current path and the photodiode can be loaded directly by $Z_{in}(\omega)$. DC coupled amplifiers with high gain and wide bandwidth can be very challenging to construct because of their exacting feedback stabilization and power supply requirements. Two techniques that have proven successful in realizing ultra-wideband DC coupled amplifiers are an embedding technique [6] or the splitting of the amplifier into two parallel paths, as shown in Figure 6.2. One path forms a low bandwidth DC coupled amplifier while the other forms a wide bandwidth AC coupled amplifier.

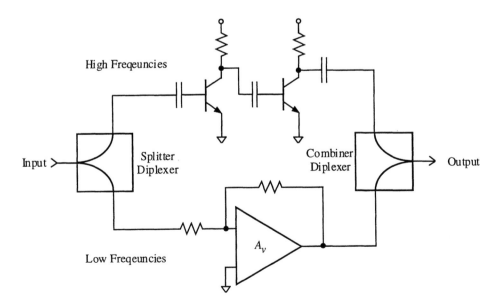

Figure 6.2 Wideband DC coupled amplifier using parallel paths.

The DC coupled path can often be realized using modest bandwidth operational amplifier technology. Care must be taken in constructing the lowpass/highpass diplexers to assure that the phase and amplitude response of the combined channels are acceptable.

The principal attractions of the low-impedance voltage amplifier front-end are its simplicity and potential for wide bandwidth. We will use this architecture to illustrate a few key points that are applicable to all front-ends. Using the techniques of Chapters 4 and 5, the equivalent circuit model for the resistor terminated voltage amplifier front-end of Fig. 6.1(a) can be constructed as illustrated in Fig. 6.3.

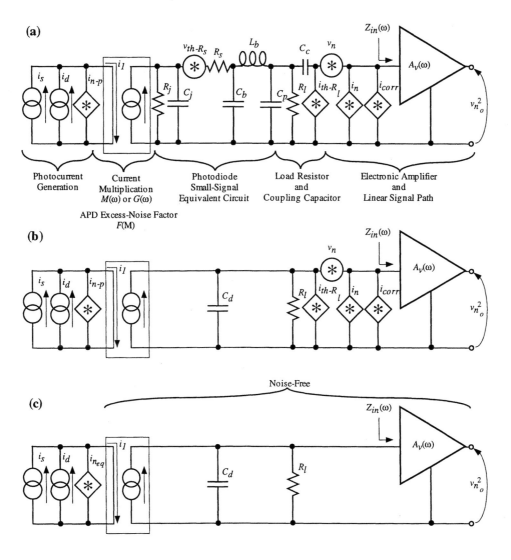

Figure 6.3 **Noise model for resistor terminated voltage amplifier front-end. (a) Complete model. (b) Simplified model. (c) Equivalent input current-noise model.**

Figure 6.3(a) illustrates the complete model, which is often best evaluated using computer aided techniques. We have explicitly included a photocurrent multiplication or gain component in the model but the model is equally applicable to photodetectors without gain. We simply set the gain and excess noise factors equal to unity. Note the presence of the thermal current-noise i_{th-R_l} that arises because of the presence of the load resistor.

In order to gain insight into the front-end's performance we will use the simplified model shown in Fig. 6.3(b). The simplified model is essentially a mid-band model in that the coupling capacitor and the series resistance and inductance in the photodiode small-signal model are ignored. We also ignore the thermal-noise in the photodiode series resistance and any correlation between amplifier noise sources. Our goal is to obtain the equivalent input current-noise model of Fig. 6.3(c). In this model we will refer all of the noise sources to the receiver's effective input. We define the effective input as being the point just after the primary photocurrent is generated but before any internal photodetector gain mechanism can occur. The single current-noise source $i_{n_{eq}}$ accounts for all of the receiver noise sources and the remaining portions of the receiver become noise-free.

We could have chosen to refer everything to an alternative location such as the photodetector's output, or even the output of the linear channel. We chose the input in this analysis so that we can directly compare the overall receiver noise to the quantum shot-noise associated with the creation of the primary photocurrent. Another benefit of using the equivalent input current-noise defined in this way is that we can obtain the signal and noise at any other point in the receiver simply by multiplying by the transimpedance that relates the primary photocurrent to the voltage at our point of interest. For example, the overall transimpedance of the circuit shown in Fig. 6.3(c) is given by

$$Z_t(\omega) = \frac{v_o(\omega)}{i_s(\omega)} = \frac{M(\omega)Z_{in}(\omega)A_v(\omega)R_l}{Z_{in}(\omega) + R_l + j\omega C_d R_l Z_{in}(\omega)} \quad (\Omega), \tag{6.1}$$

where $C_d = C_j + C_b + C_p$. In many cases we can approximate the amplifier's input impedance as the parallel combination of an input resistance R_{in} and an input capacitance C_{in}. Equation 6.1 then becomes

$$Z_t(\omega) = M(\omega)A_v(\omega)R_{eq} \frac{1}{1 + j\dfrac{\omega}{\omega_p}} \quad (\Omega), \tag{6.2}$$

where $R_{eq} = \dfrac{R_{in}R_l}{R_{in} + R_l}, \quad \omega_p = \dfrac{1}{R_{eq}C_t}, \quad C_t = C_d + C_{in}.$

Notice that the response rolls off at frequencies above the inverse of the RC time-constant set by the total capacitance C_t and the equivalent load resistance R_{eq}. In order to obtain a wideband response we would want to minimize the capacitance and use a low value for the equivalent photodetector load resistor.

Using Eq. 5.57 to describe the photodetector noise and Eq. 5.35 to describe the amplifier noise, the signal photocurrent, and dark current, the equivalent input current-noise at the receiver's input are given by

$$i_s(\omega) = \frac{\eta q}{h\nu} P_{rcvd}(\omega),$$

$$i_{n_{eq}}^2(\omega) = \frac{2qi_{du} + i_n^2(\omega) + v_n^2(\omega)|Y_s(\omega)|^2 + i_{th-R_l}^2}{|M(\omega)|^2}$$

$$+\left[2qi_{dm} + i_{shot}^2 + i_{excess}^2 + i_{nb}^2\right]F(M(\omega)), \tag{6.3}$$

$$i_{dk} = \frac{i_{du}}{M} + i_{dm} + i_b,$$

where $Y_s(\omega) = j\omega C_t + G_l$, $\quad i_{th-R_l}^2 = \frac{4kT}{R_l}$.

Note that we have not specified the characteristics of the shot-noise yet. It may be time varying (in step with the received signals as in a baseband digital system) or it may be effectively a constant, as in most analog or coherent detection systems. We are interested in determining the equivalent input current-noise density, the total equivalent input current-noise, and the signal-to-noise ratio for the front-end. For the purpose of simplicity we will ignore any optical background or optical excess-noise and we will assume that the receiver is illuminated by a signal with a constant power level P_{rcvd}. The DC dark current i_{dk}, since it amounts to an offset that is easily removed, will also be ignored. We will also assume that we are not operating near the gain-bandwidth limitation of the APD. This allows us to replace the frequency dependent multiplication transfer function $M(\omega)$ by a frequency independent value M.

With these assumptions, the equivalent input current-noise density is

$$i_{n_{eq}}^2(\omega) = \frac{2qi_{du} + i_n^2 + v_n^2 \dfrac{1 + (\omega R_l C_d)^2}{R_l^2} + \dfrac{4kT}{R_l}}{M^2} + \left[2qi_{dm} + i_{shot}^2\right]F(M). \tag{6.4}$$

Eq. 6.4 can be put in our standard form for a receiver's equivalent input current-noise density, as was given by Eq. 5.61, as

$$i_{n_{eq}}^2(f) = i_{shot}^2 + a_0 + a_2 f^2, \tag{6.5}$$

where

$$a_0 = \frac{2qi_{du} + i_n^2 + \dfrac{v_n^2}{R_l^2} + \dfrac{4kT}{R_l}}{M^2} + 2qi_{dm}F(M) + i_{shot}^2\left[F(M) - 1\right], \quad a_2 = \frac{v_n^2(2\pi C_d)^2}{M^2}.$$

The degradation from the quantum shot-noise limited current density is then given by

$$\kappa = 10\log\left(\frac{i_{shot}^2 + a_0 + a_2 f^2}{i_{shot}^2}\right). \tag{6.6}$$

The total equivalent input current-noise is found via Eq. 5.70 to be

$$I_{n_{eq}}^2 = i_{shot}^2 \int_0^\infty |H_t(f)|^2 \, df + a_0 \int_0^\infty |H_t(f)|^2 \, df + a_2 \int_0^\infty |H_t(f)|^2 f^2 df, \qquad (6.7a)$$

where $H_t(f)$ is the amplitude normalized frequency dependent portion of the overall receiver transimpedance from Eq. 6.2, given by

$$H_t(f) = \frac{M(f)A_v(f)}{M(0)A_v(0)} \frac{1}{1 + j\dfrac{f}{f_p}}. \qquad (6.7b)$$

The signal-to-noise ratio is given by

$$SNR = \frac{i_s^2}{\displaystyle\int_0^\infty i_{n_{eq}}^2(f) \, df}$$

$$= \frac{\left(\dfrac{\eta q}{h\nu} P_{rcvd}\right)^2}{2q\dfrac{\eta q}{h\nu} P_{rcvd} \displaystyle\int_0^\infty |H_t(f)|^2 \, df + a_0 \int_0^\infty |H_t(f)|^2 \, df + a_2 \int_0^\infty |H_t(f)|^2 f^2 df}, \qquad (6.8a)$$

where we have used the expression for the quantum shot-noise generated under constant illumination:

$$i_{shot}^2 = 2q\frac{\eta q}{h\nu} P_{rcvd}. \qquad (6.8b)$$

Now we analyze the receiver front-end for two cases. In the first case the photodiode is a *p-i-n* without internal gain. In this case $i_{du} = 0$, $i_{dm} = i_{dk}$, $M(\omega) = 1$, and $F(M) = 1$. The voltage amplifier is a conventional 50 Ω input impedance broadband microwave amplifier with a 1.0 dB noise figure. This would represent a relatively low load impedance to the photodiode, and a 1.0 dB noise figure is considered a comparatively low noise microwave amplifier The noise figure, which is measured in a 50 Ω environment, does not accurately predict the amplifier performance for other terminating impedances and we must use an approximation. Without any further information as to the breakdown between amplifier voltage-noise and amplifier current-noise, we can use whichever model is most convenient as long as we realize it will only be an approximation. In this case we will model all of the amplifier's noise as current-noise.

A simplified relationship between the noise figure for a microwave amplifier and the voltage-noise current-noise model is [7]

$$NF = 10\log\frac{4kTR_s + v_n^2 + i_n^2 R_s^2}{4kTR_s}. \tag{6.9}$$

Solving Eq. 6.3 for the case of $v_n = 0$ and a 1.0 dB amplifier noise figure that was measured at room temperature (290°K) with a 50 Ω source-resistance R_s, we find that the current-noise is 9.15 pA per root-Hz. Since we are assuming a p-i-n photodiode and there is no voltage-noise, the coefficients for Eqs. 6.5 through 6.8 become

$$a_0 = i_n^2 + \frac{4kT}{R_l} + 2qi_{dk}, \quad a_2 = 0. \tag{6.10}$$

Let us use a 50 Ω load resistor and a total capacitance resulting from the photodiode, amplifier input, and stray capacitance of 2.0 pF. The photodiode has a quantum efficiency of 80% at a wavelength of 1.55 microns and a dark-current of 1.0 nA. We assume that the transimpedance rolls off above the frequency given by ω_p in Eq. 6.2. This implies that the 50 Ω amplifier does not roll off first. For this example, where there are two 50 Ω impedances in parallel, $R_{eq} = 25.0\Omega$ and $C_t = 2.0$pF and the receiver rolls off above 3.185 GHz. The noise-bandwidth of this single-pole response is $\pi/2$ greater than the 3 dB bandwidth or 5.0 GHz [8].

Figure 6.4 illustrates the signal-power, total equivalent input noise-power, signal-to-noise ratio, and the degradation from the quantum shot-noise limited current density for the receiver. Figure 6.4(a) illustrates the variation in signal-current, total quantum shot-noise current, total receiver electronics current-noise, and total receiver current-noise as a function of received signal-power. Note that, as expected, the detected electrical signal-power varies as the square of the optical signal-power. Figure 6.4(a) also reveals that the receiver is electronics-noise limited for virtually all practical values of received signal-power. Quantum shot-noise is evident only for received signal-powers approaching 0 dBm. The electronics-noise is dominated by the thermal-noise in the 50Ω load resistor, which contributes a total current-noise of 1.61×10^{-12} amps^2 in our 5 GHz noise-bandwidth. In comparison the amplifier current-noise contributes only 0.42×10^{-12} amps^2 to the receiver electronics-noise.

Figure 6.4(b) plots the signal-to-noise ratio measured in the 5 GHz receiver noise-bandwidth and the degradation from the quantum shot-noise limited current-noise density. Because the receiver electronics-noise in this example does not have any frequency dependence, the degradation is the same at each frequency within the receiver bandwidth. From Fig. 6.4(b) we see that for low signal-powers our signal-to-noise ratio is negative. The SNR increases as the square of the received signal-power (i.e. a 20 dB change for every 10 dB increase in signal), a characteristic of an electronics-noise limited receiver. We can obtain a 0 dB SNR with a received signal-power of some -28 dBm and can obtain a relatively high SNR, approaching 60 dB, but only for large amounts of received signal-power. Although there are some communications applications where the received signal-power can be quite large, most systems require receivers to provide a high SNR when illuminated by much smaller received signal-powers. Figure 6.4(b) also shows that for most values of received signal-power, we are many dB away from the quantum shot-noise limit.

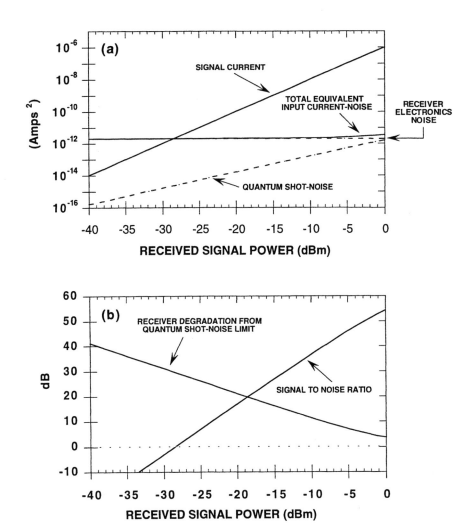

Figure 6.4   **Example performance of the resistor terminated - voltage amplifier front-end using a *p-i-n* photodiode. (a) Signal and noise currents. (b) SNR and degradation from quantum shot-noise limit.**

Now let us use an APD with the resistor terminated voltage amplifier front-end. We expect things to improve because we have already seen that the internal gain of an APD will reduce the effects of the receiver electronics-noise. Let us use an APD with a gain of 10 and an effective ionization ratio of 0.1. The unmultiplied dark-current and the multiplied dark-current are both taken to be 1.0 nA. A gain of 10 is actually quite substantial when one considers that since we have a 5 GHz noise-bandwidth, our APD must have at least a 50 GHz gain bandwidth product for our assumption of frequency independent multiplication within the 5 GHz noise-bandwidth to be realistic. Should this assumption not be true, we would include the APD avalanche multiplication transfer function $M(\omega)$ in our calculation of the equivalent noise-bandwidth and the noise performance would become frequency dependent.

When an APD is used, the coefficients for Eq. 6.8 become

$$a_0 = \frac{2qi_{du} + i_n^2 + \dfrac{4kT}{R_l}}{M^2} + 2qi_{dm}F(M) + i_{shot}^2[F(M)-1], \quad a_2 = 0, \tag{6.11}$$

where from Eq. 5.42 with $k_{eff} = 0.1$, we have $F(M) = 0.1M + 0.9(2 - 1/M)$. All other receiver parameters are the same as in the *p-i-n* example.

Figure 6.5 illustrates the signal-power, total equivalent input noise-power, signal-to-noise ratio, and the degradation from the quantum shot-noise limited current density for the receiver. Figure 6.5(a) shows the variation in current with signal-power and Fig. 6.5(b) shows the corresponding SNR and degradation from quantum shot-noise limit.

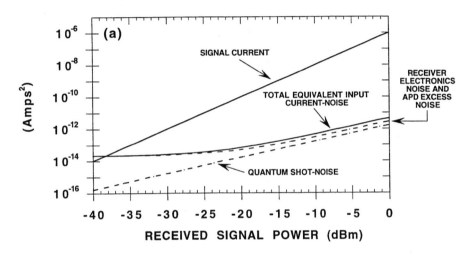

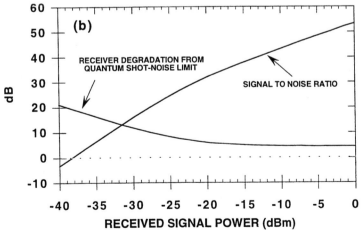

Figure 6.5 **Example performance of the resistor terminated - voltage amplifier front-end using an APD for the photodetector. (a) Signal and noise currents. (b) SNR and degradation from quantum shot-noise limit.**

Note that there is a dramatic improvement in noise performance with even this modest amount of APD gain. Quantum shot-noise is now evident for received signal-powers greater than about -20 dB. The overall receiver equivalent input current-noise also increases with signal-power for levels above -25 dBm.

From Fig. 6.5(b) we see that the 0 dB SNR point is now near -38 dBm, a 10 dB improvement over the *p-i-n* case. This is exactly what is expected since, to first order, we decrease the effect of the receiver electronics-noise by the amount of gain present in the APD. The SNR as a function of received signal-power in the APD receiver also illustrates the change in slope from square-law to linear that is expected in a receiver that is operating near the signal quantum shot-noise limited region. From Fig. 6.5(b) we can see that this receiver is operating much closer to the quantum shot-noise limit than the *p-i-n* receiver did. The excess noise in the APD ultimately prevents us from getting any closer than about 4 to 5 dB from the quantum limit.

The dramatic improvement with an APD leads us to an obvious question. What value of APD gain is correct in the sense that it provides the highest SNR for a given value of received signal-power? As APD gain increases, the impact of receiver electronics-noise decreases but the excess noise factor increases. Thus an optimum value must exist [9-12].

This can be seen in Fig 6.6 where the receiver SNR from Fig. 6.5(b) has been plotted as a function of multiplication for a received signal-power of -30 dBm and a variety of possible values for k_{eff}. For simplicity, we have ignored any gain-bandwidth limitations that could prevent us from obtaining the higher amounts of multiplication.

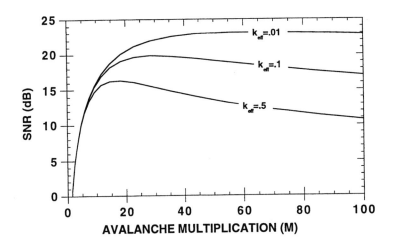

Figure 6.6   Receiver SNR as a function of avalanche multiplication.

Note that the optimum is more clearly defined for higher values of k_{eff}. For low values of k_{eff}, the APD introduces significantly less excess noise and high SNRs can be obtained for a wide variety of gains. The presence of an optimum APD gain is not restricted to just the specific front-end we have considered here. There will be an optimum gain whenever we can trade-off excess multiplication noise against receiver electronics-noise.

6.1.2 High-Impedance Amplifier

We saw that the thermal-noise associated with the load resistor dominated the receiver noise in the resistor-terminated low-impedance voltage amplifier front-end. The high-impedance amplifier is an approach that substantially reduces the effect of the thermal-noise of the load resistor, resulting in improved sensitivity. The high-impedance receiver is based on a technique that has been successfully used with other capacitive current sources such as vidicon tubes [13] and is descended from vacuum tube amplifiers. It was one of the first low-noise front-ends used in optical receivers [11, 14].

The basic design principle is to load the current-source with as large an impedance as possible. This tends to maximize the amount of voltage developed at the input of the amplifier, the idea being that since the voltage is maximized, the effects of any amplifier noise sources will be reduced. Ultimately our ability to realize a high impedance is limited by the capacitance present in the circuit. In fact, the receiver is usually constructed so that the input impedance is dominated by the total capacitance at the amplifier's input node. Since the voltage across a capacitor is the integral of the current through the capacitor, the amplifier's input voltage will be the integral of the photocurrent. This fact lends the high-impedance front-end its alternative name, the integrating front-end. Since the response is that of an integrator, which is a single-pole roll-off with frequency, we must equalize the response with a differentiator if we are to obtain a flat receiver passband.

In general, the high-impedance receiver results in the lowest noise baseband front-end that can be realized without extraordinary effort. We emphasize that the low-noise is obtained by making the load resistor as large as possible. It is not a result of the integrating nature or the high input impedance of the amplifier. A high-impedance receiver front-end is illustrated in Fig. 6.7. A large value load resistor is still usually included. It is beneficial in that it can provide a DC current path for the photodiode and it allows us to control and stabilize the low-frequency pole location ω_p so that we can accurately equalize the response.

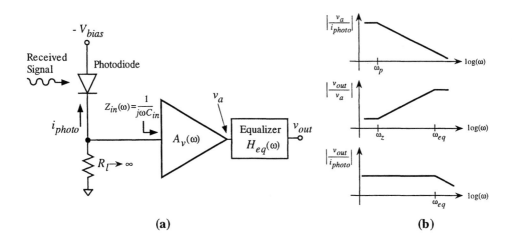

(a)   **(b)**

Figure 6.7 The high-impedance front-end. (a) Typical circuit configuration.
(b) Example transfer functions.

We can model the high-impedance amplifier using the same equivalent noise circuit from Fig 6.3 that we used for the voltage amplifier. The overall receiver transimpedance is obtained by modifying Eq. 6.2 to account for a capacitive amplifier input impedance and the presence of the equalizer stage as

$$Z_t(\omega) = M(\omega)A_v(\omega)R_l H_{eq}(\omega)\frac{1}{\left(1+j\dfrac{\omega}{\omega_p}\right)} \quad (\Omega), \tag{6.12}$$

where

$$\omega_p = \frac{1}{R_l C_t}, \quad C_t = C_d + C_{in}.$$

As an example, using the values of $R_l = 10^6\,\Omega$ and $C_t = 2.0\text{pF}$, the low frequency pole is located at 80 kHz. In order to obtain a flat response out to a high frequency we must cascade this low-pass response with an equalizing high-pass network with a zero located so that it cancels the pole at ω_p. For example if the equalizer stage has a frequency response given by

$$H_{eq}(\omega) = H\frac{\left(1+j\dfrac{\omega}{\omega_z}\right)}{\left(1+j\dfrac{\omega}{\omega_{eq}}\right)}, \tag{6.13}$$

the overall transimpedance is then

$$Z_t(\omega) = M(\omega)A_v(\omega)R_l H\frac{\left(1+j\dfrac{\omega}{\omega_z}\right)}{\left(1+j\dfrac{\omega}{\omega_p}\right)\left(1+j\dfrac{\omega}{\omega_{eq}}\right)}\frac{1}{} \quad (\Omega). \tag{6.14}$$

If we are successful in achieving pole-zero cancellation by setting $\omega_z = \omega_p$, the overall response will be flat out to ω_{eq}. The equalization process is graphically illustrated in Fig 6.7(b).

Equalization can be performed using a simple passive RC circuit, shown in Fig. 6.8(a) [15], or an active circuit, illustrated in Fig. 6.8(b) [16]. The transfer functions for both equalizers have the form of Eq. 6.13. The gain and pole-zero locations for the passive equalizer are given by

$$H = \frac{R_l}{R_s + R_{eq} + R_l}, \quad \omega_z = \frac{1}{R_{eq}C_{eq}}, \quad \omega_{eq} = \frac{R_s + R_{eq} + R_l}{R_s + R_l}\times\frac{1}{R_{eq}C_{eq}}. \tag{6.15}$$

For the active equalizer they are

$$H = -\frac{R_c}{R_e + R_{eq}}, \quad \omega_z = \frac{1}{R_{eq}C_{eq}}, \quad \omega_{eq} = \frac{R_e + R_{eq}}{R_e}\times\frac{1}{R_{eq}C_{eq}}. \tag{6.16}$$

In computing the transfer function for the active equalizer we have assumed that the transistor's current gain is large and that there is no roll-off due to the transistor.

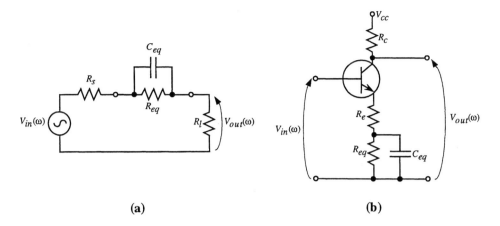

Figure 6.8 Equalizer circuits. (a) Passive. (b) Active.

In order to obtain a wideband response with a high-impedance front-end, the ratio between ω_{eq} and ω_z must be as large as possible. Thus the equalizer provides strong attenuation at low frequencies that cancels the high low-frequency gain from the high-impedance amplifier. This introduces a dynamic range constraint on the high-impedance front-end. Unless care is taken in apportioning the gain in the various stages in the receiver, the large low-frequency gain of the first stage can cause the receiver to overload with relative ease. It is generally preferable to locate a single stage equalizer near the first gain stage to immediately reduce the voltage levels within the receiver. An alternative is to distribute the equalization across several moderate gain stages so that no single high-gain stage is subject to overload. The limited dynamic range can be extended when an APD is used as the photodetector because the APD bias voltage can be used in an automatic gain-control (AGC) loop to reduce the overall receiver gain at high signal levels. In general, a high-impedance amplifier will have a dynamic range of 15-20 dB when a p-i-n is used. This can be extended up to about 35 dB when an APD in an AGC loop is used [9]. Further improvement in dynamic range can be obtained by techniques that either shunt photocurrent away from the input amplifier at high signal levels [17] or incorporate limiting amplifiers into the receiver design [18].

The need to obtain accurate pole-zero cancellation to obtain a flat wideband response can also introduce some difficulties into the receiver design. The total amount of input capacitance can vary from device to device and can be a function of temperature and photodiode and transistor bias conditions. Accurate equalization over time, temperature, and production line variations may prove difficult to obtain. Each individual equalized gain stage in the front-end could require independent manual adjustment to assure accurate pole-zero cancellation. Temperature compensation of the equalization may be required and, in addition, line-coding may be needed to reduce the low-frequency content of the received signal [1-3].

Low-noise is the high-impedance front-end's principal benefit. Limited dynamic range is generally considered to be its principal drawback. In spite of this limitation, high-impedance front-ends have been routinely used in many high sensitivity applications [19-22] and are commercially available in the form of p-i-n-FET and APD-FET receivers.

6.1.3 Transimpedance Amplifier

A front-end architecture that provides a good compromise between the low-noise characteristics of the high-impedance front-end and the wideband nature of the low-impedance voltage-amplifier front-end is known as the transimpedance amplifier, shown in Fig. 6.9(a). The amplifier is termed transimpedance because it utilizes shunt feedback around an inverting amplifier, a technique that is known to stabilize the amplifier's transimpedance [23]. Since the feedback stabilizes the transimpedance, and since the transfer function of an optical front-end is inherently a transimpedance, we might expect this amplifier to be particularly useful in optical communication receivers [24]. As with any feedback amplifier we must be careful to avoid excessive phase-shifts within the loop formed by the feedback resistor and the voltage amplifier. In general, at least a 45 degree phase-margin and a 6 dB gain-margin are needed to assure adequate stability [25].

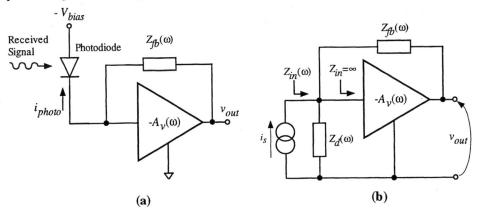

Figure 6.9   Transimpedance amplifier front-end. (a) Typical circuit configuration. (b) Simplified small-signal equivalent circuit.

In order to study the operation of the transimpedance amplifier, we will use the simplified small-signal equivalent circuit shown in Fig. 6.9(b). The photodiode is modeled as a current source and the amplifier's input impedance, any load resistance, and the photodiode's internal impedance are combined into a single impedance $Z_d(\omega)$. Thus the voltage gain-block in Fig. 6.9(b) has an infinite input impedance and a voltage-gain transfer-function $A_v(\omega)$. A straightforward circuit analysis reveals that the overall transimpedance for this circuit is

$$Z_t(\omega) = \frac{v_{out}(\omega)}{i_s(\omega)} = \frac{-Z_{fb}(\omega)}{1 + \frac{1}{A_v(\omega)}\left[1 + \frac{Z_{fb}(\omega)}{Z_d(\omega)}\right]}.$$

(6.17)

The input impedance $Z_{in}(\omega)$ is

$$Z_{in}(\omega) = \frac{\dfrac{Z_{fb}(\omega)}{1 + A_v(\omega)} \times Z_d(\omega)}{\dfrac{Z_{fb}(\omega)}{1 + A_v(\omega)} + Z_d(\omega)} = \frac{Z_{fb}(\omega)}{1 + A_v(\omega)} \bigg\| Z_d(\omega).$$

(6.18)

We have used two vertical bars to represent the parallel combination of impedances.

In the ideal case of an infinite gain-bandwidth amplifier these become

$$Z_t(\omega) \cong -Z_{fb}(\omega), \quad Z_{in}(\omega) \cong 0. \tag{6.19}$$

In such a case, the transimpedance is set entirely by the feedback impedance and the input impedance of the amplifier approaches zero. We would normally expect that the feedback impedance would be a resistor. Any realizable resistor will have a parallel stray capacitance associated with it. For this case we find that the transimpedance of the infinite gain-bandwidth is given by

$$Z_t(\omega) \cong \frac{-R_{fb}}{\left(1 + j\dfrac{\omega}{\omega_p}\right)}, \tag{6.20}$$

where $\omega_p = \dfrac{1}{R_{fb}\left(C_{fb} + \dfrac{C_d}{A_v(\omega)}\right)} \cong \dfrac{1}{R_{fb}C_{fb}}$ as $A_v(\omega) \rightarrow \infty$.

We see that for large gains, the capacitance across the amplifier's input, C_d, has no effect at all on the bandwidth. This is just the opposite of the high-impedance design, where the total capacitance at the input essentially determines the first-stage bandwidth. Since the transimpedance amplifier's input impedance approaches zero with high-gains, the input capacitance is effectively shorted out, resulting in a frequency response that is determined entirely by the feedback impedance. In reality, things are not quite this good. We have ignored the internal series resistance R_s of the photodiode in this analysis and the detector junction capacitance still plays a role, although it will be relatively insignificant if R_s is small.

A more generalized format for a feedback network is the T-network illustrated in Fig. 6.10(a) [24, 26]. For large voltage gains, the effective transimpedance for this type of feedback network is given by

$$Z_t(\omega) \cong -\frac{(Z_1(\omega)Z_2(\omega) + Z_2(\omega)Z_3(\omega) + Z_3(\omega)Z_1(\omega))}{Z_2(\omega)}. \tag{6.21}$$

The T-network can be used to compensate for the roll-off in transimpedance caused by C_{fb} by using the elements shown in Fig. 6.10(b). The transimpedance response for the circuit is now

$$Z_t(\omega) \cong -R_{fb}\frac{\left(1 + j\dfrac{\omega}{\omega_z}\right)}{\left(1 + j\dfrac{\omega}{\omega_p}\right)\left(1 + j\dfrac{\omega}{\omega_b}\right)}, \tag{6.22}$$

where $\omega_p = \dfrac{1}{R_{fb}C_{fb}}$, $\omega_z = \dfrac{1}{R_3C_2}$, $\omega_b = \dfrac{1}{R_2C_2}$, and we have assumed that $R_{fb} \gg R_3 \gg R_2$, and $C_2 \gg C_{fb}$.

The additional components introduce phase-lag into the feedback to compensate for the phase-lead produced by C_{fb}. The T-network has primarily found use in low to moderate bandwidth receivers that utilize large values of R_{fb}, in the range of hundreds of kΩ to hundreds of MΩ [27, 28]. The bandwidth of the receiver that can be obtained using this technique is limited because of the need to maintain a relatively large voltage gain throughout the receiver's bandwidth.

If the location of the low-frequency zero ω_z is set exactly equal to the location of the feedback resistor pole ω_p, pole-zero cancellation occurs and the response becomes determined by the $C_2 R_2$ time-constant. In theory, if R_2 is made zero, the response is frequency independent. In reality, finite gain-bandwidths in the amplifier and additional parasitic reactances conspire to prevent a truly frequency independent response. In many cases it is desirable to include a fixed value for R_2 so that we can accurately set the bandwidth.

Perfect pole-zero cancellation will generally require manual adjustment of the feedback components, which limits this technique's usefulness in mass-produced receivers. When manual adjustment is allowable, it is frequently best done in the time-domain. If the pulse response of the receiver is monitored, R_3 can be adjusted to obtain the minimum amount of overshot or drop in the output voltage. This is similar to the techniques used to compensate high-impedance oscilloscope probes.

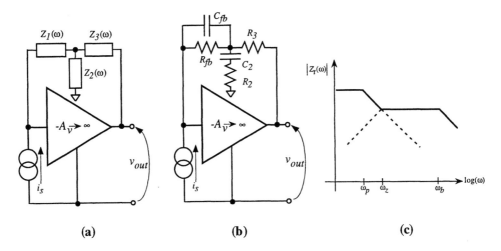

(a)   **(b)**   **(c)**

Figure 6.10 T-feedback network. (a) Generic configuration. (b) Network to compensate for effects of C_{fb}. (c) Transimpedance response.

Our assumption of an infinite gain-bandwidth in our previous analyses is clearly impractical. A more realistic form for an amplifier's transfer function is

$$A_v(\omega) = -\frac{A_o}{\left(1 + j\dfrac{\omega}{\omega_a}\right)}. \tag{6.23}$$

This is a good approximation to the type of transfer function obtained in many single stage transistor amplifiers and operational amplifiers. Substituting Eq. 6.23 into Eq. 6.17 and Eq. 6.18 results in a transimpedance and an input impedance for this circuit that follow classic second-order transfer-functions [25], given by

$$Z_t(\omega) = -R_t \frac{1}{\left(1 + j\dfrac{\omega}{Q\omega_o} - \dfrac{\omega^2}{\omega_o^2}\right)},$$

$$Z_{in}(\omega) = R_{in} \frac{\left(1 + j\dfrac{\omega}{\omega_a}\right)}{\left(1 + j\dfrac{\omega}{Q\omega_o} - \dfrac{\omega^2}{\omega_o^2}\right)}, \qquad\qquad (6.24)$$

where

$$R_t = R_{fb}\left[\frac{A_o R_d}{(A_o+1)R_d + R_{fb}}\right], \qquad R_{in} = \frac{\dfrac{R_{fb}}{(A_o+1)}R_d}{R_d + \dfrac{R_{fb}}{(A_o+1)}} = \frac{R_{fb}}{(A_o+1)}\bigg\| R_d,$$

$$Q = \sqrt{\frac{R_d R_{fb} C_t}{\omega_a\left[(A_o+1)R_d + R_{fb}\right]}} \times \frac{\omega_a\left[(A_o+1)R_d + R_{fb}\right]}{R_d + R_{fb} + \omega_a R_d R_{fb}\left(C_t + C_{fb}A_o\right)},$$

$$\omega_o = \sqrt{\frac{\omega_a\left[(A_o+1)R_d + R_{fb}\right]}{R_d R_{fb} C_t}}, \qquad C_t = C_{fb} + C_d.$$

A second order circuit is characterized by its natural resonance frequency ω_o and its Q. At resonance, the response is equal to Q times the low-frequency value. A high Q results in a narrow, highly peaked response with an under-damped pulse response in the time-domain. A low Q results in an over-damped response. A critically-damped pulse response is often desirable and corresponds to a Q of 0.5

As with any second-order response, proper selection of the circuit's natural frequency and Q allows us to obtain a variety of desirable transfer functions such as the maximally-flat amplitude response of the Butterworth, the equal-ripple Chebyshev, or the maximally-flat in group-delay Bessel [29, 30]. In particular, for the Butterworth response, we want $Q = 1/\sqrt{2}$. Given a desired transimpedance bandwidth ω_o and the values for the total input capacitance, feedback resistance, and feedback capacitance, we can solve Eq. 6.24 for the required values of voltage amplifier gain and bandwidth needed to obtain a Butterworth response. Because of the coupling between Q and ω_o in Eq. 6.24, the design parameters are best found via computer-aided techniques. Using an approach similar to that of Manterola-Barros [28] we can obtain the design curves illustrated in Fig 6.11.

Figure 6.11 plots the required amplifier bandwidth f_a and DC gain A_o needed to obtain a Butterworth response for the values $R_d = 1.0\text{M}\Omega$, $C_t = 1.0\text{pF}$, $R_{fb} = 1\text{k}\Omega$, and $C_{fb} = 0.2\text{pF}$. This corresponds to a reasonably low-level of total capacitance. Also shown in the figure are the associated gain-bandwidth of the amplifier, $A_o \times f_a$, and the ratio of the effective DC transimpedance realized by the amplifier, R_t, to the value of the feedback resistor, R_{fb}. The design curves are exhaustive for the chosen parameters, that is, there are no other realizable solutions that will allow a Butterworth response. For example, with the specified parameters, we cannot obtain a 100 MHz bandwidth Butterworth response because the required amplifier bandwidth would become negative.

As the bandwidth increases the voltage amplifier must have an increasingly large gain-bandwidth product. Bandwidths of several hundred MHz require gain-bandwidths of several GHz. Ultimately, bandwidths in excess of 1 GHz cannot be obtained unless we either relax our requirements for a Butterworth response or use a more sophisticated voltage amplifier and feedback network.

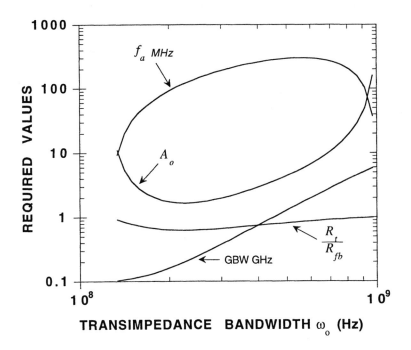

Figure 6.11 Design curves for a Butterworth response.

Figure 6.11 provides voltage amplifier parameters given known feedback and photodetector parameters. In some instances we may have a certain type of amplifier available and need to know the type of performance obtainable. We can get a simple estimate for the overall-bandwidth and the circuit Q resulting from the use of an amplifier with a known gain-bandwidth (GBW) product (in Hz) via

$$\omega_o \cong \sqrt{2\pi GBW \times \omega_p},$$

$$Q \cong \frac{1}{\sqrt{2\pi GBW \times \omega_p}} \times \frac{2\pi GBW \times \omega_{fb}}{2\pi GBW + \omega_{fb}},$$

$$(6.25)$$

where $\omega_p = \dfrac{1}{R_{fb}C_t}$, $\omega_{fb} = \dfrac{1}{R_{fb}C_{fb}}$, and we have assumed that $R_{in} \gg R_{fb}$, $A_o \gg 1$,

and $\dfrac{C_d}{A_o} \ll C_{fb}$.

Using these approximations we can determine that for a 1.0 pF total capacitance and a 1000 Ω feedback resistor, a voltage amplifier with a 1 GHz gain-bandwidth product would allow us to realize a receiver with a bandwidth of approximately 400 MHz. If the feedback capacitance was 0.1 pF, the circuit would have a Q of approximately 1.5 which may represent an unacceptable amount of peaking. If we increase the feedback capacitance to 0.5 pF, we would obtain a flatter response at the expense of a slightly narrower bandwidth of 340 MHz.

Another case of particular interest is when the gain-bandwidth of the amplifier is much larger than the bandwidth we are interested in obtaining. In this case the transimpedance and input impedance are approximately

$$Z_t(\omega) \cong -R_{fb} \frac{1}{\left(1+j\dfrac{\omega}{\omega_{fb}}\right)},$$

$$Z_{in}(\omega) \cong \frac{R_{in}}{A_o} \frac{1}{\left(1+j\dfrac{\omega}{\omega_{fb}}\right)}, \tag{6.26}$$

where $R_d \gg R_{fb}$, $\quad A_o \gg 1$, $\quad A_o\omega_a \gg \omega_{fb}$.

The input impedance of the transimpedance amplifier is essentially the feedback resistor divided by the open loop gain of the amplifier. The input impedance can therefore reach a low value for high gain. The low impedance effectively short-circuits the photodetector, which minimizes the influence of photodetector capacitance. This can dramatically increase the bandwidth of the amplifier over that of an unequalized high-impedance front-end. The other benefit of the transimpedance amplifier is a high dynamic range. Unlike the high-impedance design, the response can be flat from DC to the design bandwidth. There is no integration of the signal photocurrent that can lead to overload. The dynamic range of transimpedance designs using *p-i-n* photodiodes can exceed 30 dB while those using APDs and AGC loops can exceed 40 dB [9].

In addition to the small-signal and dynamic range response of the transimpedance amplifier we are also interested in its noise performance. This has been extensively studied [9, 12, 24] and we will summarize the results. The principal effect is that the thermal-noise of the feedback resistor appears as an additional current-noise source that appears across the input of the amplifier in parallel with the amplifier's current-noise $i_n(\omega)$. This is a very important point. The transimpedance amplifier allows us to obtain the noise performance of a relatively large value feedback resistor simultaneously with the bandwidth benefits obtained by loading the photodetector with a relatively low value of load impedance. Thus the amplifier represents a compromise of sorts between the low-noise, low-bandwidth, high-impedance front-end and the high-noise, high-bandwidth, resistor terminated voltage amplifier.

To illustrate the noise analysis of a transimpedance amplifier front-end we will use the noise equivalent circuit illustrated in Fig. 6.12. We have simplified the photodiode equivalent circuit and are neglecting correlation between the amplifier's noise sources. Our analysis procedure follows those outlined in Chapter 5. We use superposition to combine the uncorrelated noise sources to obtain an expression for the voltage-noise-power at the output of the amplifier. We then divide by the magnitude squared of the overall transimpedance to obtain the equivalent input current-noise.

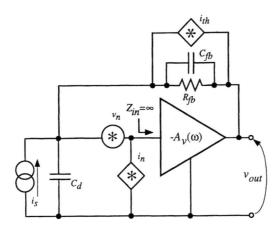

Figure 6.12 Transimpedance amplifier noise equivalent circuit.

The output voltage-noise-power is given by

$$v_{n_o}^2(\omega) \cong |Z_t(\omega)|^2 \left[i_n^2 + \frac{4kT}{R_{fb}} + \frac{v_n^2}{R_t^2}\left(1 + \left(\frac{\omega}{\omega_z}\right)^2\right) \right], \qquad (6.27)$$

where $\omega_z = 1/R_{fb}\,C_t$ and we have assumed that $A_o \gg 1$. We are also ignoring the capacitive shunting from C_{fb} that will reduce the thermal-noise-current contribution from R_{fb} at high frequencies. Using Eq. 6.24, the overall transimpedance, under the same assumptions as we made to obtain Eq. 6.27, is given by

$$Z_t(\omega) \cong -R_{fb}\,\frac{1}{1 + j\omega\left[\dfrac{1}{\omega_a A_o} + \dfrac{R_{fb}C_d}{A_o} + R_{fb}C_{fb}\right] - \omega^2\,\dfrac{R_{fb}C_t}{\omega_a A_o}}. \qquad (6.28)$$

The equivalent input current-noise is then

$$i_{n_{eq}}^2(\omega) \cong i_n^2 + \frac{4kT}{R_{fb}} + \frac{v_n^2}{R_t^2}\left(1 + \left(\frac{\omega}{\omega_z}\right)^2\right). \qquad (6.29)$$

Figure 6.13 plots the performance of the transimpedance front-end illustrated in Fig. 6.12 for the case where $C_d = 1.0\,\mathrm{pF}$, $R_{fb} = 1\mathrm{k}\Omega$, $C_{fb} = 0.18\,\mathrm{pF}$, $A_o = 100$, $f_a = 100\mathrm{MHz}$, $i_n = 1.0\,\mathrm{pA}/\sqrt{\mathrm{Hz}}$, and $v_n = 1.0\,\mathrm{nV}/\sqrt{\mathrm{Hz}}$. Figure 6.13(a) shows the magnitudes of the open and closed loop voltage gains for the circuit. The open-loop voltage gain is simply the transfer function of the amplifier, $A_v(f)$. The gain is flat out to f_a, beyond which point the response follows a single-pole roll-off. The gain crosses through unity at the gain bandwidth product of 10 GHz. The closed-loop gain is important because it relates the voltage-noise source to the output voltage-noise. The closed-loop is very close to unity up to the frequency f_{fb}, above which the gain increases because the amount of feedback decreases. The reduction of open-loop gain at high frequencies eventually affects the closed-loop gain and the closed loop-gain rolls-off above f_o.

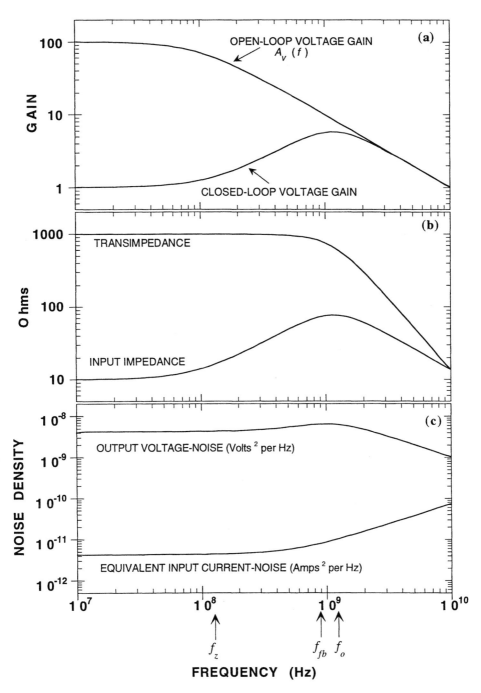

Figure 6.13 Transimpedance front-end. (a) Open and closed loop voltage gains. (b) Input impedance and transimpedance. (c) Total output voltage-noise and equivalent input current-noise.

$$f_z = \frac{1}{2\pi R_{fb} C_t}, \quad f_{fb} = \frac{1}{2\pi R_{fb} C_{fb}}, \quad f_o \cong \sqrt{GBW \frac{1}{2\pi R_{fb} C_t}}.$$

The magnitudes of the effective input receiver impedance and the overall transimpedance are shown in Figure 6.13(b). The input impedance at low frequencies is equal to the feedback resistor divided by one plus the open-loop gain. At frequencies above f_z the amount of feedback decreases and the input impedance begins to rise, reaching a value of nearly 80 Ω near f_o. The input impedance continues to rise until the effects of capacitance at the input node begin to dominate. At high frequencies the capacitance effectively shorts out the input, reducing both the input impedance and the transimpedance. At low frequencies we obtain an overall transimpedance of 990 Ω. The Q of the circuit is just under 0.67 and we would expect a response that is nearly a maximally flat Butterworth response. The receiver's bandwidth, defined as the point at which the transimpedance is approximately 70% of its low frequency value (i.e. the 3 dB or half power point) is just over 950 MHz.

Figure 6.13(c) shows the spectral densities of the output voltage-noise and equivalent input current-noise of the receiver. The two curves are related by the overall transimpedance. The output voltage-noise is flat until the variation in closed-loop voltage gain causes a slight peak to occur in the vicinity of f_o. At high frequencies the output voltage-noise drops because the capacitance in the circuit is effectively short-circuiting the noise (and the signal for that matter) to ground. The equivalent input current-noise is flat at low frequencies but increases at high frequencies because of the differentiating action of the input capacitance on the amplifier's voltage-noise. The low frequency level is set by the combination of the 1.0 pA per root Hz amplifier current-noise density and the 4 pA per root Hz thermal-noise current contributed by the 1000 Ω feedback resistor. Note that the equivalent input current-noise density *does not* in general have the same shape as the output voltage-noise density. This difference can become important when designing the following demodulator stages since the demodulator observes the signal plus output voltage-noise, not the input current-noise.

Before concluding our discussion of transimpedance front-ends, we highlight some non-traditional transimpedance designs. The first is an approach in which the feedback resistor is replaced by an "artificial" resistor constructed using active components. Since an active resistor can be constructed to have less than the "standard" $4kT/R$ thermal-noise current of a conventional dissipative resistor, an improvement in receiver noise performance can be obtained [31]. A variation on this approach uses an active feedback path to obtain high dynamic range [32]. Another approach uses optical feedback to obtain the transimpedance feedback. The output of the amplifier stages drives a laser or light-emitting diode that is detected with an additional photodetector that is in parallel with the signal detector. The photodetector biases are arranged such that the two photocurrents subtract so that the fed-back signal is removed from the received signal. High dynamic range and high sensitivity for moderate bandwidths have been demonstrated using this approach [33].

In summary, the transimpedance design represents a compromise between the wideband, but comparatively noisy, resistor terminated low-impedance voltage-amplifier and the low-noise, but equalization dependent, high-impedance approach. A drawback to the three receiver front-ends that we have examined is that, in general, the low-frequency noise performance will be relatively flat (neglecting any 1/f effects) before rising dramatically at high frequencies because of the differentiating effect that the input capacitance has on the amplifier's voltage-noise. Capacitance at the input node thus ultimately determines the receiver noise performance.

6.1.4   Noise-Matched or Resonant Amplifiers

Noise matching techniques allow us to significantly reduce the influence of the input node capacitance. This is often the preferred approach for designing receivers with multi-GHz bandwidths because noise matching makes use of conventional microwave noise parameters and lends itself to use in commercially available computer-aided microwave design routines. A noise-matched architecture is in some sense the most general case of a receiver front-end in that it is applicable to a wide variety of amplifier configurations, including high-impedance and transimpedance. The design principles are based on low-noise microwave amplifier design techniques [34-36] that can be applied to both resonant narrow-band [37-39] and broad-band receiver designs [40, 41].

We have already seen that the overall noise performance of an optical receiver is generally a strong function of the photodetector's impedance. The basic principle behind the noise-matched receiver is the use of an impedance matching network located between the photodetector and the first amplifier stage to convert the highly capacitive source to an impedance that results in a better noise match for the transistor. The topic of broadband impedance matching can be very complex and entire books have been written on the subject [42]. For our purposes we will use the simplified circuit illustrated in Fig 6.14(a). Our goal will be to obtain the equivalent input model of Fig. 6.14(b).

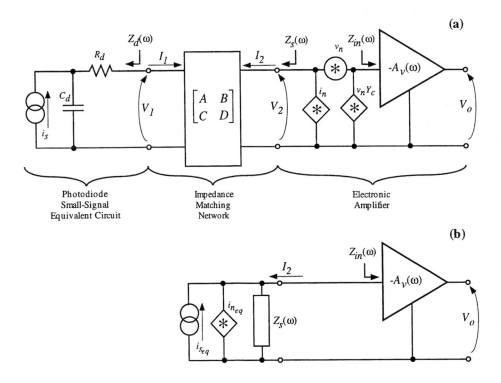

Figure 6.14 Noise-matched or resonant front-end. (a) Simplified model. (b) Model using equivalent inputs.

The photodetector is represented by a simplified small-signal model consisting of a parallel capacitance C_d and a series resistance R_d. Between the photodetector and the first amplifier stage there is an impedance matching network that can be represented by its transmission matrix:

$$\begin{bmatrix} V_1(\omega) \\ I_1(\omega) \end{bmatrix} = \begin{bmatrix} A(\omega) & B(\omega) \\ C(\omega) & D(\omega) \end{bmatrix} \times \begin{bmatrix} V_2(\omega) \\ -I_2(\omega) \end{bmatrix}. \qquad (6.30)$$

The amplifier's noise is represented by a voltage-noise source, a current-noise source, and a correlated current-noise source. In many cases, the individual noise sources of an amplifier or transistor are not specified at microwave frequencies. Instead, the specification sheet will usually have three noise parameters specified. Three noise parameters are required because a measurement of just the amplifier noise figure is inadequate to specify noise performance for a reactive source-impedance. The first parameter is the minimum noise figure that is obtainable $F_{min}(f)$, the second is the noise resistance $R_n(f)$, and the third is the optimum source admittance $Y_{opt}(f)$. Note that they are all generally functions of frequency. This approach is used with microwave amplifiers because each parameter can be conveniently measured using standard microwave test equipment [43], whereas the individual noise sources are difficult to measure at high frequencies.

The noise figure $F(\omega)$ for the front-end amplifier can be expressed in three equivalent ways [35, 36, 44]:

$$F(\omega) = 1 + \frac{i_n^2(\omega) + v_n^2(\omega)|Y_s(\omega) + Y_c(\omega)|^2}{4kTG_s(\omega)},$$

$$F(\omega) = F_{min}(\omega) + \frac{R_n(\omega)}{G_s(\omega)}|Y_s(\omega) - Y_{opt}(\omega)|^2, \qquad (6.31)$$

$$F(\omega) = F_{min}(\omega) + \frac{G_n(\omega)}{R_s(\omega)}|Z_s(\omega) - Z_{opt}(\omega)|^2,$$

where $Z_s(\omega) = R_s\omega + jX_s(\omega), \quad Y_s(\omega) = \dfrac{1}{Z_s(\omega)}, \quad Y_s(\omega) = G_s(\omega) + jB_s(\omega),$

$$R_n(\omega) = \frac{v_n^2(\omega)}{4kT}, \quad G_n(\omega) = \frac{i_n^2(\omega)}{4kT}, \quad Y_c(\omega) = G_c(\omega) + jB_c(\omega).$$

Note that Eq. 6.31 shows that the noise figure is a direct function of the source admittance and that a minimum possible noise figure can be defined. The variation of noise figure with source admittance at a specific frequency can then be plotted, as in Fig. 6.15(a).

If the optimum source admittance is presented to the device over all frequencies, that is, $Y_s(\omega) = Y_{opt}(\omega)$, the minimum noise figure $F_{min}(\omega)$ is obtained as [36]

$$F_{min}(\omega) = 1 + 2R_n(\omega)\left[G_c(\omega) + \sqrt{G_c^2(\omega) + \frac{G_n(\omega)}{R_n(\omega)}} \right]. \qquad (6.32)$$

The actual noise figure obtained at a given frequency can be plotted as a function of the complex source admittance Y_s. The resulting graph is a series of contours of constant noise figure, as illustrated in Fig. 6.15 (b).

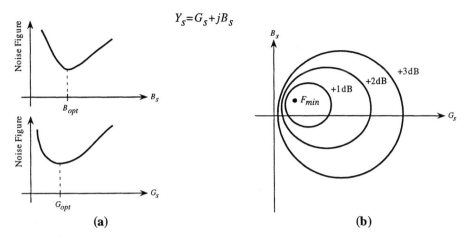

$$Y_s = G_s + jB_s$$

Figure 6.15 Noise figure variations. (a) Variations with the individual components of the source admittance. (b) Contours of constant noise-figure.

It is possible to solve Eqs. 6.31 and 6.32 to obtain the individual noise sources but this is not necessary. By calculating the short-circuit current appearing at the amplifier input we can compute the equivalent input current-noise generator $i_{n_{eq}}$ that appears in parallel with i_s. This is equivalent to determining the Norton equivalent circuit. Expressed in terms of the conventional microwave noise parameters $i_{n_{eq}}$ gives [45]

$$i_{n_{eq}}^2(\omega) = 4kTG_s(\omega)F(\omega)[\omega C_d]^2 \times |B(\omega) + D(\omega)Z_d(\omega)|^2, \qquad (6.33)$$

where $Y_s(\omega) = G_s(\omega) + jB_s(\omega) = \dfrac{A(\omega) + C(\omega)Z_d(\omega)}{B(\omega) + D(\omega)Z_d(\omega)}, \quad Z_d(\omega) = R_d + \dfrac{1}{j\omega C_d}.$

The noise figure in terms of just the individual amplifier noise generators is

$$i_{n_{eq}}^2(\omega) = |C(\omega) + j\omega C_d[C(\omega)R_d + A(\omega)]|^2 v_n^2(\omega)$$
$$+ |D(\omega) + j\omega C_d[D(\omega)R_d + B(\omega)]|^2 [i_n^2(\omega) + Y_c^2(\omega)v_n^2(\omega)]. \qquad (6.34)$$

By selecting an appropriate matching network it is possible to significantly reduce the effects of C_d on the overall noise performance of the receiver. Reducing the noise is clearly not enough. We must also preserve the signal at the output of the linear channel. In selecting the matching network we must also take into account the equivalent signal component that appears at the input to the amplifier. This is given by

$$i_{s_{eq}}(\omega) = \frac{i_s(\omega)}{D(\omega) + j\omega C_d[B(\omega) + D(\omega)R_d]}. \qquad (6.35)$$

The overall transimpedance for the receiver is then

$$Z_t(\omega) = \frac{v_o(\omega)}{i_s(\omega)}$$

$$= \frac{Z_{in}(\omega)A_v(\omega)}{D(\omega) + Z_{in}(\omega)[j\omega C_d[C(\omega)R_d + A(\omega)] + C(\omega)] + j\omega C_d[D(\omega)R_d + B(\omega)]}. \qquad (6.36)$$

There are a wide variety of possible impedance matching circuits. Some popular ones are illustrated in Fig. 6.16. Many impedance matching circuits have the form of a T-circuit, as shown in Fig. 6.16(a). The transmission matrix for this is given by

$$
\begin{bmatrix} A(\omega) & B(\omega) \\ C(\omega) & D(\omega) \end{bmatrix} = \begin{bmatrix} 1 + \dfrac{Z_1(\omega)}{Z_3(\omega)} & \dfrac{Z_1(\omega)Z_2(\omega) + Z_2(\omega)Z_3(\omega) + Z_1(\omega)Z_3(\omega)}{Z_3(\omega)} \\ \dfrac{1}{Z_3(\omega)} & 1 + \dfrac{Z_2(\omega)}{Z_3(\omega)} \end{bmatrix}. \quad (6.37)
$$

Figure 6.16(b) illustrates a lossy low-pass filter topology that has been used to provide noise matching over bandwidths of several GHz [45-47]. The resistors spoil the Q of the circuit, resulting in a flatter broad-band response. Resistors can also be used to set an intermediate impedance level for impedance-matching purposes. The additional thermal-noise the resistors introduce into the circuit is compensated for by an improvement in the impedance match between the photodiode and amplifier.

A common component used in broadband matching is known as the inductive L-section transformer, illustrated in Fig. 6.16(c) [48-50]. This can be replaced with the T-circuit illustrated in Fig. 6.16(d) via the transformation

$$
L_1 = L_s + (1-n)L_p, \qquad L_2 = nL_p, \qquad L_3 = \left(n^2 - n\right)L_p, \quad (6.38)
$$

where there is a requirement of $1 \le n \le 1 + \dfrac{L_s}{L_p}$ for realizability.

As with the low-pass filter of Fig. 6.16(b), loss in the form of R_3 can be added to the circuit to improve the overall match. This approach has been used to realize receivers with bandwidths of some 16 GHz [40].

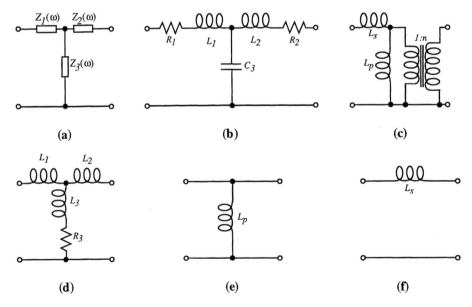

Figure 6.16 Impedance matching circuits. (a) General T-circuit. (b) Lossy low-pass filter (c) Transformer. (d) Equivalent circuit for a lossy transformer. (e) Parallel inductor. (f) Series inductor.

Simple yet effective noise matching circuits are the parallel and series inductors shown in Figs. 6.15(e) and 6.15(f). To demonstrate the benefit of the parallel inductor, we first compute the transmission matrix using Eq. 6.35 as

$$\begin{bmatrix} A(\omega) & B(\omega) \\ C(\omega) & D(\omega) \end{bmatrix} = \begin{bmatrix} 1 & 0 \\ 1/j\omega L_p & 1 \end{bmatrix}.$$

(6.39)

We will simplify the analysis by assuming that the series resistance R_s and the correlation admittance Y_c can be ignored. We note that more extensive analytical expressions than those we will use in this simple example exist [51, 52] and that in general, computer-aided techniques are required to obtain adequate accuracy at microwave frequencies. In our simplified example, Eq. 6.34 reduces to

$$i_{n_{eq}}^2(\omega) = \left| C(\omega) + j\omega C_d[A(\omega)] \right|^2 v_n^2(\omega) + \left| D(\omega) + j\omega C_d[B(\omega)] \right|^2 i_n^2(\omega).$$

(6.40)

Substituting the matrix elements from Eq. 6.39 into Eq. 6.40 yields

$$i_{n_{eq}}^2(\omega) = \left| \frac{1}{j\omega L_p} + j\omega C_d \right|^2 v_n^2(\omega) + i_n^2(\omega) = \left| \frac{1 - \left(\dfrac{\omega}{\omega_n} \right)^2}{j\omega L_p} \right|^2 v_n^2(\omega) + i_n^2(\omega),$$

(6.41)

where $\omega_n = \dfrac{1}{\sqrt{C_d L_p}}$.

At the noise resonance frequency ω_n the voltage-noise has no effect on the circuit. The price we pay for this is a relatively narrow band-pass response. If we model the amplifier input impedance as a resistor R_{in} in parallel with a capacitor C_{in}, the overall transimpedance has the form of a classic parallel resonance that peaks at ω_o and can be written as

$$Z_t(\omega) = \frac{j\omega L_p A_v(\omega)}{\left(1 + j\omega \dfrac{\omega}{\omega_o Q} - \dfrac{\omega^2}{\omega_o^2} \right)},$$

(6.42)

where $\omega_o = \dfrac{1}{\sqrt{L_p C_t}}$, $Q = R_{in} \sqrt{\dfrac{C_t}{L_p}}$, $C_t = C_d + C_{in}$.

Note that the resonant frequency ω_o that is associated with the peak of the transimpedance is not the same as the noise resonance frequency ω_n that results in tuning-out of the voltage-noise contribution. In this simple model they differ by the amplifier input capacitance. Some difference in the frequencies for minimum noise and maximum response will generally occur because the noise tuning is only concerned with the source-impedance while the overall transimpedance involves maximizing the transducer power gain, which is subject to both source *and* amplifier impedance.

If a series inductor had been used instead of a parallel inductor, we would observe a reduction in the effects of the current-noise term at resonance, instead of the voltage-noise term. We would also obtain a third-order low-pass transfer function for the transimpedance. This makes the series inductor particularly attractive for low-noise wideband amplifiers [37, 52, 53].

The choice of parallel inductance or series inductance is also influenced by the relative size of the voltage and current-noise sources and the ratio of detector capacitance to input capacitance. In general, when the photodetector capacitance is less than or equal to the amplifier's input capacitance, series tuning is used to advantage. When the detector capacitance is large it emphasizes the voltage-noise, and a parallel inductor is often preferred.

Figure 6.17 illustrates the general shapes for the overall receiver transimpedance and equivalent input current-noise obtained for various matching circuits. The effects of using the parallel inductor from Fig 6.15(e) are illustrated in Fig. 6.17(a). Note that a relatively broad-reduction in noise is obtained even though the response may have significant peaking. This implies that an additional equalization network could be used to further flatten the response without accumulating a severe noise penalty.

In Fig. 6.17(b) the effects of a series inductor or a low-pass filter are illustrated, while in Fig 6.17(c) the effects of the transformer T-circuit equivalent are illustrated. The T-circuit transformer results in a double-tuned response that exhibits two distinct peaks in transimpedance and two distinct nulls in equivalent input current-noise.

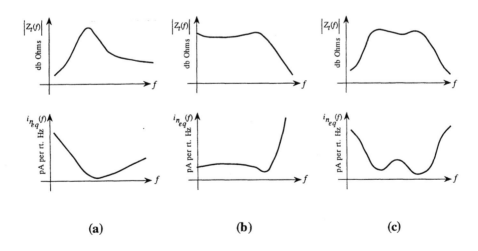

(a)   (b)   (c)

Figure 6.17   Effect of noise matching circuits on overall receiver transimpedance and equivalent input current-noise. (a) Parallel inductor. (b) Series inductor or low-pass filter. (c) Transformer or T-circuit transformer equivalent (double-tuned response).

The benefits obtained with noise-matching are not without limits. As with any broadband impedance match there are fundamental limits on the amount of power that can be transferred without loss from a reactive source such as a photodiode to a reactive load such as an amplifier [42]. This constraint can be written as [54, 55]

$$\int_0^\infty \frac{1}{\omega^2} \ln\left[\frac{1}{|\rho(\omega)|}\right] d\omega \le \pi \times \min\{R_d C_d, R_{in} C_{in}\}, \qquad (6.43)$$

where the amount of power delivered to the amplifier by the photodetector is given by

$$P_l(\omega) = \left[1 - |\rho(\omega)|^2\right] \frac{i_s^2(\omega)}{8\omega^2 C_d^2 R_d}. \qquad (6.44)$$

This constraint gives the bound for lossless matching using a passive network consisting of inductors, capacitors, and transformers. In theory, a lossless match is preferred because it maximizes the signal-power and can be particularly desirable in amplifiers working with high level signals. A lossless match is not always desirable in a receiver front-end. By introducing small resistors into the matching network we can trade-off signal loss for a flatter overall receiver response. In many practical cases a small amount of loss can dramatically improve the overall response function by spoiling the relatively high Q resonances formed between the highly reactive photodetector source-impedance and the amplifier input impedance. A flatter, more well behaved response will often result in better overall receiver performance than that obtained by simply maximizing power transfer.

6.2 Amplifier Circuit Design

We have now investigated four receiver front-end architectures and have expressed their performance in terms of electronic amplifier transfer functions and noise sources. In this section we will examine the performance of the two principal types of device used to construct electronic amplifiers for receiver front-ends, the bipolar junction transistor (BJT) and the field effect transistor (FET). Either can be used to obtain low-noise performance, although the FET can have certain advantages at high frequencies.

6.2.1 Bipolar Transistor Amplifiers

Bipolar junction transistors are one of the key building blocks of electronic amplifiers. The noise sources present in a bipolar junction transistor have been extensively studied [9, 12, 56-59]. The noise model most suitable for use in optical receivers is based on the hybrid-π model of a transistor and is illustrated in Fig. 6.18.

The hybrid-π model shown in Fig. 6.18(b) replaces the BJT with a combination of resistors and capacitors, and the current amplification that occurs between the base terminal and collector terminal is modeled by a controlled current source. The base-spreading resistance r_x accounts for any resistance between the base terminal contact and the actual active base region of the device. This is a true dissipative resistance and its associated thermal-noise is accounted for by a voltage-noise term v_{nx}. The base current and collector current shot-noises are accounted for by two noise current generators i_{nb} and i_{nc}. Any 1/f noise effects are accounted for by a noise current generator i_{nf}. The remaining resistances r_π, r_μ, and r_o are dynamic resistances; they do not dissipate energy and do not contribute thermal-noise.

(a)   **(b)**

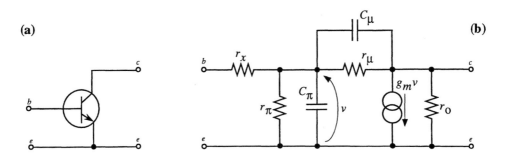

(c)

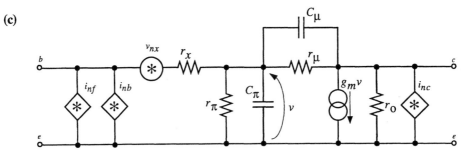

Figure 6.18 Bipolar junction transistor models. (a) Schematic symbol of the common emitter configuration. (b) Small-signal hybrid-π model. (c) Corresponding noise model.

Neglecting second-order effects that are introduced by the internal feedback path comprised of r_μ and C_μ, the equivalent noise current and noise voltage generators for a BJT operating in the common emitter configuration are given by [9, 12]

$$v_n^2(f) \cong \frac{1}{2} r_x q I_b \left(\frac{f_c}{f} \right) + 4kT r_x + 2q I_c r_e^2 \quad (\frac{\text{volts}^2}{\text{Hz}}),$$

$$i_n^2(f) \cong 2q I_b \left(\frac{f_c}{f} \right) + 2q I_b + 2q I_c \left(\frac{f}{f_t} \right)^2 \quad (\frac{\text{amps}^2}{\text{Hz}}), \qquad (6.45)$$

where

$I_c = $ DC collector current, $I_b = $ DC base current,

$r_e = $ dynamic resistance of the emitter $= \dfrac{1}{g_m}$,

$g_m = $ transistor transconductance $= \dfrac{q}{kT} I_c = \dfrac{I_c}{V_t} = 40 I_c$ @ $T = 290^\circ$ K,

$V_t = \dfrac{kT}{q} = 0.025$ volts @ $T = 290^\circ$ K

$f_c = $ 1/f corner frequency, $f_t = $ transition frequency $\approx \dfrac{1}{2\pi r_e C_\pi}$.

The capacitance C_π is composed of two contributions $C_\pi \cong C_{je} + C_{de}$. The first term is the space-charge in the emitter junction. The second term accounts for the diffusion capacitance of the emitter junction that increases with emitter current.

The voltage-noise term has contributions from the thermal-noise in the base spreading resistance and the collector shot-noise appearing across the dynamic emitter resistance. The current-noise term corresponds to the shot-noise in the base current and the shot-noise in the collector current. Note that both the voltage-noise and the current-noise have a 1/f component and a component that is independent of frequency. The current-noise also has a component that increases with frequency. This type of frequency dependence is characteristic of the BJT and is illustrated in Fig. 6.19(a).

It is not necessary to restrict ourselves to just the common-emitter configuration. Noise sources for the common-base and common-collector configuration can be obtained in a similar manner. To first order, the ultimate noise performance obtained in each configuration is the same [57]. All configurations can provide power gain but the voltage and current gains available in each configuration are very different. The common-base (CB) provides voltage gain but not current gain. The common-collector (CC) provides current gain but not voltage gain. The common-emitter (CE) configuration is usually preferred because it provides both current and voltage gain. It also usually provides the highest amount of power gain.

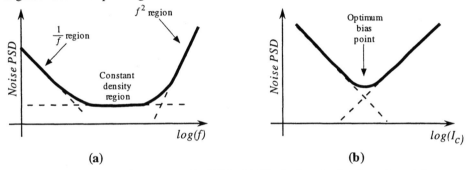

(a)   (b)

Figure 6.19 Noise performance of a BJT. (a) Typical variation in noise with frequency. (b) Variation with bias current.

We can substitute Eq. 6.45 directly into any of the expressions that we have developed for the equivalent input current-noise of a receiver. Using Eq. 6.3 with $Y_s(f) = 2\pi f C_d$ for the combination of the photodetector junction capacitance and any stray capacitance that may be present across the input allows the equivalent input current-noise density for a BJT based front-end to be written as [9, 12, 16]

$$i_{n_i}^2(f) = 2qI_b + 4kTr_x(2\pi C_d)^2 f^2 + 2qI_c r_e^2 (2\pi(C_\pi + C_d))^2 f^2 \quad (\frac{\text{amps}^2}{\text{Hz}}). \quad (6.46)$$

Note that Eq. 6.46 neglects any noise associated with the internal photodiode resistance or any load or feedback resistors. If desired these additional terms can be accounted for by adding them to Eq. 6.46 as uncorrelated current-noise generators.

The collector current is related to the base current by the DC current gain β_o as $I_c = \beta_o I_b$ and the dynamic emitter resistance is the inverse of the transistor transconductance. The transconductance is also related to the collector current as shown in Eq. 6.45. Substituting these relationships into Eq. 6.46 yields

$$i_{n_i}^2(f) = 4kTr_x(2\pi f C_d)^2 + \frac{2q}{\beta_o}I_c + 2q(2\pi f C_t)^2 \frac{V_t^2}{I_c} \quad (\frac{\text{amps}^2}{\text{Hz}}) \quad (6.47)$$

where $V_t = kT/q$, $C_t = C_d + C_\pi$. For room temperature operation, $V_t = 0.025$ volts.

The total capacitance C_t can be approximated by a first-order linear relationship $C_\pi = C_o + \alpha I_c$, where α is an experimentally determined coefficient that is typically between 0.25 pF/mA and 0.75 pF/mA [9, 60].

Note that the first term of Eq. 6.47 is independent of collector current. The second term *increases* with increasing collector current while the third term *decreases* with increasing collector current. This relationship is plotted in Fig. 6.19(b). The optimum current will depend on both the noise-bandwidth of the receiver and the total capacitance at the input. For a baseband digital receiver of bit-rate B an approximate form for the optimum collector current is [60]

$$I_{c_{opt}} \cong 2\pi f C_o V_t B \sqrt{\beta_o \frac{I_3}{I_2}} \quad \text{(amps)}, \tag{6.48}$$

where I_2 and I_3 are the normalized noise-bandwidth integrals of Eq. 5.81.

Eq. 6.48 must be used with some care at both very low and very high data-rates. At rates below a few hundred Mbps, it tends to predict too small a value of collector current to maintain adequate gain, and a practical lower limit of a few hundred microamps is used. At rates above a few Gbps, the optimum collector current approaches a constant value that is independent of frequency [9, 60].

Using values of $C_d = 0.5\text{pF}$, $C_t = 2.3\text{pF}$, $I_c = 3.0\text{mA}$, $\beta_o = 100$ and $r_x = 15.0\Omega$ for the base spreading resistance, Eq. 6.46 can be plotted as shown in Fig. 6.20.

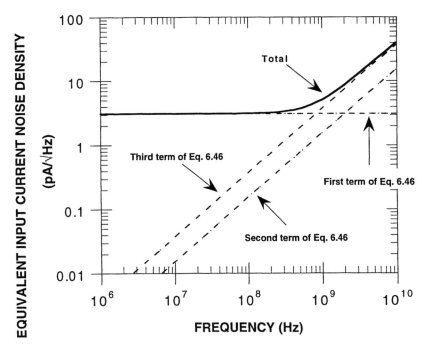

Figure 6.20 Example equivalent input current-noise density for a common emitter BJT amplifier.

The low frequency noise performance is determined by the first term in Eq. 6.46, which accounts for the shot-noise in the base current. The noise at high frequencies is determined by the second and third terms. This accounts for the base spreading resistance thermal-noise and collector shot-noise, respectively.

Equation 6.46 and Fig. 6.20 provide insights into the type of BJT device parameters that are desirable for low noise operation. In general, the higher the DC current gain, the lower the low-frequency noise since less base current is then required to achieve a given collector current. This reduces the amount of base current shot-noise. Similarly the smaller the base spreading resistance, the lower the high frequency noise since less thermal-noise is generated in the device.

6.2.2 FET Amplifiers

The field-effect transistor is another key building block of electronic amplifiers. The noise sources present in a field-effect transistor are well known [9, 12, 59, 61-65]. The schematic symbol, small-signal model, and noise equivalent models for a FET are illustrated in Fig. 6.21.

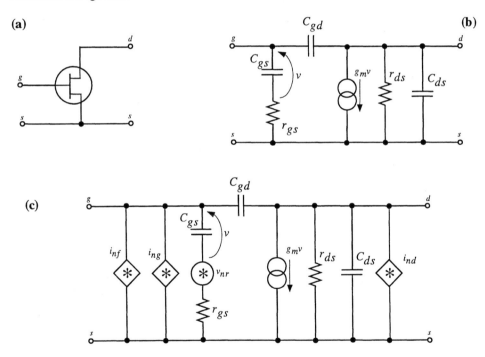

Figure 6.21 FET noise. (a) Schematic symbol for a FET in the common-source configuration. (b) Small-signal model. (c) Corresponding noise equivalent circuit.

The structure and materials used in fabricating the FET and the frequency at which the FET is operated determine which components in the small-signal model play significant roles. In the silicon junction FET (JFET) and in the metal-oxide-semiconductor FET (MOSFET), the gate-to-source resistance r_{gs} is usually ignored. The FET input then looks simply like a pair of capacitors, one from gate to source C_{gs}, the second from gate to drain C_{gd}. For the GaAs metal-semiconductor FET (MESFET), most heterojunction FET (HFET) structures, and for high electron mobility transistors (HEMTs) that are fabricated from III-V semiconductors, the gate-to-source resistance is usually included [64-67].

In FETs there is inherent correlation between the noise at the gate and the noise at the drain. This is particularly evident at high frequencies. In order to simplify our analysis we will ignore any direct effects from the gate-to-drain feedback capacitor . This is usually valid since in most devices $C_{gs} \cong 10 \times C_{gd}$. The expressions for voltage and current-noise vary, depending on the exact model used to represent the device. We can combine the results from several authors to obtain the equivalent noise current and voltage-noise generators,

$$v_n^2(f) \cong 4kTr_n + 4kTr_n\left(\frac{f_c}{f}\right) \quad (\frac{\text{Volts}^2}{\text{Hz}}),$$

$$i_n^2(f) \cong 2qI_g + 4kTr_n\left(2\pi C_{gs}\right)^2 f^2 \quad (\frac{\text{Amps}^2}{\text{Hz}}), \qquad (6.49)$$

and a correlated noise current given by

$$v_n^2(f)Y_{corr}(f)Y_s(f) = 2 \times 4kTr_n C_{gs} 2\pi f Y_s(f) \quad (\frac{\text{Amps}^2}{\text{Hz}}), \qquad (6.50)$$

where

I_g = DC gate leakage current, g_m = transistor transconductance,

$f_c = 1/f$ corner frequency, r_n = effective noise resistance,

$= \dfrac{T_{eff}}{T}\dfrac{1}{r_{ds}g_m^2}$ for HEMTs with T_{eff} being the effective temperature of the drain [21],

$\cong \dfrac{0.7}{g_m}$ for Si JFETs [19, 20], $\qquad \cong \dfrac{1.03}{g_m}$ for Si MOSFETs [19, 24],

$\cong \dfrac{1.75}{g_m}$ for GaAs MESFETs [19, 22, 23].

The 1/f component in a FET's noise is accounted for in the voltage-noise. With the exception of the gate leakage current shot-noise, FET noise is essentially a thermal-noise that originates in the FET's channel.

When analyzing FET noise we cannot just restrict ourselves to a midband region as we did with the BJT. The 1/f noise corner can be significantly higher (tens to hundreds of MHz) and the current-noise term containing the channel thermal-noise contribution is a direct function of frequency.

Equations 6.49 and 6.50 can be used in any of our expressions for equivalent input current-noise. Following the procedure we used with the BJT we will use Eq. 6.3 but ignore photodetector and load resistor effects. Using $Y_s(f) = 2\pi f C_d$ for the source admittance allows the equivalent input current-noise spectral-density for an optical receiver using an FET to be obtained as

$$i_n^2(f) = 2qI_g + 4kTr_n\left(2\pi f_c\left(C_{gs} + C_d\right)^2\right)f + 4kTr_n\left(2\pi\left(C_{gs} + C_d\right)\right)^2 f^2. \qquad (6.51)$$

The first term in Eq. 6.51 is the shot-noise due to the base leakage current. The second term is the 1/f noise of the device, while the third is thermal-noise of the channel reflected into the gate. The low frequency noise performance is set by the combination of the 1/f term and the gate current shot-noise. The coefficients used in Eq. 6.51 vary depending on the material and structure of the FET. Typical values for the coefficients are listed in Table 6.1

Table 6.1 Typical parameters of FETs at room temperature.

Parameter	Si JFET	Si MOSFET	GaAs MESFET	HEMT
I_g (nA)	<0.1	< 0.01	1-1000	1-1000
r_n (Ohms)	35-140	25-50	35-125	10-40
g_m (mS)	5-20	20-40	15-50	40-60
T_{eff} (°K)	n/a	n/a	n/a	1500-2500
r_{ds} (Ohms)	n/a	n/a	n/a	150-500
C_{gs} (pF)	1.0-5.0	0.25-1.0	0.15-0.50	0.15-0.50
f_c (MHz)	<<1	1-10	10-100	10-100

Equation 6.51 also indicates the type of FET device parameters that are desirable for low noise operation. At low frequencies we want a device to have a low 1/f corner frequency and as low a gate leakage current as possible. At high frequencies, the noise is proportional to $r_n C_t^2$, where $C_t = C_{gs} + C$ is the total capacitance across the input node. Since the effective noise resistance is reduced as the transconductance increases, the noise can be reduced by making the term C_t^2/g_m as small as possible. The ratio g_m/C_t^2 is frequently used as a figure of merit for an FET and should be maximized for lowest noise performance. This can be accomplished by selecting a FET with a short gate length to maximize high frequency operation and then selecting the gate width to optimize the noise performance [68].

If we substitute values for a low noise GaAs MESFET of $I_g = 10.0$nA, $R_n = 40.0\Omega$, $f_c = 50.0$MHz, $C_{gs} = 0.5$pF, and $C = 0.5$pF into Eq. 6.51 and plot the result as a function of frequency, Fig. 6.22 is the result.

The low frequency performance is determined by the gate current shot-noise and the 1/f noise terms. For a GaAs MESFET these can be comparatively high. Consequently, low and moderate bandwidth receivers are frequently constructed using a Si MOSFET in the first stage to take advantage of the MOSFET's much lower 1/f corner frequency and gate leakage current. The high-frequency performance for all FETs is determined by the channel thermal-noise, and cooling can play a role in reducing the total receiver noise.

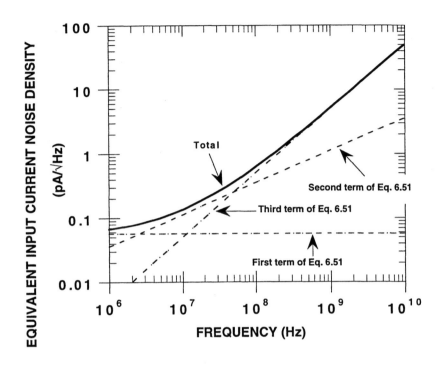

Figure 6.22 Example of the equivalent input current-noise density for a typical common source GaAs MESFET first stage amplifier.

If we replot Eq. 6.51 using parameters for the various FET structures of Table 6.1 and also add an optimized BJT/heterojunction bipolar transistor (HBT) current-noise density, the curves shown in Fig. 6.23 result. We emphasize that these curves are by no means definitive and that the curves use values typical of contemporary devices with f_t's of a few tens of GHz. In addition the curve plotted for the BJT/HBT is obtained by optimizing the noise performance of the front-end for each frequency. This is a feat that no realistic amplifier could manage. Noise matching has not been employed, thermal noise from either a detector load resistor or a feedback resistor has been neglected, and any additional noise from the following gain stages has been ignored. The final decision between device types should only be based on calculations for the specific devices and amplifier designs in question.

In spite of these limitations the curves can be used to provide insight into the trade-off between devices that are indicative of the *general* trends among the available devices. The quietest device for frequencies between the audio region and a few hundred MHz is often the Si MOSFET. This is because of the relatively high 1/f noise corner frequencies associated with HEMTs and MESFETs. Their relatively high corner frequencies usually preclude the use of MESFETs or HEMTs in receivers with bandwidths of less than 5 to 10 MHz.

At very low frequencies, in the kHz region and below, the JFET can sometimes be quieter than the MOSFET because of its lower 1/f corner frequency. The final trade-off often depends on the amount of gate bias-current required by each device. JFETs and MOSFETs are not generally used beyond a few GHz because of their comparatively high capacitance and low f_t's. The BJT/HBT is considered comparatively noisy at low frequencies because of the base bias-current shot-noise. This can be reduced by operating the device at lower levels of collector current if reduced bandwidths and lower transconductance are tolerable.

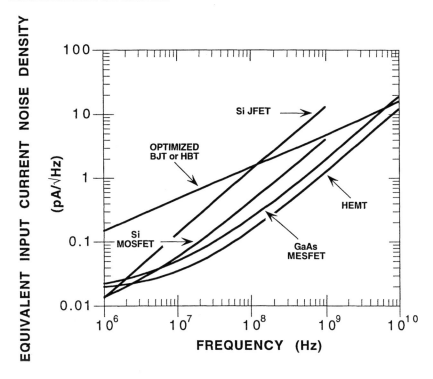

Figure 6.23 Examples of the equivalent input current-noise densities obtainable for a variety of semiconductor devices.

At high frequencies, contemporary GaAs MESFETs and HEMTs have similar performance, with BJTs/HBTs being marginally noisier. In many cases where the detector capacitance is relatively large the BJT/HBT can be used to advantage if the transistor's r_x is small because a BJT/HBT can have as much as an order of magnitude larger transconductance than a FET.

The HEMT may ultimately prove to be the device of choice for low-noise receivers as fabrication techniques and material parameters mature. However, the characteristics of the HEMT, MESFET, and BJT/HBT all continue to improve, and f_t's in excess of 100 GHz are expected in the future. For bandwidths of many GHz, the noise performance of the various device types is so dependent on the exact nature of the specific transistor and the associated photodiode and noise-matching circuitry that the trade-offs in performance must be obtained from a more detailed analysis that is typically aided by microwave CAD tools.

6.2.3 Transimpedance of a Single-Stage Amplifier

The simplified small-signal model illustrated in Fig 6.24(a) can be used to estimate the transimpedance that can be obtained from a single stage transistor amplifier. For a BJT amplifier the input capacitance C_{in} and input resistance R_{in} are the parallel combinations of the photodiode capacitance C_d and parallel resistance with C_π and R_π. For an FET, R_π is essentially infinite and C_{in} is the combination of the photodiode capacitance and C_{gs}. For either transistor type, the feedback elements consist of the parallel combination of any internal feedback from collector to base and any external feedback resistor.

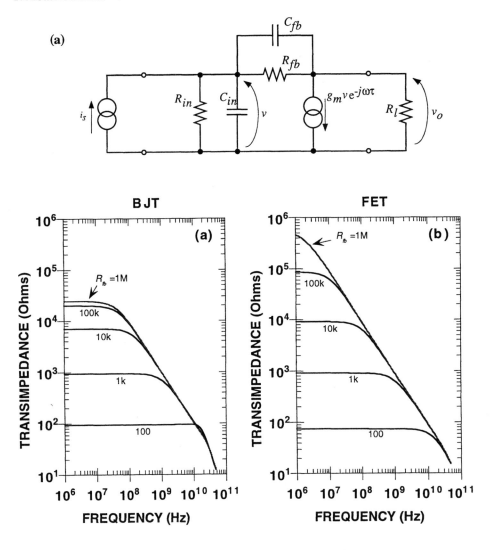

Figure 6.24 Transimpedance of a single stage transistor amplifier. (a) Small-signal model. (b) Calculated values of a BJT amplifier. (c) Calculated values for a FET amplifier.

The transimpedance for the model shown in Fig 6.24(a) is given by

$$Z_t(\omega) = \frac{v_o}{i_s} = \frac{Y_{fb} - g_m}{Y_{fb}[Y_{in} + Y_l + g_m] + Y_l Y_{in}}, \tag{6.52}$$

where $Y_{fb} = \frac{1}{R_{fb}} + j\omega C_{fb}$, $\quad Y_{in} = \frac{1}{R_{in}} + j\omega C_{in}$, $\quad Y_l = \frac{1}{R_l}$, $\quad g_m = g_{m_o} e^{-j\omega\tau}$,

and τ accounts for the transistor's transit-time delay.

Using parameters for a BJT amplifier of $C_{fb} = 0.125\text{pF}$, $C_{in} = 3\text{pF}$, $R_{in} = 250\Omega$, $g_m = 0.4$, $\tau = 10\text{ps}$, and $R_l = 250\Omega$, the plot in Fig 6.24(b) can be obtained for a variety of possible values of feedback resistance. In comparison, using FET amplifier parameters of $C_{fb} = 0.125\text{pF}$, $C_{in} = 0.5\text{pF}$, $R_{in} = 1\text{M}\Omega$, $g_m = 0.050$, $\tau = 7\text{ps}$, and $R_l = 250\Omega$, the curves plotted in Fig 6.24(c) are obtained.

The effective transimpedance of the BJT saturates for large values of feedback resistance because of the finite value for R_π. This limits the BJT's usefulness in realizing a high-impedance amplifier design. For transimpedance feedback amplifiers with moderate values of feedback resistance, both devices perform similarly and gains of a few thousand ohms with bandwidths of a few GHz can be obtained.

A finite transit-time is usually considered to be a detriment to a broadband receiver response but in certain situations it can actually be beneficial. When the gain stage has a relatively high transconductance and a low circuit capacitance it is possible to improve the frequency response of the receiver by utilizing a finite amount of transit-time delay to introduce phase-lag into the closed loop gain of a transimpedance amplifier [69]. This causes a peak in the overall transimpedance to form, as illustrated in Fig. 6.26.

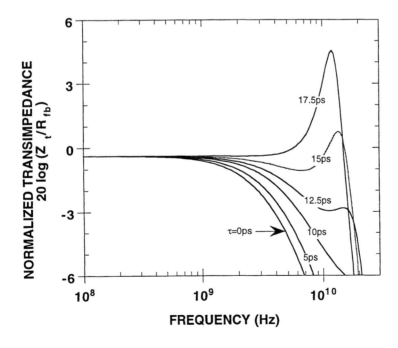

Figure 6.25 Effects of transistor transit-time on receiver response.

Figure 6.25 was obtained using the amplifier values of $R_{fb} = 1\mathrm{k}\Omega$, $C_{fb} = 0.030\mathrm{pF}$, $C_{in} = 0.750\mathrm{pF}$, $R_{in} = 500\Omega$, $g_m = 0.2$, and $R_l = 500\Omega$. Note that for a zero transit-time delay the bandwidth is dominated by the time-constant of the feedback resistor and capacitor and is equal to just over 4 GHz. By introducing a 15 ps transit-time delay into the loop-gain a peak is introduced into the response that extends the bandwidth out to some 20 GHz. The amplitude of the peak is quite sensitive to the total delay, and circuit instability can occur for large transit-time delays.

6.3 Multistage Amplifiers

Receiver front-end amplifiers are by no means limited to using a single gain-stage. There are a wide variety of multiple stage amplifiers that can be used to advantage in optical receivers. Some commonly used amplifiers are shown in Fig. 6.26. The circuits illustrated are all small-signal AC circuits. The required DC bias circuitry and any AGC or monitoring circuitry has been omitted for clarity.

Figure 6.26(a) illustrates a popular transimpedance amplifier using bipolar transistors [9, 70]. The bandwidth for this design can be relatively large since the transimpedance feedback is developed around a single transistor gain stage. The voltage gain that can be developed in a single stage is relatively modest, and consequently the feedback resistor is often limited to no more than a kΩ or so if a low impedance load is to be presented to the photodiode. Emitter bypassing is used in the second stage to introduce a peak in the voltage gain at the inverse of the $R_e C_e$ time-constant. This peak is used to compensate for the roll-off in response caused by the $R_{fb} C_{fb}$ time-constant. A transimpedance front-end amplifier where the feedback is developed around two stages is shown in Fig. 6.26(b) [71]. The first stage can either be an FET as illustrated or a BJT [72]. This amplifier is more limited in bandwidth because of the extra phase-shift that accumulates in two gain stages instead of one, but the result is a particularly stable feedback amplifier.

An example of a high-impedance front-end is shown in Fig. 6.26(c) [73-75]. This circuit can also be used as the gain block for a transimpedance front-end. The high output impedance of the first stage FET drives the relatively low input impedance of a second stage shunt-feedback amplifier. A third stage is used to provide equalization via emitter bypassing. Alternating high output-impedance stages with low input-impedance stages tends to maximize the impedance mismatch between amplifier stages. This is a technique that minimizes the effects of stray reactances and can result in a broadband response [76]. An alternative to the shunt-feedback stage is the use of a common-base stage as illustrated in Fig. 6.26(d) [73]. This forms a classic cascode combination that minimizes the Miller effect capacitance of the first stage [77]. A third stage, which in this example is a shunt feedback stage, is then used as an output driver. A variation on this high-impedance design that is particularly suitable for monolithic integration into an opto-electronic integrated circuit (OEIC) is shown in Fig. 6.26(e) [73, 78]. The *pnp* transistor tends to have a lower bandwidth than an *npn* but provides a level-shifting function that allows the amplifier to be DC coupled from input to output. This makes it particularly attractive for use as the gain block for a transimpedance amplifier [67, 79, 80]. Three stages will introduce substantial loop delay, but since the forward voltage gain is large, higher value feedback resistors can be used while still presenting a low impedance load to the photodiode. Variations of this circuit have also shown promise when used with heterojunction bipolar transistors [81]. An *npn* second stage is used to maximize the achievable bandwidth, and additional level-shifting circuitry is included.

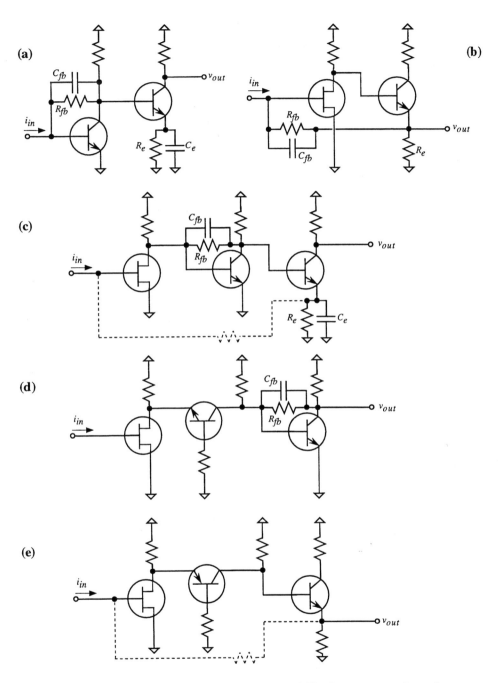

Figure 6.26 Multiple stage receiver front-ends. (a) Single-stage transimpedance feedback with second stage equalization. (b) Two-stage transimpedance feedback. (c) High-impedance with shunt feedback second stage. (d) High-impedance with common base second stage and shunt-feedback third stage. (e) High-impedance with *pnp* second stage.

The circuits illustrated in Fig 6.26 used multiple stages but did not use impedance matching between stages or noise-matching between the photodiode and the first stage amplifier. Figure 6.27 illustrates multiple-stage receiver front-ends that use both interstage impedance-matching and input noise-matching to obtain both wideband and low-noise performance. The circuits illustrated are again all small-signal AC circuits, where the DC bias circuitry has been omitted for clarity. All of the circuits shown utilize GaAs MESFETs or HEMTs because these have historically been the devices with the highest value cutoff frequencies and have been most suitable for receiver bandwidths in excess of 10 GHz. Similar circuit design techniques can, however, be used for junction transistors such as BJTs and HBTs.

The circuits illustrated are usually constructed as chip-and-wire hybrid circuits on low-loss dielectric substrates, although the techniques can also be applied to microwave integrated circuit (MMIC) realizations. Successful construction will generally require accurate models for all of the active devices, CAD optimization, and a mature microwave thin-film hybrid circuit fabrication capability. The comparatively long path-lengths and the correspondingly large amount of phase-shift that accumulates in a chip-and-wire hybrid circuit design generally preclude the use of feedback around more than a single gain-stage. MMIC and OEIC techniques dramatically reduce path-lengths and can use combinations of noise-matching, interstage impedance matching, and multi-stage feedback amplifiers as necessary. In many cases the ultimate bandwidth limitation is either the photodetector or the packaging of the front-end amplifier.

Figure 6.27(a) is an example of a multi-GHz bandwidth FET based receiver front-end that uses series inductors L_1, L_2, L_3 for input noise-matching and interstage impedance matching [52]. The resistor R_{in} that provides a bias path for the first stage gate is typically a fairly large value so that the contributions to the thermal current-noise in the circuit are minimized. The RC equalizer between the second and third stages compensates for the integrating behavior of the amplifier from the few hundred MHz region to a few GHz. Above a few GHz, series inductors in combination with the bonding pads found on active devices are used as impedance matching networks that produce a flat overall response.

A receiver that has achieved a 16 GHz bandwidth via interstage matching is shown in Fig 6.27(b) [40]. Lossy inductive T-networks are used for both the input noise match and the interstage impedance match. The input resistor R_{in} is a few hundred ohms as a compromise between low-noise and improved high-frequency response. The interstage networks again consist of the combination of the device bond pads and bond wire inductance.

Another approach for realizing wide bandwidths is the use of inductive peaking in both the drain L_d and gate L_g of a HEMT, as shown in Fig 6.27(c) [41]. This approach has been used to construct a front-end amplifier with a 20 GHz 3 dB bandwidth and less than 10 pA per root Hz of equivalent input current-noise density [53]. The overall topology is a cascade of inductively peaked shunt-feedback amplifiers. The final stage can also be used as an equalizer to compensate for a limited amount of roll-off in the photodetector.

It is possible to add additional peaking into the receiver by including an inductor L_{fb} in the feedback path, as shown in Fig 6.27(d) [41]. The feedback inductor reduces the amount of feedback at high frequencies, which tends to increase the apparent transimpedance of each individual gain-stage.

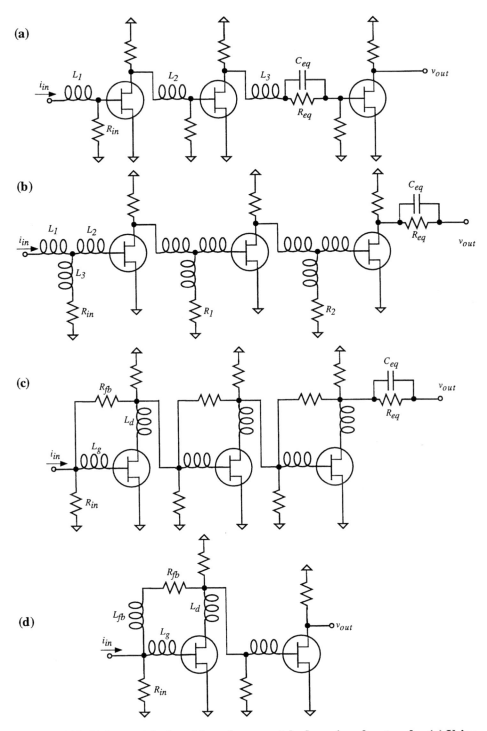

Figure 6.27 Noise-matched and impedance matched receiver front-ends. (a) Using series inductors. (b) Using lossy T-circuits. (c) Inductive peaking. (d) Inductive peaking with feedback reduction at high frequencies.

Table 6.2 concludes this chapter by summarizing the principal features of the four optical receiver front-end architectures that we have explored. The features are grouped according to the generally applicable advantages and disadvantages.

Table 6.2 Summary of Front-end Architectures.

Front-End Type	*Advantages*	*Disadvantages*
Low-impedance voltage amplifier.	Simplicity. Uses commercial microwave amplifiers. Good dynamic range.	High noise level. Photodetector is usually in a separate package.
High-impedance.	Minimizes the thermal noise in the receiver. Often the lowest noise obtainable for a moderate bandwidth baseband receiver design.	Output voltage is the integral of the photodetector current. Response rolls-off at 20 dB/Decade. Flat-response requires equalization via pole-zero cancellation. Poor dynamic range.
Transimpedance.	High dynamic range. Equalization generally not required. Receiver response can be very stable over time, temperature, etc. due to desensitizing nature of negative feedback.	Noisier than high-impedance but not as noisy as low-impedance voltage-amp. Stray capacitance from feedback resistor often determines achievable bandwidth. Accumulating phase-shifts in multi-stage amplifiers limits achievable open-loop gains and sets upper limit on size of feedback resistor. This limits the noise performance that can be achieved.
Noise-matched or resonant receiver.	Best performance at multi-GHz frequencies. Techniques can be used with high-impedance or transimpedance designs.	More complicated microwave circuit design. Optimum performance may require CAD modeling and individual circuit tuning.

6.4 References

1. R. Brooks, "7B8B Balanced Code with Simple Error Detecting Capability," Electronics Letters, 1980, vol. 16, pp. 458-459.

2. M. Rousseau, "Block Codes for Optical Fibre Communication," Electronics Letters, 1976, vol. 12, pp. 478-479.

3. Y. Takasaki, *et al.*, "Optical Pulse Formats for Fiber Optic Digital Communications," IEEE Transactions on Communications, April 1976, vol. COM-24, no. 4, pp. 404-413.

4. T. Darcie, "Subcarrier Multiplexing for Lightwave Multiple-Access Lightwave Networks," Journal of Lightwave Technology, August 1987, vol. LT-5, pp. 1103-1110.

5. R. Olshansky, *et al.* "Design and Performance of Wideband Subcarrier Multiplexed Lightwave Systems," in *European Conference on Optical Communication.* 1988, Brighton U.K., IEE:London.

6. G. White and G. Chin, "A DC-2.3 GHz Amplifier Using an 'Embedding' Scheme," Bell System Technical Journal, January 1973, vol. 52, no. 1, pp. 53-63.

7. C. D. Motchenbacher and F. C. Fitchen, *Chap. 2 - Amplifier Noise* in *Low-Noise Electronic Design,* 1973, John Wiley and Sons: New York.

8. H. Ott, *Chap. 8 - Intrinsic Noise Sources* in *Noise Reduction Techniques in Electronic Systems,* 1976, John Wiley & Sons: New York.

9. T.V. Muoi, "Receiver Design for Optical-Fiber Systems," Journal of Lightwave Technology, 1984, vol. LT-2, no. 3, pp. 243-265.

10. R.J. McIntyre, "Multiplication Noise in Uniform Avalanche Diodes," IEEE Transactions on Electron Devices, January 1966, vol. ED-13, no. 1, pp. 164-169.

11. S.D. Personick, "Receiver Design for Digital Fiber Optic Communication Systems, I & II," Bell System Technical Journal, July-August 1973, vol. 52, no. 6, pp. 843-886.

12. R.G. Smith and S.D. Personick, *Receiver Design for Optical Fiber Communication Systems*, in *Semiconductor Devices for Optical Communication,* 2nd, H. Kressel ed. Vol. 39, 1982, Springer Verlag: New York, pp. 89-160.

13. O.H. Schade, "A Solid-State Low-Noise Preamplifier and Picture-Tube Drive Amplifier for a 60 MHz Video System," RCA Review, March 1968, vol. 29, no. 1, pp. 3-11.

14. J.E. Goell, "An Optical Repeater with High-Impedance Input Amplifier," Bell System Technical Journal, April 1974, vol. 53, no. 4, pp. 629-643.

15. B.L. Kasper, *Chap. 18 - Receiver Design* in *Optical Fiber Telecommunications II*, S.E. Miller and I.P. Kaminow, Editors. 1988, Academic Press: New York.

16. T.V. Muoi, *Chap. 12 - Receiver Design of Optical-Fiber Systems*, in *Optical Fiber Transmission*, E.E.B. Basch ed. 1987, H. W. Sams & Co. Div. of Macmillan, Inc.: Indianapolis, .

17. T.V. Muoi, "Optical Receiver with Improved Dynamic Range," 1983, U.S. Patent No. 4415803.

18. D.W. Faulkner, "A Wideband Limiting Amplifier for Optical-Fiber Receivers," IEEE Journal of Solid-State Circuits, June 1983, vol. SC-18, no. 3, pp. 333-340.

19. P.K. Runge, "An Experimental 50 Mb/s Fiber-Optic PCM Repeater," IEEE Transactions on Communications, 1976, vol. COM-24, pp. 413-418.

20. B.L. Kasper and J.C. Campbell, "Multigigabit-per-second Avalanche Photodiode Lightwave Receivers," Journal of Lightwave Technology, 1987, vol. LT-5, pp. 1351-1364.

21. M. Shikada, *et al.* "1.5 µ High Bit Rate Long Span Transmission Experiments Employing a High Power DFB-DC-PBH Laser Diode," in *European Conference on Optical Communication*. 1985, Istituto Internazionale delle Communicazioni: Geneva. Post Deadline Paper.

22. M. Brain, *et al.*, "PIN-FET Hybrid Optical Receivers for 1.2 Gbit/s Transmission Systems at 1.3 and 1.55 µm Wavelength," Electronics Letters, 1984, vol. 20, pp. 894-896.

23. C.A. Holt, *Chapter 20 - Feedback Amplifiers* in *Electronic Circuits: Digital and Analog*, 1978, John Wiley and Sons: New York.

24. J.L. Hullett and T.V. Muoi, "A Feedback Receive Amplifier for Optical Transmission Systems," IEEE Transactions on Communications, October 1976, vol. COM-24, no. 10, pp. 1180-1185.

25. C.A. Holt, *Chapter 21 - Frequency and Transient Response of Feedback Amplifiers* in *Electronic Circuits: Digital and Analog*, 1978, John Wiley and Sons: New York.

26. J. Wait, *et al.*, *Section 1.5 - The General Inverting Amplifier* in *Introduction to Operational Amplifier Theory and Applications*, 1975, McGraw Hill, Inc.: New York.

27. Y. Netzer, "Simplify Fiber-Optic Receivers with a High-Quality Preamp," Electronics Design News, September 20, 1980, pp. 161-164.

28. M.A. Manterola-Barros, "Low-Noise InSb Photodetector Preamp for the Infrared," IEEE Journal of Solid-State Circuits, August 1982, vol. SC-17, no. 4, pp. 761-766.

29. M. Abraham, "Design of Butterworth-Type Transimpedance and Bootstrap-Transimpedance Preamplifiers for Fiber-Optic Receivers," IEEE Transactions on Circuits and Systems, June 1982, vol. CAS-29, no. 6, pp. 375-382.

30. L. Weinberg, *Network Analysis and Synthesis,* 1962, McGraw-Hill, Inc.: New York.

31. R.L. Forward, "Reducing Noise in Photoamplifier Circuits with Artificial Low-Noise Resistors," Journal of Applied Physics, May 1982, vol. 53, no. 5, pp. 3365-3371.

32. G. Williams and H. LeBlanc, "Active Feedback Lightwave Receivers," Journal of Lightwave Technology, October 1986, vol. LT-4, no. 10, pp. 1502-1507.

33. B.L. Kasper, *et al.*, "An Optical-Feedback Transimpedance Receiver for High Sensitivity and Wide Dynamic Range at Low Bit Rates," Journal of Lightwave Technology, 1988, vol. 6, no. 2, pp. 329-338.

34. T.T. Ha, *Chap. 5 - Microwave Amplifiers: Broadband Design* in *Solid-State Microwave Amplifier Design,* 1981, John Wiley & Sons: New York.

35. G.F. Vendelin, *Section 4.4 - Low Noise Design* in *Design of Amplifiers and Oscillators by the S-Parameter Method,* 1982, John Wiley & Sons: New York.

36. T.T. Ha, *Appendix A8 - Theory of Noisy Two-Port Networks* in *Solid-State Microwave Amplifier Design,* 1981, John Wiley & Sons: New York.

37. T. Darcie, *et al.*, "Resonant p-i-n-FET Receivers for Lightwave Subcarrier Systems," Journal of Lightwave Technology, April 1988, vol. 6, no. 4, pp. 582-589.

38. K. Alameh and R. Minasian, "Tuned Optical Receivers for Microwave Subcarrier Multiplexed Systems," IEEE Transactions on Microwave Theory and Techniques, May 1990, vol. 38, no. 5, pp. 546-551.

39. A. Olomo-Ngongo, *et al.*, "Network Synthesis Applied to the Design of Tuned Optical Front-Ends for Microwave Lightwave Systems," Electronics Letters, January 1994, vol. 30, no. 2, pp. 164-165.

40. J.L. Gimlett, "Ultrawide Bandwidth Optical Receivers," Journal of Lightwave Technology, October 1989, vol. 7, no. 10, pp. 1432-1437.

41.   N. Ohkawa, "Fiber-Optic Multigigabit GaAs MIC Front-End Circuit with Inductor Peaking," Journal of Lightwave Technology, November 1988, vol. 6, no. 11, pp. 1665-1671.

42.   W. Chen, *Broadband Matching: Theory and Implementation*, 2nd ed. Advanced Series in Electrical and Computer Engineering, Vol. 1. 1988, Teaneck, NJ: World Scientific Publishing Co.

43.   A. Cappy, "Noise Modeling and Measurement Techniques," IEEE Transactions on Microwave Theory and Techniques, January 1988, vol. 36, no. 1, pp. 1-10.

44.   Y. Netzer, "The Design of Low-Noise Amplifiers," Proceedings of the IEEE, 1981, vol. 69, no. 6, pp. 728-741.

45.   M. Park and R. Minasian, "Ultralow Noise 10 Gb/s p-i-n-HEMT Optical Receiver," IEEE Photonics Technology Letters, February 1993, vol. 5, no. 2, pp. 161-162.

46.   M.S. Park and R.A. Minasian, "Synthesis of Lossy Noise Matching Network for Flat-Gain and Low-Noise Tuned Optical Receiver Design," IEEE Photonics Technology Letters, February 1994, vol. 6, no. 2, pp. 285-287.

47.   M.S. Park and R.A. Minasian, "Ultra-Low-Noise and Wideband-Tuned Optical Receiver Synthesis and Design," Journal of Lightwave Technology, February 1994, vol. 12, no. 2, pp. 254-259.

48.   G. Jacobsen, *et al.*, "Improved Design of Tuned Optical Receivers," Electronics Letters, July 1987, vol. 23, no. 15, pp. 787-788.

49.   J. Kan, *et al.*, "Transformer Tuned Front-Ends for Heterodyne Optical Receivers," Electronics Letters, July 1987, vol. 23, no. 15, pp. 785-786.

50.   A. Peterson, *et al.*, "MMIC Tuned Front-End for a Coherent Optical Receiver," IEEE Photonics Technology Letters, June 1993, vol. 5, no. 6, pp. 679-681.

51.   Q. Liu, "Unified Analytical Expressions for Calculating Resonant Frequencies, Transimpedances, and Equivalent Input Noise Current Densities of Tuned Optical Receiver Front Ends," IEEE Transactions on Microwave Theory and Techniques, February 1992, vol. 40, no. 2, pp. 329-336.

52.   E. Kimber, *et al.*, "High Performance 10 Gbit/s pin-FET Optical Receiver," Electronics Letters, 1992, vol. 28, no. 2, pp. 120-122.

53.   N. Ohkawa, "20 GHz Bandwidth Low-Noise HEMT Preamplifier for Optical Receivers," Electronics Letters, August 1988, vol. 24, no. 17, pp. 1061-1062.

54. A. Czylwik. "Theoretical Limitations of Broadband Matching Networks in PIN-FET Receivers," in *Sources and Detectors for Fiber Communications*. 1992, SPIE Vol. 1788.

55. V. Prabhu, "Theory of Broadband Matching for Optical Heterodyne Receivers," Applied Optics, April 1968, vol. 7, no. 4, pp. 657-666.

56. M.J. Buckingham, *Chap. 4 - Inherent Noise in Junction Diodes and Bipolar Transistors* in *Noise in Electronic Devices and Systems*, 1983, Halsted Press Division of John Wiley & Sons: New York.

57. C. D. Motchenbacher and F. C. Fitchen, *Chap. 4 - Bipolar Transistor Noise Mechanisms* in *Low-Noise Electronic Design*, 1973, John Wiley and Sons: New York.

58. S.M. Sze, *Chap. 3 - Bipolar Transistor* in *Physics of Semiconductor Devices*, 1981, John Wiley & Sons: New York.

59. H. Ott, *Chap. 9 - Active Device Noise* in *Noise Reduction Techniques in Electronic Systems*, 1976, John Wiley & Sons: New York.

60. T.V. Muoi, *Chap. 16 - Optical Receivers* in *Optoelectronic Technology and Lightwave Communications Systems*, C. Lin, Editor. 1989, Van Nostrand Reinhold: New York.

61. M.J. Buckingham, *Chap. 5 - Noise in JFETs and MOSFETs* in *Noise in Electronic Devices and Systems*, 1983, Halsted Press Division of John Wiley & Sons: New York.

62. C. D. Motchenbacher and F. C. Fitchen, *Chap. 6 - Noise in Field-Effect Transistors* in *Low-Noise Electronic Design*, 1973, John Wiley and Sons: New York.

63. S.M. Sze, *Chap. 6 - JFET and MESFET* in *Physics of Semiconductor Devices*, 1981, John Wiley & Sons: New York.

64. M.V. Schneider, "Reduction of Spectral Noise Density in PIN-HEMT Lightwave Receivers," Journal of Lightwave Technology, July 1991, vol. 9, no. 7, pp. 887-892.

65. K. Ogawa, "Noise caused by GaAs MESFETs in Optical Receivers," Bell System Technical Journal, July-August 1981, vol. 60, pp. 923-928.

66. R.A. Pucel, *et al.*, *Signal and Noise Properties of GaAs Microwave Field-Effect Transistors* in *Advances in Electronics and Electron Physics*, 1975, Academic Press: New York.

67.   K. Ogawa, "A Long-Wavelength Optical Receiver Using a Short-Channel Si-MOSFET," Bell System Technical Journal, May-June 1983, vol. 62.

68.   A. Abidi, "On the Choice of Optimum FET Size in Wideband Transimpedance Amplifiers," Journal of Lightwave Technology, 1988, vol. LT-6, pp. 64-66.

69.   M. Govindarajan and S.R. Forrest, "Transit-Time Broad-Banding of Very High Bandwidth Monolithic p-i-n/HBT Optical Receivers," IEEE Photonics Technology Letters, September 1992, vol. 4, no. 9, pp. 1015-1017.

70.   W. Albrecht and C. Baack, "A Broad-Band Opto-Electronic Receiver with Bipolar Transistors," Journal of Optical Communication, 1981, vol. 2, no. 1, pp. 24-25.

71.   S.A. Siegel and D.J. Channin, "PIN-FET Receiver for Fiber Optics," RCA Review, March 1984, vol. 45, pp. 3-22.

72.   R.G. Smith, et al., "An Optical Detector Package," Bell System Technical Journal, 1978, vol. 57, pp. 1809-1821.

73.   K. Ogawa, "Considerations for Optical Receiver Design," IEEE Journal of Selected Areas in Communications, April 1983, vol. SAC-1, no. 3, pp. 524-532.

74.   W. Albrecht and C. Baack, *Chap. 8 - Optoelectronic Receivers* in *Optical Wideband Transmission Systems,* C. Baack, Editor. 1986, CRC Press: Boca Raton, Florida.

75.   V.K. Jain, et al., "Design of an Optimum Optical Receiver," Journal of Optical Communication, 1985, vol. 6, no. 3, pp. 106-112.

76.   E.M. Cherry and D.E. Hooper, "The Design of Wideband Transistor Feedback Amplifiers," Proceedings of the IEEE, 1963, vol. 110, no. 2, pp. 375-379.

77.   C.A. Holt, *Chapter 19 - Wideband Multistage Amplifiers* in *Electronic Circuits: Digital and Analog,* 1978, John Wiley and Sons: New York.

78.   D.R. Smith, et al., "pin/FET Hybrid Optical Receiver for Longer Wavelength Optical Communication Systems," Electronics Letters, 1980, vol. 16, pp. 69-71.

79.   K. Ogawa and E.L. Chinnock, "GaAs FET Transimpedance Front-end Design for a Wideband Optical Receiver," Electronics Letters, August 1979, vol. 15, pp. 650-652.

80.   J.L. Hullett and S. Moustakas, "Optimum Transimpedance Broadband Optical Preamplifier Design," Optical and Quantum Electronics, 1981, vol. 13, pp. 65-69.

81.   K.D. Pedrotti, et al., "High-Bandwidth OEIC Receivers Using Heterojunction Bipolar Transistors: Design and Demonstration," Journal of Lightwave Technology, October 1992, vol. 11, no. 10, pp. 1601-1614.

Chapter 7

RECEIVER PERFORMANCE ANALYSIS

In our concluding chapter we will combine our photodetector and receiver-noise modeling techniques with front-end and demodulator designs to construct complete receiver structures. Our goal is to predict the overall performance of an optical receiver. We begin this chapter with a short introduction to the demodulation techniques used to recover digital information from the electrical signals that are present at the output of the receiver front-end.

7.1 Digital Demodulation

The functional components of a digital receiver are illustrated in Fig. 7.1. The optical signal received from the channel may be optically processed prior to photodetection. The electrical signal resulting from photodetection is then amplified and demodulated to recover the digital information. The electrical amplifier is usually configured as either an automatic-gain-control amplifier or as a limiting amplifier [1-3]. This provides a relatively constant electrical signal level to the demodulator regardless of the amount of optical power received.

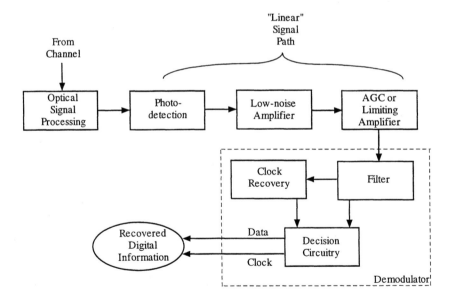

Figure 7.1 Digital receiver subsystems.

The digital receiver must recover both the digital information present in the received signal and a corresponding digital clock waveform. The clock must have a stable, well defined timing relationship with the boundaries between adjacent symbols in the information waveform because it is used by the decision circuitry to examine the information carrying waveform at precise times to determine the symbol being received.

Accurate recovery of the transmitted information usually requires that the receiver be built so that the electrical signal-to-noise ratio is as large as possible given the expected received optical power levels. In an analog system, we typically try to maximize the SNR at the receiver's output port. In a digital receiver the SNR is usually maximized at the input to the data demodulator's decision circuitry. Fig. 7.1 reveals that a digital receiver contains a linear signal path similar to that found in an analog receiver. Maximizing SNR and minimizing waveform distortion within this linear signal path will be important in obtaining a low bit-error-rate.

When analyzing the performance of a demodulator, a digital communication system can be modeled as shown in Fig. 7.2.

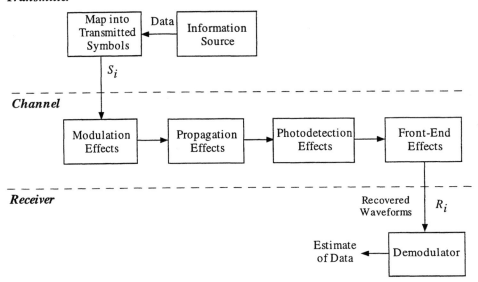

Figure 7.2 The digital demodulator's view of the link.

The digital data from the information source is first mapped into a set of ideal transmitted symbols S_i. The symbols usually correspond to discrete states of the transmitter such as *on* or *off*, *0°-phase* or *180°-phase*, *frequency* f_1 or *frequency* f_2. In reality, the transmitter may not completely shut-off, may ring when turned on, or may change phase or frequency slowly. Consequently, the symbols generated are not ideal. Modulation effects corrupt the symbols as soon as they are generated.

The transmitted symbols are sent through either a fiber or free-space channel that introduces additional effects. The symbols may be reduced in energy, spread in time, disturbed by other signals, or subject to random fluctuations. Upon reception, the symbols are further corrupted by noise and distortion effects in the photodetection process. The receiver front-end and amplifier chain introduce additional noise and distortion effects. Ultimately, a set of receiver waveforms R_i are produced. The demodulator uses these waveforms to determine which symbol was transmitted.

In effect, the demodulator sees a "channel" that consists of the combined effects of modulation, propagation, photodetection, and front-end amplification. Since the identity of the symbol transmitted at a specific time is unknown to the receiver and the recovered waveforms are corrupted versions of the ideal symbols, the demodulator can only estimate which one of the original symbols S_i was transmitted. Estimation is inherently a statistical process that will have an associated probability of success and probability of failure. The data at the demodulator's output is generally characterized in terms of the probability of a bit being in error. In some systems the bit error rate may be so low that "rate" is essentially meaningless and error-free minutes or error-free hours may be a more appropriate measure of performance.

Let us model the demodulator as an observer of the recovered electrical waveforms present at the output of the receiver front-end. At the end of each symbol-time the demodulator assigns a unique value R_i to the waveform based on what is observed. One example of R_i is a voltage that corresponds to the total energy in the waveform. Another is a current corresponding to the frequency of the waveform. Once the value of R_i is known, the demodulator computes the conditional probability

$$P(S_i|R_i) = \text{probability that } S_i \text{ was sent given that } R_i \text{ was received,} \qquad (7.1)$$

for each one of the known possible transmitted symbols and picks the symbol with the highest probability as the estimate.

In a binary communication system the transmitter sends symbol S_{zero} for a data *zero* and S_{one} for a data *one*. The demodulator observes the recovered waveform, assigns the value R_i, and then computes two probabilities. The demodulator decides a *one* was transmitted if

$$P(S_{one}|R_i) > P(S_{zero}|R_i), \qquad (7.2)$$

and decides a *zero* was transmitted if

$$P(S_{zero}|R_i) > P(S_{one}|R_i). \qquad (7.3)$$

If both probabilities are equal then a random choice can be made. This simple comparison test is frequently written as

$$P(S_{zero}|R_i) \lessgtr P(S_{one}|R_i). \qquad (7.4)$$

The probability that a symbol S_i was transmitted given that R_i was observed is known as an *a posteriori* probability since it is determined once the information moves through the channel to the receiver and R_i is generated.

It can be shown that for a communication system with N possible transmitted symbols, the optimum demodulator, the one with the minimum probability of making an error in estimating the symbol transmitted, is a *maximum a posteriori,* or MAP, demodulator [4-6]. The MAP demodulator observes R_i, computes the a posteriori probabilities for each of the N possible transmitted symbols, and declares the symbol with the maximum a posteriori probability to be the correct choice.

The MAP decision criteria given by Eq. 7.4 can be rewritten as a ratio as

$$k = \frac{P(S_{one}|R_i)}{P(S_{zero}|R_i)} \gtrless 1. \qquad (7.5)$$

The MAP decision criteria is now to decide *one* for $k > 1$, *zero* for $k < 1$, and pick *one* or *zero* at random if $k=1$.

Equation 7.5 requires computing a probability that a specific symbol was transmitted given that the waveform R_i was observed. Alternatively, we can calculate the probability that R_i would be observed given that a known symbol S_i was transmitted. This can be accomplished by applying Bayes's rule [7],

$$P(A|B) = \frac{P(B|A)P(A)}{P(B)}, \quad P(B) \neq 0, \tag{7.6}$$

to Eq. 7.5. This allows it to be rewritten as

$$k = \frac{P(S_{one}|R_i)}{P(S_{zero}|R_i)} = \frac{\dfrac{P(R_i|S_{one})P(S_{one})}{P(R_i)}}{\dfrac{P(R_i|S_{zero})P(S_{zero})}{P(R_i)}} = \frac{P(S_{one})}{P(S_{zero})} \frac{P(R_i|S_{one})}{P(R_i|S_{zero})} \gtrless 1, \tag{7.7}$$

where

$P(R_i|S_i)$ = probability of receiving R_i given that S_i was sent,

$P(S_i)$ = probability that S_i was sent,

$P(R_i)$ = probability that R_i was received.

The probability that a symbol S_i was sent is known as an *a priori* probability. In most systems the probability of any given symbol being transmitted is equal to the probability of any other symbol (*ones* are as equally likely as *zeroes*) and Eq. 7.7 reduces to

$$k = \frac{P(R_i|S_{one})}{P(R_i|S_{zero})} \gtrless 1. \tag{7.8}$$

This ratio is termed a likelihood ratio. A demodulator that uses likelihoods to estimate the received data is known as a *maximum likelihood* (ML) demodulator [4, 5]. It is equivalent to a MAP demodulator when the transmitted symbols are equally likely. A ML demodulator observing R_i computes the likelihood for each of the possible transmitter symbols S_i and picks the symbol with the maximum likelihood as the estimate of the transmitted symbol. The ML demodulator usually represents the optimum type of demodulator for a digital communications receiver in that it will have the lowest probability of making an error. With some signaling formats it is possible to construct a demodulator that is essentially optimum, while with others the ML demodulator may be unacceptably complex and approximations must be made.

7.1.1 Digital Signaling Formats

Since a laser can be either amplitude, frequency, phase, or polarization modulated, there is a wide variety of possible transmitter symbol sets that are used in optical communications [8, 9]. Some popular formats are illustrated in Fig 7.3. The formats shown in Fig. 7.3(a)-(d) are typically associated with direct detection systems, while the formats in Fig. 7.3(e)-(g) are most often associated with coherent detection systems.

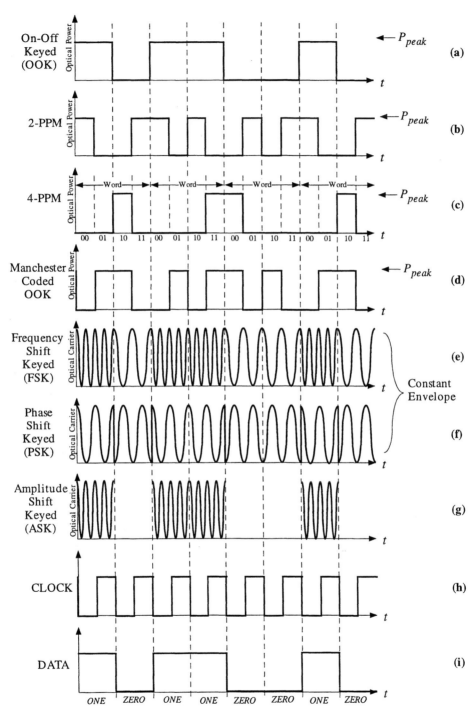

Figure 7.3 Common modulation formats for digital signaling. (a) On-off-keying (OOK). (b) Binary PPM. (c) 4-ary PPM. (d) Manchester coded OOK. (e) FSK. (f) PSK. (g) ASK. (h) Clock waveform. (i) Data sequence.

The choice of a specific transmitted symbol set or modulation format is dependent on the type of laser used, the target cost for the system, the type of channel present, and the performance goals of the system. Some signal sets convey information with a higher efficiency than others. One way of specifying system efficiency is the number of bits carried per Hz of occupied bandwidth. Increasing the bandwidth or spectral efficiency of a system often requires an increase in the SNR needed at the receiver. Another way of specifying system efficiency is in terms of the number of photons required to transmit a bit of information at a specified probability of error, typically 10^{-9}. An increased power efficiency is often accompanied by an increase in modulator and demodulator complexity.

A modulation scheme that is popular because of its simplicity is on-off keying illustrated in Fig. 7.3(a). The laser output is simply turned on and off in response to whether the data is a *one* or a *zero*. This format is particularly popular with direct detection fiber systems. One problem with this format is that with long strings of contiguous *ones* or *zeroes* there are no transitions present that a clock recovery system can use to recover the bit clock. The OOK illustrated is an example of non-return-to-zero signaling. The laser is turned on when transmitting a *one* and stays on for the entire bit-time. A variation on OOK is return-to-zero signaling, where the laser is on during a *one* but for a time that is less than the bit-time. This format is commonly found in short-pulse and soliton systems.

Another variation on OOK is binary pulse-position-modulation (PPM), shown in Fig. 7.3(b) [10, 11]. In this scheme a bit time is divided into two slots. When the data is a *one* a pulse is transmitted in the first slot; when the data is a *zero*, the pulse is sent in the second time-slot (or vice-versa). PPM can be readily expanded to larger alphabet sizes. The general case is M-ary PPM, where the data is divided up into words that are $\log_2 M$ bits long. Each word then corresponds to a specific transmitted symbol. Time is divided into M time-slots, with an optical pulse sent in only one of the possible slots. An 8-ary system will transmit 3 bits of data per optical pulse while a 256-ary system would transmit 8 bits per pulse. Since multiple bits of information can be transmitted using a single pulse, PPM can be more efficient than simple OOK in terms of the number of photons required to transmit an individual bit of information. PPM systems are also attractive when there is a substantial amount of optical background present.

4-ary PPM is illustrated in Fig. 7.3(c). The time required for two bits is divided into four time-slots. Each time slot has a corresponding two-bit address starting at 00 and counting through 01 and 10 up to 11. For each two-bit word, a light pulse is transmitted in the corresponding time-slot. 4-ary PPM has been demonstrated in some moderate data-rate free-space systems [12, 13]. The signaling format is attractive because it becomes increasingly power efficient as the alphabet size increases. Large alphabets have been proposed for power efficient optical communications with space probes [14]. There are three principal drawbacks to higher order PPM systems. The first is the need for accurate clock synchronization across all of the slot-times [15, 16]. This becomes increasingly difficult as the number of slots increases. The second is that semiconductor lasers are typically peak power limited and as M increases, the pulse width decreases and the peak power increases. Ultimately a limit of a relatively modest M is reached [17]. Q-switched solid-state lasers may be used to advantage at the expense of a more complicated transmitter. The third drawback with M-ary PPM is that the receiver bandwidth increases with increasing M. It is harder to achieve low-noise performance in a wider bandwidth, and M-ary PPM is typically restricted to moderate data-rates [18].

Manchester coded OOK [19], as shown in Fig. 7.3(d), is equivalent to binary PPM. In the example illustrated we are sending a low-high transition to indicate a *one* and a high-low transition to indicate a *zero*. If this mapping was reversed, our examples of the 2-PPM and Manchester formats would be identical. The Manchester coded waveform can be derived by exclusive-OR'ing the data and clock waveforms. Manchester coding has the benefit of producing a transition every bit. This aids clock acquisition. The penalty paid is that the bandwidth required is twice that of a simple OOK system.

The formats in Fig. 7.3(a)-(d) correspond to variations in the power in the received optical field. Once photodetected, the electrical signal is recovered directly at baseband. When the receiver is sensitive to the frequency or phase of the optical carrier, frequency-shift keying (FSK) shown in Fig. 7.3(e), or phase-shift-keying (PSK) shown in Fig. 7.3(f), can be used [20]. This is usually implemented using a coherent receiver.

In an FSK system the transmitter generates a specific frequency for a *one* and a different frequency for a *zero*. An FSK system can also be expanded to *M*-ary signaling by utilizing additional transmitter frequencies [21]. In PSK, the transmitter sends an in-phase signal for a *one* and an out-of-phase signal for a *zero* (or vice-versa). A PSK receiver requires an absolute phase reference and is typically implemented using a homodyne receiver [22].

A variation on PSK is differential phase-shift-keying (DPSK), where a phase transition is sent to indicate transition from a *zero* to a *one* or a *one* to a *zero*. A DPSK receiver need only compare the phase between bit-times to determine if a transition has occurred. This reduces its complexity when compared to a PSK homodyne system. Another variation is quadrature-phase-shift-keying (QPSK), where the phase between symbols differs by 90 degrees [23].

Coherent receivers can also utilize the amplitude-shift-keying (ASK) format illustrated in Fig. 7.3(g). ASK is similar to OOK except that the detected signal is obtained at an intermediate frequency instead of directly at baseband. The ASK receiver, being a coherent receiver, is also sensitive to the frequency of the received signal but this sensitivity is typically minimized by operating with a wide IF bandwidth or by precision frequency tracking. With any coherent receiver the spectral purity of the optical signals [24, 25] and the stability of the received polarization [26] will strongly influence the receiver's performance [27].

7.2 Direct Detection Digital Receivers

The simplest form of a classical direct detection communications receiver is one that performs simple energy detection. If the receiver detects light of sufficient intensity it outputs a logical *one*; otherwise the receiver output is a logical *zero*. Direct detection receivers are typically thought of as being able to respond only to intensity fluctuations in the received optical signal. Phase, frequency, and polarization information is often considered lost in a direct detection communication system. However, by utilizing additional optical elements, it is possible to construct direct detection receivers that respond to polarization, frequency, or phase modulated signals. For example, it is possible to build a direct detection receiver that could demodulate a FSK signal by including an optical frequency discriminator such as a Fabry-Perot or Mach-Zehnder interferometer in front of the photodetector [28]. A direct detection receiver using a Mach-Zehnder can also be used to demodulate DPSK [9, 29].

7.2.1   Ideal Photon Counting On-Off Keying

An OOK system utilizing an ideal photon counting receiver is shown in Fig. 7.4. In a binary OOK system, the transmitter sends a pulse of light into the channel to represent a *one* and does not send any light for a *zero*. The channel is assumed to be linear and there are no sources of background illumination. Any photons in the received signal must therefore originate in the transmitter. We assume that the receiver is noiseless and that the photodetector converts photons to photoelectrons with a 100% quantum efficiency.

The receiver counts the number of photoelectrons generated during the time interval T. For simplicity, the counter in our example is completely noiseless and generates an output voltage equal to the number of photons counted during the interval. The decision circuit is a simple clocked voltage comparator. Comparisons are made on the falling edge of the clock waveform. If the input voltage is greater than or equal to the threshold, a *one* is generated at the comparator's output. If the input voltage is less than the threshold, a *zero* is generated. The counter and comparator are controlled by a clock that is perfectly aligned with the bit timing boundaries present in the received signal.

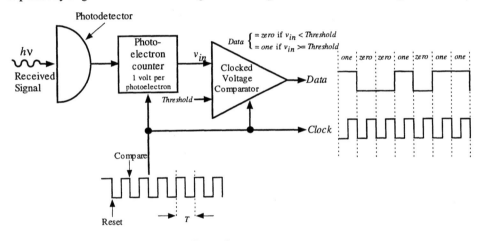

Counter is reset every T seconds

Figure 7.4 An ideal photon counting receiver for OOK.

When the received signal is a *one*, the photoelectrons exhibit Poisson statistics as described in Chapter 3. Using Eq. 3.1, the probability that n photoelectrons are counted during our observation-time T is given by

$$P(n \text{ photons counted given a } \textit{one} \text{ is transmitted}) = \frac{(rT)^n e^{-rT}}{n!}, \qquad (7.9)$$

where r is the mean value of the photon arrival rate in photons per second

Once we know the probability of detecting a given number of photons during our observation-time, we can derive the probability of making an error in estimating whether the transmitter sent a *one* or a *zero*. Fig. 7.5 is a plot of the probability density functions for the voltage at the input of the comparator. The probability density for the comparator's input voltage when a *zero* is received is an impulse function, while the probability density when a *one* is received is a Poisson distribution.

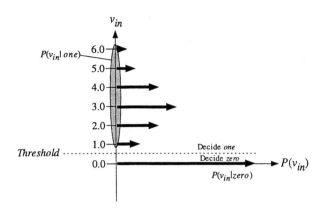

Figure 7.5 Example probability density functions for the digital comparator's input voltage in a photon counting OOK receiver.

The probability of making a bit-error in a binary decision system is composed of two parts. The receiver may decide a *one* was transmitted when a *zero* was actually sent. This is sometimes termed an insertion error since a *one* was inserted into the data stream. The receiver may also decide a *zero* was sent when in fact a *one* was transmitted. This is sometimes called an erasure or deletion error since a *one* has been removed from the data.

The probability that a *one* or *zero* was sent also enters into the determination of error probability because if it was known a-priori that a *one* is more likely to have been transmitted than a *zero*, this additional information could be used to help determine the received bit. The complete expression for error probability is then

$$P_e = \text{(Probability of deciding } zero \text{ when a } one \text{ is sent)}$$
$$\times \text{(Probability that a } one \text{ is being sent)}$$
$$+ \text{(Probability of deciding } one \text{ when a } zero \text{ is sent)} \quad (7.10)$$
$$\times \text{(Probability that a } zero \text{ is being sent)}.$$

Using conventional notation, this probability is written as

$$P_e = P(zero|one)P(one) + P(one|zero)P(zero)$$
$$or$$
$$P_e = P(0|1)P(1) + P(1|0)P(0). \quad (7.11)$$

Assuming that the transmitter is sending *ones* and *zeroes* with equal probability, each has probability equal to one-half, so the probability of a bit error is then given by

$$P_e = \frac{1}{2}P(0|1) + \frac{1}{2}P(1|0). \quad (7.12)$$

We have excluded the possibility of any background illumination in our formulation of this OOK system. The probability of receiving a photon when a *zero* is transmitted must therefore be exactly zero. Since all photons must originate in the transmitter, the only way photoelectrons can be counted is if the transmitter was sending a *one*. Even if only a single photoelectron is counted we should therefore decide that a *one* has been received. Conversely, if no photoelectrons are counted we should decide a *zero* has been received. The maximum likelihood decision rule that results in optimum receiver performance is then to simply set a threshold between 0.0 and 1.0 volt and decide a *zero* if the input voltage falls below 1.0 volt and a *one* if the input is 1.0 volt or higher.

Since the probability of receiving a photon when a *zero* is sent is equal to zero, the only way this photon counting OOK system can make an error is if a *one* is transmitted and no photoelectrons are counted. Setting $n=0$ in Eq. 7.9 to determine the probability of not counting any photons when a *one* is received yields

$$P_e = \left[\frac{1}{2}e^{-rT} + \frac{1}{2}(0)\right] = \frac{1}{2}\exp\left\{-\frac{P_{rcvd_{one}}T}{h\nu}\right\}, \qquad (7.13)$$

where $P_{rcvd_{one}}$ is the power in the received signal during the *one* and T is the bit time.

Equation 7.13 represents a fundamental limit on the performance of an OOK direct detection system. To achieve a probability of error of 10^{-9}, an OOK system requires 20 photons every time it transmits a bit with value *one*. Since we are assuming equally probable *ones* and *zeroes* and since a *zero* requires no photons from the transmitter, the ultimate sensitivity for OOK direct detection is said to be an average of 10 photons per bit. This sensitivity is called the quantum-limit for OOK signaling since quantum effects associated with photodetection are the only limiting factor in determining the sensitivity. A plot of the variation in BER as a function of received signal for a quantum limited OOK direct detection system is shown in Fig. 7.6.

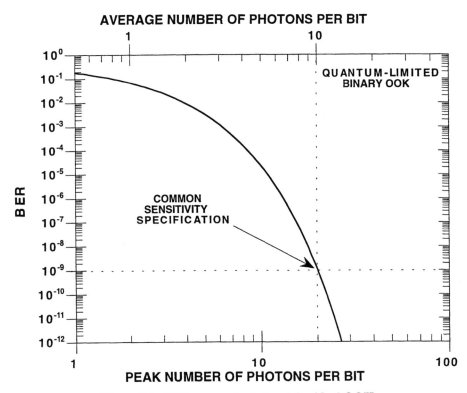

Figure 7.6 BER vs. received signal for ideal OOK.

Figure 7.6 plots BER versus both the peak and the average number of photons per bit required. Because lasers are usually peak-power, as opposed to average-power limited, the peak value of photons-per-bit is of particular interest in some systems. The difference between peak and average is simply a factor of 2 for this 50% duty-cycle OOK system. Photons-per-bit is a particularly attractive way of specifying sensitivity because it is wavelength and data-rate independent. The overall sensitivity is often specified at the 10^{-9} BER point. The quantum limited sensitivity for OOK is therefore commonly quoted as being 20 photons-per-bit (peak) or 10 photons-per-bit (average).

7.2.2 Receiver Noise, Pulse Shaping Filters, and the Eye Diagram

An ideal photon counting receiver is unrealistic. A practical photodetector is not perfectly efficient and any realizable receiver front-end introduces noise. A practical receiver for OOK signaling is illustrated in Fig. 7.7. The output of the photodiode passes through a low-noise front-end amplifier. A transimpedance configuration is illustrated, but this discussion is generally applicable to any front-end configuration.

The output of the front-end is split into two paths. One portion is used to derive a clock signal that is aligned to the symbol transitions. The other portion is filtered and passed to a digital decision circuit. In this example, the falling edge of the recovered clock causes the decision circuit to sample the recovered waveform and compare the input voltage v_{in} to the threshold voltage. In practice, the decision circuit can be realized with a continuously acting voltage comparator and a clocked flip-flop or a sample-and-hold circuit [30].

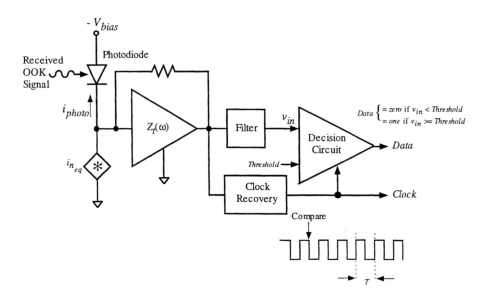

Figure 7.7 Practical OOK receiver.

If the received waveform's voltage is below the threshold, the decision circuit produces a *zero* at its output. If the voltage exceeds the threshold, a *one* is produced. Examples of the signal that appears at the input to the decision circuit are illustrated in Fig. 7.8. The recovered waveform is corrupted by noise, as illustrated in Fig. 7.8(a). There is a probability density associated with the reception of a *zero* and a with a *one*. In simple binary OOK with a high extinction ratio, signal shot-noise is present only when a *one* is received. If the receiver is low-noise, so that the signal shot-noise is a major contributor to the overall noise observed in the receiver, *ones* will appear to be "noisier" than *zeroes*. The area of each probability density is the same since we are assuming equally probable *ones* and *zeroes*, but the exact shapes of the densities will represent the actual statistics of the noise present when either a *one* or a *zero* is being received.

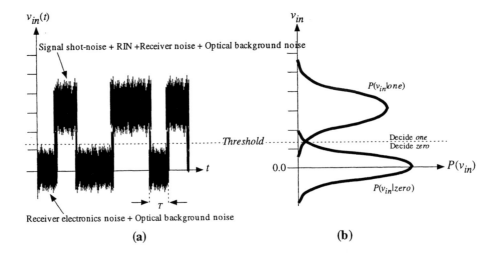

Figure 7.8   Signals at the decision circuit's input in a practical OOK receiver.
(a) Example time-domain waveforms. (b) Example probability density functions.

The decision operation performed in Fig. 7.8 is essentially the detection of a pulse waveform in noise. Both the shape of the pulse waveform and the amount of noise present at the input to the decision circuitry influence the receiver's ability to correctly identify the symbol that has been received. We would like to select the waveform and noise level so that we maximize our ability to make correct decisions. This is a common problem in data communications that has been extensively studied, and techniques for obtaining maximum likelihood detection are well-known [31-35].

A simple view of the known results for the optimization of a conventional digital receiver is that is involves two complementary steps. We want to design the waveform that appears at the input to the decision circuit such that any interference between adjacent received symbols is minimized. Intersymbol-interference (ISI) occurs when the detection of one symbol interferes with our ability to determine the symbols that follow and can be a severe source of degradation in digital systems. In addition to using a waveform with minimal ISI, we want to maximize the signal-to-noise ratio at the decision time.

The primary method a digital receiver uses to control the waveform shape and signal-to-noise ratio at the input to the decision circuit is via post-detection electrical filtering. In practice, each stage of the receiver contributes something to the overall post-detection filter. The combination of the photodetector and receiver front-end transfer functions can be used to develop the filter function, or a separate electrical filter stage may be used. The decision circuit itself may have a low-pass filtering effect that contributes to the overall filter function. *Pre*-detection *optical* filtering will also play a role if the bandwidth of the optical filter is comparable to the bandwidth of the received signal.

The design of minimal ISI pulse waveforms and pulse shaping filters for digital communication has been extensively developed since Nyquist's original work on waveforms for use in telegraph transmission [36, 37]. Nyquist identified criteria that a waveform must meet in order to avoid interactions between consecutive symbols, and a wide variety of practical pulse waveforms have been shown to satisfy the Nyquist criteria [38]. He also showed that to avoid ISI, the minimum bandwidth that an ideal binary waveform of data rate R bits-per-second could pass through was $B_{min} = R/2$. This is effectively a restatement of the sampling theorem that requires the sampling rate to be at least twice the maximum frequency that is to be reconstructed. It is in fact possible to signal faster than that set by the Nyquist criteria if we are willing to accept a controlled amount of ISI. This technique is known as partial response signaling [33]. The duobinary format is an example of this type of signaling that has been used to extend the range of high-speed systems by reducing the amount of bandwidth required [39]. In practice, most systems do not signal at even the Nyquist rate. It is more common for a system transmitting x symbols per second to require x Hz of bandwidth.

Techniques that maximize SNR in a digital receiver are also well known. The matched-filter demodulator is equivalent to the maximum-likelihood demodulator in many practical cases, allowing optimum detection of a known waveform in the presence of additive noise [31]. We can now restate our simple view of the conventional digital receiver optimization criteria as involving two criteria. We select a received pulse waveform that satisfies Nyquist's criteria for minimizing ISI and we use a matched filter to maximize the SNR at the decision time. It is also possible to separate the matched filter into several parts, with some filtering occurring at the transmitter, some in the channel, and the remainder in the receiver. In particular, in the case of a channel that does not contribute a significant amount of dispersion, the filter can be equally divided between the transmitter and the receiver while still obtaining near-optimum performance [33, 40, 41]. This tends to minimize the overall bandwidth requirements of the opto-electronic components used in the transmitter and receiver.

Practical optical communication receivers complicate our simple view of receiver optimization because they combine bandlimited waveforms having signal dependent Poisson photodetection statistics with additive Gaussian-noise from the receiver electronics. The fiber-optic channel may be dispersive or suffer from optical nonlinearities, polarization dependent fading, or channel-to-channel crosstalk. The free-space channel may introduce fading due to atmospheric effects. Avalanche multiplication-noise from an APD may also be present, and the waveforms may need to be designed to be tolerant of jitter in the receiver decision sampling or transmitter pulse timing. We also saw in Chap. 5 that even though the receiver electronics-noise is Gaussian, the power spectral density of the electronics-noise in wide bandwidth receivers is often frequency dependent instead of "white." This may require that an additional "whitening" filter be included in the receiver's design if a true matched filter is to be constructed [41, 42]. The whitening filter can itself introduce unacceptable amounts of ISI, and an additional zero-forcing equalization stage, similar to those used to combat dispersion, may be required [35, 43-45].

Several approaches have been taken to rigorously determine the optimum receiver filter for an optical communication receiver. The optimization techniques usually involve various combinations of specific criteria such as minimum mean-square-error, maximum signal-to-noise ratio, minimum ISI, or maximum tolerance to jitter. Optimum filter schemes for the case of Poisson signaling [46] and Poisson signaling in additive white-Gaussian-noise have also been determined [47].

In many cases the mathematically optimum receiver filter is either unrealizable or impractical. Fortunately there are several approximation techniques available and it is usually possible to obtain nearly ideal filter performance with relatively simple filter structures [48-50]. In particular, a third order all-pole low-pass filter with a bandwidth of approximately 70% of the symbol rate performs within a few dB of optimum in many circumstances [51, 52]. More complicated higher-order filters can result in performance within a few tenths of a dB of the optimum filter [50, 52-61].

Filters that produce output waveforms that approximate members of the raised-cosine family have been used in many digital communication systems [62-64]. These waveforms have also proven useful in optical communication systems, and practical direct detection receivers can be fabricated with filters that realize near-optimum raised-cosine pulse waveforms at the input to the decision circuitry. This will be true whether the system uses NRZ pulses that occupy the full symbol time, RZ pulses that occupy only a portion of the symbol time, or Manchester coded signals [19].

A raised-cosine pulse is nearly optimum for the case where the receiver is accurately modeled as a signal in additive white or colored Gaussian-noise. This is the case in many broadband *p-i-n* based receivers [60]. It is somewhat worse than optimum when there is significant signal dependent noise such as found in an APD or optically preamplified receiver. In these cases it is possible to modify the pulse waveform and receiver filter and again achieve performance within approximately 1 dB of optimum [50, 52, 59].

The time-domain waveform and the corresponding spectrum for the raised-cosine pulse family of symbol duration T are given by

$$h_{out}(t) = \frac{\sin\left(\dfrac{\pi t}{T}\right)}{\dfrac{\pi t}{T}} \times \frac{\cos\left(\dfrac{\alpha \pi t}{T}\right)}{1-\left(\dfrac{2\alpha t}{T}\right)^2},$$

$$H(f) = \begin{cases} T; & 0 \le |f| \le \left(\dfrac{1-\alpha}{2T}\right), \\ \dfrac{T}{2}\left[1-\sin\left(\pi T\left[\dfrac{|f|}{\alpha}-\dfrac{1}{2T\alpha}\right]\right)\right]; & \left(\dfrac{1-\alpha}{2T}\right) \le |f| \le \left(\dfrac{1+\alpha}{2T}\right), \\ 0; & |f| > \left(\dfrac{1+\alpha}{2T}\right). \end{cases} \qquad (7.14)$$

Examples of the raised-cosine family are illustrated in Fig 7.9. A raised-cosine pulse is described by the parameter α, which is known as the roll-off factor. The limits on the range of α are $0 \le \alpha \le 1$. When $\alpha = 0$, the pulse has a perfectly rectangular spectrum with a single sided bandwidth of $1/2T$. T is the interval between symbol decisions and is equal to the time between bits for a binary system. Fig. 7.9(a) illustrates the pulse waveforms in the time-domain. The decision between a *zero* and a *one* for the current symbol would ideally occur at the point $t = 0$. The decision on the next symbol would occur at $t = T$, while the decision on the previous symbol occurred at $t = -T$. Note that all waveforms in the raised-cosine family have no ISI because the waveform always equals zero at all sample times $\pm nT$ except for $n = 0$.

Figure 7.9(b) illustrates the corresponding frequency-domain spectra. The case of $\alpha = 0$ occupies the minimum bandwidth and would therefore have the smallest noise equivalent bandwidth. Unfortunately, the tails of the minimum bandwidth pulse roll-off as $1/t$. This gives rise to relatively large amplitude tails that cause any error in the timing of the decision samples to cause significant degradation because of the resulting ISI.

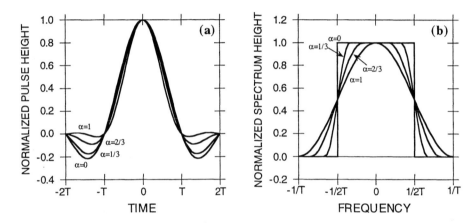

Fig. 7.9 Raised-cosine pulse waveforms. (a) Time-domain. (b) Frequency-domain.

An important feature of the raised-cosine family is that they allow us to trade an increase in bandwidth for a more rapid pulse roll-off. For cases where $\alpha \approx 1$, the tails of the pulse waveform roll-off as $1/t^3$. An error in the decision circuit timing is therefore significantly less important than for a pulse shape that rolls-off as $1/t$. The case of $\alpha = 1$ is termed the 100% raised-cosine waveform since there is 100% more bandwidth occupied than would be required by the minimum bandwidth pulse obtained when $\alpha = 0$. We note that the term bandwidth as used to describe the pulse is the total width of the spectrum. This is neither the receiver's 3 dB bandwidth nor the noise equivalent bandwidth. Both the 3 dB electrical bandwidth (i.e. the half-power width) and the noise equivalent bandwidth of an electrical filter that a receiver would use to convert rectangular NRZ pulses appearing at its input into 100% raised-cosine pulses at the decision circuit are actually approximately 60% of the bit-rate [65]. It is also possible to use a raised-cosine pulse instead of a rectangular on-off-keyed pulse as the actual channel symbols [33, 40, 66]. This approach has been investigated as a technique to reduce the bandwidth required for the opto-electronic components in free-space systems [67].

A convenient visual method that is often used to qualitatively measure the properties of a recovered data waveform is illustrated in Fig. 7.10. Figure 7.10(a) is an example of a recovered data waveform. If we superimpose all of the various symbol transitions at their respective decision times, we obtain a waveform as shown in Fig. 7.10(b). This is called an eye-diagram because of its similarity to a human eye.

An eye diagram is easily generated using an oscilloscope that is triggered by the symbol timing clock. Long pseudo-random data patterns are often used when generating eye-diagrams to guarantee that the eye-diagram is representative of virtually all possible symbol transitions. By measuring the width of the opening of the eye in both the vertical and horizontal directions we obtain information about the system's ISI, noise, and jitter. Jitter, whether due to variations in the received pulse duration or the accuracy of the recovered symbol clock, will cause the eye to close in the horizontal direction. Noise and ISI cause the eye to close in the vertical dimension. The ideal decision sampling point occurs at the time of maximum vertical opening. This point corresponds to the time when the signal-to-noise ratio is at its maximum.

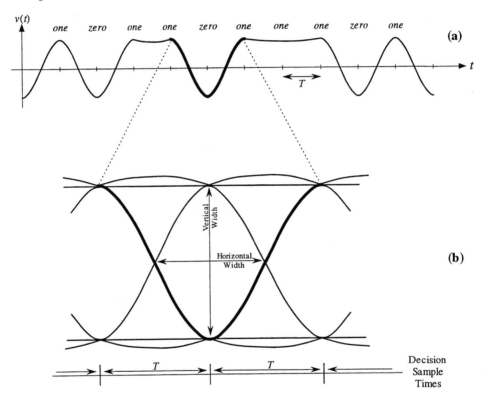

Fig. 7.10 Origin of the eye-diagram.   (a) Example data waveform. (b) Superposition of transitions, yielding an eye-diagram.

In addition to the qualitative measurements obtained with an oscilloscope it is possible to obtain more detailed information using eye-diagrams. More quantitative results can be obtained by either accurately sampling the eye-pattern and constructing histograms of jitter and noise or by combining the eye-diagram measurements with bit-error-rate measurements taken with an accurate decision sample gate. This allows contours of constant error-rate to be obtained for the eye diagram that can sometimes reveal subtle performance degradation mechanisms [68].

7.2.3   OOK with a *p-i-n* Photodetector Front-End

A simple example of a *p-i-n* based receiver is illustrated in Fig. 7.11. The electrical schematic of the receiver is shown in Fig. 7.11(a). A *p-i-n* photodetector with an 80% quantum efficiency is being used to detect a 2.0 Gbps OOK modulated signal at a wavelength of 1550 nm. The receiver front-end is a transimpedance type with a 10 kΩ feedback resistor. There is an additional voltage amplifier with a gain of 100 and a low-pass filter that is used to set the receiver's noise-bandwidth and shape the signal waveform. The demodulator is a decision circuit that samples the waveform that appears at v_{in} and compares the sample to a threshold voltage V_T.

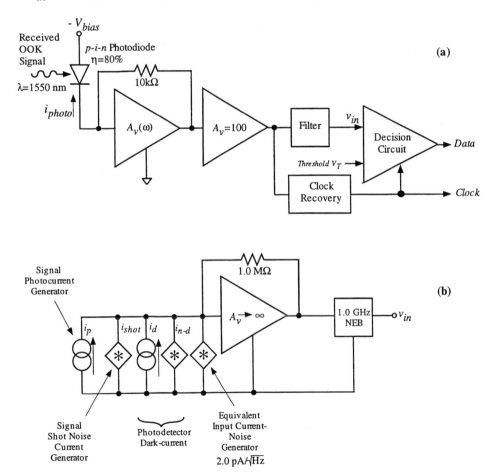

Fig. 7.11   Example of a *p-i-n* receiver. (a) Electrical schematic. (b) Equivalent circuit for determining the decision circuit's input voltage v_{in}.

We wish to determine how accurately the *p-i-n* receiver is able to determine the information being received. Our primary measure of a digital receiver's accuracy is the probability of error P_e. We therefore need to understand the statistics of the signal at the time at which we sample v_{in}.

The equivalent circuit used to determine the characteristics of the voltage at the decision circuit's input is shown in Fig. 7.11(b). We first need to determine the equivalent input current-noise for the receiver. This is done using Eq. 5.81 or Eq. 5.83. In this example we will use a noise density of $2.0\text{pA}/\sqrt{\text{Hz}}$. The thermal-noise of the $10\,\text{k}\Omega$ feedback resistor contributes $1.2\,\text{pA}/\sqrt{\text{Hz}}$ to the receiver-noise. Electronic shot-noise and thermal-noise from the transistors and biasing circuitry used in the front-end amplifier contribute the remainder. The front-end amplifiers, in combination with the filter, produce a noise equivalent bandwidth (the BI_2 term in Eq. 5.83) of $1.0\,\text{GHz}$. The overall transimpedance is $1.0\,\text{M}\Omega$.

Electronics-noise is well modeled by the Gaussian probability distribution [69]. The distribution describing the total equivalent input current-noise from the receiver electronics can then be written as

$$i_n(i) = \frac{1}{\sqrt{2\pi}I_{elec}} \exp\left\{-\frac{i^2}{2I_{elec}^2}\right\}, \tag{7.15}$$

where I_{elec} is the total equivalent input current-noise from the receiver electronics. For our receiver with a $1.0\,\text{GHz}$ NEB and an equivalent input current-noise density i_{elec} of $2.0\text{pA}/\sqrt{\text{Hz}}$, the total input current-noise I_{elec} is 63nA.

The appropriate statistics for the signal photocurrent are not immediately obvious. We have seen that the statistics of the signal photocurrent are exactly described by a Poisson distribution. Unfortunately, the Poisson distribution does not lend itself to convenient analytical manipulation, and the use of the exact Poisson photodetection statistics in addition to the Gaussian-noise from the receiver electronics will usually require numerical evaluation techniques [70, 71].

However, for cases where the mean photocarrier counting rate is high, exceeding about one hundred generated photocarriers per observation-time, it is possible for the statistics of photodetection to be reasonably well approximated by a Gaussian distribution, given by [72]

$$i_{photo}(i) = \frac{1}{\sqrt{2\pi}I_{shot}} \exp\left\{-\frac{(i - i_{DC})^2}{2I_{shot}^2}\right\}, \tag{7.16}$$

where $i_{DC} = \frac{\eta q}{h\nu}\overline{P}_{rcvd}$, $\overline{P}_{rcvd}$ = average received signal power, and $I_{shot} = \sqrt{2qi_{DC}B}$.

In order to use the Gaussian approximation we need to verify that the mean photocarrier count rate is high. Since at least a SNR>1 is required for the receiver to be able to make reliable decisions, we can determine a conservative lower bound on the number of signal photons received in each symbol simply by determining what photon arrival rate n_s is required during a symbol to obtain a zero dB SNR. The actual photons per symbol must then be greater than the equivalent number of "noise" counts, or

$$n_s > \frac{I_{elec}}{q \times NEB}. \tag{7.17}$$

For our example *p-i-n* receiver, the receiver-noise level of $63\,\text{nA}$ and NEB of $1.0\,\text{GHz}$ correspond to at least 400 photons in each symbol. The Gaussian distribution can then be used to describe the signal photocurrent with minimal loss in accuracy.

If we neglect optical excess-noise and any potential optical background fields, the total equivalent input current-noise is obtained from $i_{n_{eq}}^2 = i_{elec}^2 + i_{shot}^2$. Since the sum of two Gaussians is another Gaussian [73], we can express the voltage at the input to the decision circuit at the instant we sample the waveform at v_{in} as a pair of Gaussian distributions, given by

$$P(v) = \frac{1}{\sqrt{2\pi}\sigma_{one}} \exp\left\{-\frac{(v - v_{one})^2}{2\sigma_{one}^2}\right\} \quad \text{when a } one \text{ is received}$$

and $\hspace{8cm}$ (7.18)

$$P(v) = \frac{1}{\sqrt{2\pi}\sigma_{zero}} \exp\left\{-\frac{(v - v_{zero})^2}{2\sigma_{zero}^2}\right\} \quad \text{when a } zero \text{ is received}$$

where

$$\sigma_{one}^2 = Z_t^2 I_{n_{one}}^2, \hspace{5cm} \sigma_{zero}^2 = Z_t^2 I_{n_{zero}}^2,$$

$$I_{n_{one}}^2 = \left(i_{elec}^2 + 2qi_d\right)BI_2 + 2qRP_{rcvd_{one}}BI_1, \hspace{2cm} R = \frac{\eta q}{h\nu},$$

$$I_{n_{zero}}^2 = \left(i_{elec}^2 + 2qi_d\right)BI_2 + 2qRP_{rcvd_{zero}}BI_1,$$

$$v_{one} = Z_t\left(i_d + RP_{rcvd_{one}}\right), \hspace{3cm} v_{zero} = Z_t\left(i_d + RP_{rcvd_{zero}}\right),$$

and BI_1 and BI_2 are the noise-equivalent bandwidths from Eq. 5.81.

In many cases the transmitter has a high extinction ratio and the power received during a *one* is much larger than the power received during a *zero*. If in addition to a high extinction ratio, there is no significant source of background illumination, the photodiode dark-current i_d is negligible, and there is no noticeable DC offset in the receiver, it is usually safe to assume that $v_{zero} \cong 0.0$. Making this assumption allows the probability density functions shown in Fig. 7.12 to be obtained for the receiver from Fig. 7.11.

The heights of the distributions in Fig. 7.12 have all been normalized to unity for simplicity of presentation. Fig. 7.12(a) illustrates the probability density functions $P(v|zero)$ and $P(v|one)$ for the case when the received signal power during a *one* is 0.5μW. The normalized density for the signal shot-noise is also shown. Note that the receiver electronics-noise completely dominates the signal shot-noise, and for all practical purposes the noise level of the recovered waveform is the same for a *zero* or a *one* so that $i_{n_{zero}} \cong i_{n_{one}}$, $i_{n_{eq}} = i_{elec}$, and $\sigma_0 = \sigma_1 = \sigma_n$.

Fig. 7.12(b) shows the probability densities from Fig. 7.12(a) plotted on a logarithmic scale. This type of plot emphasizes the tails of the densities, which dominate the statistics at low probabilities. The area under $P(v|zero)$ above the threshold voltage V_t is the probability that we will decide a *one* when a *zero* was received. The area under the $P(v|one)$ curve below the threshold voltage is the probability that we will decide a *zero* when a *one* has been received. Fig. 7.12(c) illustrates the change in the distributions when the received signal level during a *one* is increased to 1.0μW.

The general expression for the probability of error in a binary system using a threshold detector demodulator is then obtained using Eq. 7.11 as

$$P_e = P(one) \int_{-\infty}^{V_t} P(v|one)dv + P(zero) \int_{V_t}^{\infty} P(v|zero)dv. \qquad (7.19)$$

By substituting the actual distributions from Eq. 7.18 into Eq. 7.19 we can obtain an expression for the probability of error in our *p-i-n* receiver, assuming equally likely *ones* and *zeroes,* as

$$P_e = \frac{1}{2} \int_{-\infty}^{V_t} \frac{1}{\sqrt{2\pi}\sigma_n} \exp\left\{-\frac{(v-v_{one})^2}{2\sigma_n^2}\right\} dv + \frac{1}{2} \int_{V_t}^{\infty} \frac{1}{\sqrt{2\pi}\sigma_n} \exp\left\{-\frac{(v-v_{zero})^2}{2\sigma_n^2}\right\} dv, (7.20)$$

where

$$\sigma_n^2 = Z_t^2 I_{n_{eq}}^2, \quad v_{one} = Z_t \frac{\eta q}{h v} P_{rcvd_{one}}, \quad v_{zero} = Z_t \frac{\eta q}{h v} P_{rcvd_{zero}} \approx 0.0.$$

The choice of a value for the threshold voltage V_t will obviously have a major effect on our ability to accurately distinguish a *zero* from a *one*. This is similar to the problem of selecting a threshold to establish the desired probability of detection, probability of miss, and probability of false-alarm that occurs in radar and sonar systems [6].

In a communications application we generally wish to select the threshold voltage such that we minimize the overall probability of error P_e. The optimum threshold is found by differentiating the expression for P_e with respect to the threshold value and then solving for the threshold value that sets the derivative equal to zero. This can be done using Leibnitz's rule for differentiating integrals [6, 31]. In the most common case, where the cost of mistaking a *zero* for a *one* is the same as the cost of mistaking a *one* for a *zero*, the expression for the optimum threshold is [6]

$$\frac{dP_e}{dV_t} = P(one)P(V_t|one) - P(zero)P(V_t|zero) = 0. \qquad (7.21)$$

For the case of equally probable *ones* and *zeroes*, the optimum threshold occurs when

$$P(V_t|one) = P(V_t|zero). \qquad (7.22)$$

That is, we set the threshold to the point where the probability densities intersect.

For this *p-i-n* receiver, the probability densities are symmetric and the densities intersect halfway between the peaks. The optimum decision criteria is therefore to chose a threshold that is half-way between the peaks of the $P(v|zero)$ and $P(v|one)$ curves as

$$V_{t_{opt}} = \frac{v_{zero} + v_{one}}{2}. \qquad (7.23)$$

This criteria is essentially the minimax criteria that is frequently used in the detection of binary signals in additive white Gaussian-noise [6, 31]. It can be shown that the minimax criteria implements a maximum likelihood decision when the probability densities are symmetric [31].

We are able to use this simple criteria in a *p-i-n* based optical receiver because the noise performance is dominated by receiver electronics-noise instead of signal shot-noise. Note that the optimum threshold depends on the received signal level. This can also be seen by comparing Fig. 7.12(b) and 7.12(c). The threshold value has increased from 0.25 V to 0.50 V as the received signal power during a *one* increased from $0.5\mu W$ to $1.0\mu W$.

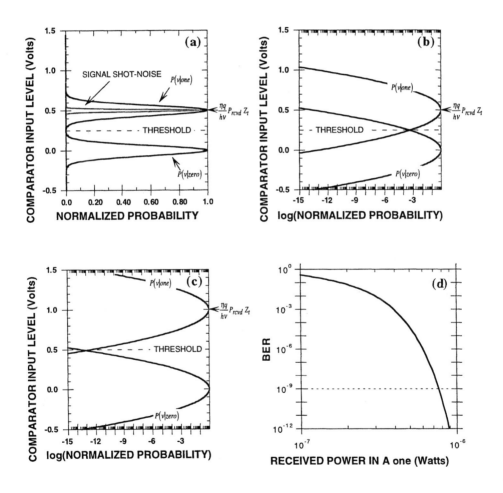

Figure 7.12 Distributions for the *p-i-n* receiver of Fig. 7.11. (a) Densities for 0.5 μW received power. (b) Densities plotted on a log scale. (c) Log densities for 1.0 μW received power. (d) Bit-error-rate plot.

The integrals in Eq. 7.20 cannot be obtained in closed form. Instead, they are usually expressed in terms of tabulated functions. Both the Q-function and the complimentary error function are used. These functions are defined as

$$Q(z) = \frac{1}{\sqrt{2\pi}} \int_z^\infty e^{-x^2/2}\, dx, \qquad Q(z) \approx \frac{1}{\sqrt{2\pi} z} e^{-z^2/2} \quad \text{for } z > 3,$$

$$\text{erfc}(z) = \frac{2}{\sqrt{\pi}} \int_z^\infty e^{-x^2}\, dx, \qquad Q(z) = \frac{1}{2}\text{erfc}\left(\frac{z}{\sqrt{2}}\right). \tag{7.24}$$

We will use the Q-function representation because it requires us to keep track of fewer factors of 2 and $\sqrt{2}$. The Q-function is plotted in Fig. 7.13.

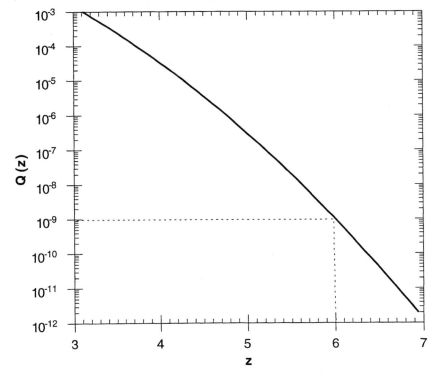

Figure 7.13 The Q-function for determining the probability of error when the noise is Gaussian distributed.

When the argument of the Q-function is approximately equal to 6, the probability of error is 10^{-9}. In terms of the Q-function, Eq. 7.20 can be rewritten using a simple change of variables as

$$P_e = \frac{1}{2}Q\left(\frac{-V_t + v_{one}}{\sigma_n}\right) + \frac{1}{2}Q\left(\frac{V_t - v_{zero}}{\sigma_n}\right). \tag{7.25}$$

Substituting the expression for the optimum threshold from Eq. 7.23 into Eq. 7.25 and assuming the source has a high extinction ratio and that there is no background or DC bias so that $v_{zero} = 0.0$ yields

$$P_e = Q\left(\frac{v_{one}}{2\sigma_n}\right). \tag{7.26}$$

We can express this in terms of the received signal power and the receiver's total equivalent input current-noise as

$$P_e = Q\left(\frac{\frac{\eta q}{hv} P_{rcvd_{one}}}{2I_{n_{eq}}}\right) \tag{7.27}$$

where $P_{rcvd_{one}}$ is the power received during a *one* and $I_{n_{eq}}$ is the total equivalent input current-noise.

Figure 7.12(d) plots the probability of error as a function of the received signal power during a *one* for our example receiver. We can also express the probability of error in terms of an electrical signal-to-noise ratio as

$$P_e = Q(SNR_e), \tag{7.28}$$

where the expression for the electrical signal-to-noise ratio is

$$SNR_e = \frac{\frac{\eta q}{hv}\left(\frac{1}{2}\left(P_{rcvd_{one}} - P_{rcvd_{zero}}\right)\right)}{I_{n_{eq}}} = \frac{\frac{1}{2}\left(i_{one} - i_{zero}\right)}{I_{n_{eq}}} \cong \frac{\frac{\eta q}{hv}\overline{P}_{rcvd}}{I_{n_{eq}}} = \frac{\overline{I}_{sig}}{I_{n_{eq}}} \tag{7.29}$$

and we are assuming 50% duty cycle in computing the average SNR value from the peak SNR. We require that the average electrical SNR be a factor of six or the peak voltage SNR be a factor of twelve to obtain a 10^{-9} BER. Note that the term average in this context refers to the fact that the SNR is given in terms of the average received optical signal power instead of the peak optical received power.

The voltage (or current) SNR is often termed the Q-factor [62, 63]. The Q-factor is a popular means for analyzing receiver performance because it is easily observed on an oscilloscope. Contemporary digitizing oscilloscopes are capable of creating histograms of the voltages present in an eye-diagram and can be used to calculate the Q-factor directly or can compute estimates for $v_{zero}, \sigma_{zero}, v_{one},$ and σ_{one}.

The Q-factor generally gives a good estimate of receiver performance when the noise statistics are nearly Gaussian. In practice, this is often a good assumption when the receiver is electronics-noise limited or there are several receiver degradation mechanisms acting in combination. It can also be used to predict the BER at high SNR when the time required to accurately measure the BER would be prohibitively long [74]. By verifying that the Q-factor continues to increase with received signal power we can predict that a BER floor will not be observed.

Solving Eq. 7.29 for the amount of received signal power needed to obtain a specified Q-factor, assuming a high extinction-ratio signal, yields

$$P_{rcvd_{one}} = 2Q \frac{hv}{\eta q} \sqrt{I_{n_{eq}}^2} .$$ (7.30)

To obtain a 10^{-9} BER we need an average SNR (Q-factor) of 6, or

$$P_{rcvd_{one}} = 12.0 \frac{hv}{\eta q} \sqrt{I_{n_{eq}}^2} , \quad \overline{P}_{rcvd} = 6.0 \frac{hv}{\eta q} \sqrt{I_{n_{eq}}^2} .$$ (7.31)

Our example receiver has a 63 nA equivalent input current-noise. The 10^{-9} sensitivity of this receiver is therefore 0.76 μW for a *one* or an average of -34.2 dBm of received signal power. Note that all of the BER expressions in Eqs. 7.26 - 7.31 assume the use of the optimum threshold given by Eq. 7.23.

If the extinction ratio is not perfect, some power is received even during a *zero*. For the case where $P_{rcvd_{zero}} = \varepsilon P_{rcvd_{one}}$, we can modify Eq. 7.30 to be [63]

$$\overline{P}_{rcvd} = \frac{1+\varepsilon}{1-\varepsilon} Q \frac{hv}{\eta q} \sqrt{I_{n_{eq}}^2} .$$ (7.32)

We can then define a power loss penalty due to a finite extinction ratio as

$$P_{loss} = \frac{1+\varepsilon}{1-\varepsilon} .$$ (7.33)

The penalty is 1 dB when the ratio between the power in a received *one* and a received *zero* is about 8.7:1. This is equivalent to about a 9.4 dB extinction ratio.

All of the analysis in this section has relied on the use of Gaussian statistics. This was done in order to simplify the receiver analysis problem and to gain insight into the statistical operation of an optical receiver [75]. In particular, the use of Gaussian statistics allowed us to utilize a simple signal-to-noise ratio expression when determining the receiver's BER. In the example receiver considered in Fig. 7.11, Gaussian statistics are easily justified because receiver electronics-noise is by far the dominant noise source, and the exact form for the distribution for the signal shot-noise is effectively irrelevant.

The goal of a low-noise receiver, however, is for the electronics-noise to *not* be the dominant noise source. In some cases we must use the exact Poisson statistics and numerical evaluation techniques when analyzing sensitive optical communication receivers. As an example of the difference between a Gaussian approximation and an exact distribution, Fig. 7.14 compares the Gaussian approximation to the exact Poisson distribution for an ideal noise-free receiver with a mean-count of 20 photocarriers per observation-time. The Gaussian distribution is symmetric, while the Poisson distribution is skewed to the right. Note that both distributions accurately predict the mean-count of 20 photons and are in reasonably good agreement for counts with probabilities greater than about 10^{-3}. For counts with lower probabilities the agreement degrades.

From Fig. 7.14 we see two potential problems with the Gaussian approximation. The first is that the Gaussian predicts a finite probability of obtaining a negative count. This is clearly incorrect since we can only count upward from zero. Another problem is that the Gaussian approximation overestimates the receiver's probability of error.

Using our photon counting decision rule of declaring a *zero* only if no photons are counted, the Gaussian approximation would predict a probability of error for choosing a *zero* when a *one* was received of some 2×10^{-6}. This can be compared to the 10^{-9} error probability with 20 photons that we obtained in Eq. 7.13 using the exact Poisson statistics. If we wish to obtain a 10^{-9} error probability using the Gaussian approximation, 36 photons are apparently required. This is an error in determining the receiver sensitivity of ~2.5 dB.

This example somewhat exaggerates the actual error in using a Gaussian distribution since some Gaussian distributed receiver electronics-noise is inevitable in any practical receiver. This example does, however, illustrate two important points. The first is that the Gaussian approximation only holds when many photons are received per symbol. A sensitive receiver operates with a small number of photons per symbol. The second point is that it is the *tails* of the distributions that determine the low probabilities of error.

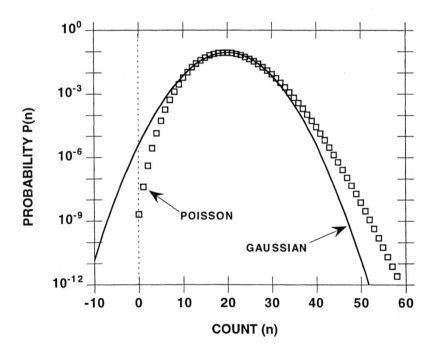

Figure 7.14 Gaussian and Poisson distributions for $\bar{n} = 20$.

Substituting a Gaussian for another distribution when analyzing an optical receiver is a valid option that can dramatically simplify a receiver's analysis [62, 63, 76]. In many cases, accuracies within a few dB of the exact solution can be achieved when careful simplifying assumptions are made. It is important, however, that the inherent limitations of this approach on the achievable accuracy be understood.

7.2.4   OOK with an APD Front-end

A simple example of an APD based receiver is illustrated in Fig. 7.15. The electrical schematic of the receiver is shown in Fig. 7.15(a), while the equivalent circuit used to determine the characteristics of the voltage at the decision circuit's input is shown in Fig. 7.15(b). The APD is characterized by a quantum efficiency η, average multiplication gain $\overline{M}$, an excess-noise factor $F(\overline{M})$, and multiplied and unmultiplied dark-currents. The remainder of the receiver is the same as that used in the *p-i-n* example, with the exception of the intermediate gain stage. In this example it has a gain of 50 instead of 100. The overall transimpedance Z_t is then $M \times 500\text{k}\Omega$. The APD receiver will also be used with the same received signal as we used for the *p-i-n* receiver, a 2.0 Gbps OOK modulated signal at wavelength of 1550 nm.

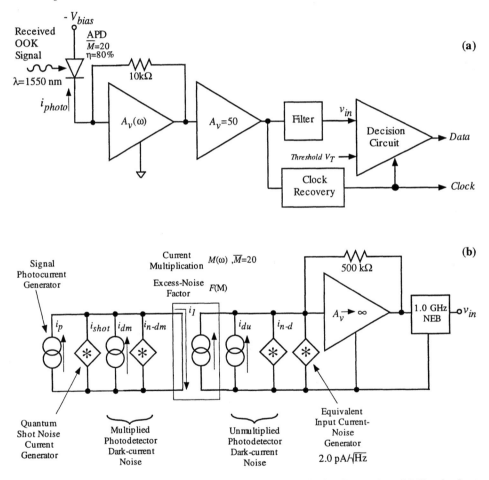

Fig. 7.15   Example of an APD receiver. (a) Electrical schematic. (b) Equivalent circuit for determining the decision circuit's input voltage v_{in}.

Since an APD involves a random avalanche multiplication between the primary and secondary photocarriers, the APD output must be described in terms of a probability distribution. For a given time interval, the probability of obtaining m secondary photocarriers after n primary photocarriers are generated has been both theoretically derived and experimentally verified to be given by [77-79]

$$P(m|n) = \sum_{i=1}^{\infty} \frac{i\,\Gamma\left(\dfrac{m}{1-k_{eff}}+1\right)}{m(m-i)!\,\Gamma\left(\dfrac{k_{eff}m}{1-k_{eff}}+1+i\right)} \times \left[\frac{1+k_{eff}\left(\overline{M}-1\right)}{\overline{M}}\right]^{\frac{i+k_{eff}m}{1-k_{eff}}}$$

$$\times \left[\frac{\left(1-k_{eff}\right)\left(\overline{M}-1\right)}{\overline{M}}\right]^{m-i} \times \left(\frac{n^{i}}{i!}\right)e^{-n}, \tag{7.34}$$

where $\overline{M}$ = average APD multiplication gain, k_{eff} = effective ionization ratio, and $\Gamma(z)$ is the gamma function.

Equation 7.34 is far too complex for anything but numerical analysis. Several approaches have been taken to determining the $P(v|zero)$ and $P(v|one)$ probability densities needed to compute the corresponding probability of error. The techniques used have included Monte Carlo simulations with importance sampling, Gram-Charlier expansion, the Gauss quadrature rule, the Chernoff bound, and the method of characteristic functions [70, 71, 80-82]. A good approximation to Eq. 7.34 has also been provided by Webb [83]:

$$P(m|n) \cong \frac{1}{\sqrt{2\pi n \overline{M}^{2} F}} \cdot \frac{1}{(1+\Psi)^{3/2}} \exp\left\{-\frac{\left(m-n\overline{M}\right)^{2}}{2n\overline{M}^{2}F(1+\Psi)}\right\}, \tag{7.35}$$

where

$$\Psi = \frac{\left(m-n\overline{M}\right)}{n\overline{M}\dfrac{F}{F-1}},$$

$$F = F(M) = k_{eff}\overline{M} + \left(1-k_{eff}\right)\left(2-\frac{1}{\overline{M}}\right).$$

Even with this approximation, the mathematics are relatively complex and do not allow easy manipulation or interpretation. Fortunately, when $nF \gg 1$ the Ψ term in Eq. 7.35 is nearly zero and it is possible to approximate Eq. 7.34 with a Gaussian probability density as

$$P(m|n) \cong \frac{1}{\sqrt{2\pi n \overline{M}^{2} F}} \exp\left\{-\frac{\left(m-n\overline{M}\right)^{2}}{2n\overline{M}^{2}F}\right\}. \tag{7.36}$$

The $nF \gg 1$ condition requires that the receiver be illuminated by a strong received signal, causing the generation of a large number of primary signal photoelectrons during each symbol-time. It can also be satisfied when there are comparatively high dark-currents, a large number of background photoelectrons are being generated, or the APD has a high excess-noise factor. The actual magnitude of the errors are strongly dependent on the characteristics of the APD used and the details of the communication system the receiver is used in. Errors generally range from less than 1 dB in APD's with high excess-noise factors to several dB in low-noise APDs used in high sensitivity, low background applications [71, 79, 84, 85].

Figure 7.16 plots the Gaussian approximation of Eq. 7.36 and the more accurate Webb approximation from Eq. 7.35 for an APD when the number of primary photoelectrons is equal to either 1 or 100. This would be the case for a received signal with a 100:1 or 20 dB extinction ratio.

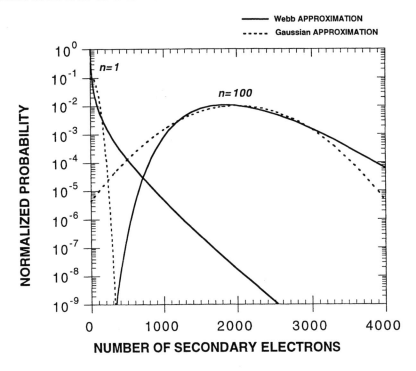

Figure 7.16 The Gaussian and Webb approximations for an APD with M=20, k_{eff}=0.25.

The Gaussian approximation is accurate only near the peaks of the distributions. The approximation becomes much poorer in the tails. The optimum threshold corresponding to the locations where the distributions intersect is also different for the Gaussian and Webb approximations. In general, the Gaussian approximation is known to underestimate the achievable receiver sensitivity, overestimate the optimum avalanche gain, and underestimate the optimum threshold [71, 84, 86].

In spite of these acknowledged inaccuracies, the Gaussian approximation is widely used because it does not require numerical techniques for evaluation, provides reasonably accurate results, and provides good insight into the receiver's operation. Since it tends to predict a less sensitive receiver than can actually be obtained using the optimum threshold, it is considered a conservative technique for estimating receiver performance.

Using the Gaussian approximation in combination with the signal and noise currents of an APD given by Eq. 5.44, we can express the probability densities for a *one* and a *zero* in the APD receiver as

$$P(v) \cong \frac{1}{\sqrt{2\pi}\sigma_{one}} \exp\left\{-\frac{\left(v - v_{one}\right)^2}{2\sigma_{one}^2}\right\} \quad \text{when a } one \text{ is received,}$$

and   (7.37)

$$P(v) \cong \frac{1}{\sqrt{2\pi}\sigma_{zero}} \exp\left\{-\frac{\left(v - v_{zero}\right)^2}{2\sigma_{zero}^2}\right\} \quad \text{when a } zero \text{ is received,}$$

where

$$\sigma_{one}^2 = Z_t^2 I_{n_{one}}^2, \quad \sigma_{zero}^2 = Z_t^2 I_{n_{zero}}^2, \quad R = \frac{\eta q}{h v},$$

$$I_{n_{one}}^2 = I_{elec}^2 + 2q\left(i_{du} + i_{dm}M^2 F(M)\right)BI_2 + 2qM^2 F(M)RP_{rcvd_{one}}BI_1,$$

$$I_{n_{zero}}^2 = I_{elec}^2 + 2q\left(i_{du} + i_{dm}M^2 F(M)\right)BI_2 + 2qM^2 F(M)RP_{rcvd_{zero}}BI_1,$$

$$v_{one} = Z_t\left(i_{du} + M\left(i_{dm} + RP_{rcvd_{one}}\right)\right),$$

$$v_{zero} = Z_t\left(i_{du} + M\left(i_{dm} + RP_{rcvd_{zero}}\right)\right),$$

and BI_1 and BI_2 are the noise-equivalent bandwidths from Eq. 5.81.

We will assume that the introduction of the transfer-function $M(\omega)$ associated with the APD's avalanche multiplication does not significantly alter the 1.0 GHz receiver noise-bandwidth, and we will also use the same total electronics-noise I_{elec} of 63nA that we used in the *p-i-n* example. Representative values for a long-wavelength APD would be $k_{eff} = 0.25$, $i_{dm} = i_{du} = 1.0$nA. Substituting these values into Eq. 7.37 and operating the APD with a gain of 20 to produce an overall transimpedance gain Z_t of $10^7 \, \Omega$, allows us to obtain the probability densities shown in Fig. 7.17.

The height of the distributions in Fig. 7.17 have again been normalized for simplicity of presentation and we are again assuming a high extinction ratio so that the power received during a *zero* is assumed to be negligible. Fig. 7.17(a) illustrates the probability density functions $P(v|one)$ and $P(v|zero)$ for the case when the received signal power during a *one* is 0.05μW. Fig. 7.17(b) shows the probability densities from Fig. 7.17(a) plotted on a logarithmic scale. Fig 7.17(c) illustrates the change in the distributions when the received signal level during a *one* is increased to 0.10μW. Note that in both Figs. 17(b) and 7.17(c), the $P(v|one)$ curve is not centered around zero. It is slightly offset because of the multiplied portion of the dark-current i_{dm}. As with the *p-i-n* receiver the optimum threshold varies with received signal power. In the APD receiver, the threshold will also vary with multiplication gain M and excess-noise factor $F(M)$.

As with the *p-i-n* receiver, the area under $P(v|zero)$ above the threshold voltage V_t is the probability that we will mistake a *one* for a *zero*, while the area under the $P(v|one)$ curve below the threshold voltage is the probability that we will mistake a *zero* for a *one*. Using a simple threshold based decision circuit, we would decide a *one* for voltages greater than about 0.150 volts in Fig. 7.17(b) and about 0.250 volts in Fig. 7.17(c).

There is an unexpected feature associated with the plots in Fig. 7.17(b). Note that the $P(v|one)$ curve intersects the $P(v|zero)$ curve in two locations. This indicates that the demodulator would actually have a better chance of making a correct decision for voltages that are less negative than about -0.25 volts by picking a *one* instead of a *zero*! Strictly speaking, the maximum-likelihood demodulator for this case is then not a single threshold. It would instead divide the decision space into three regions by introducing a second threshold test. This feature is largely a result of our use of the Gaussian approximation and can be safely ignored in a practical receiver.

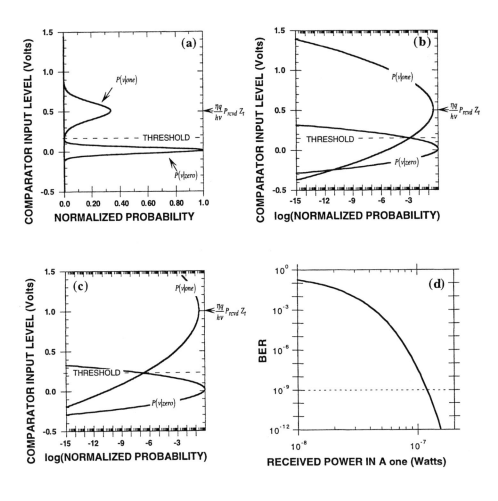

Figure 7.17 Distributions for the APD receiver of Fig. 7.15 using the Gaussian approximation. (a) Densities for 0.05 μW received power. (b) Same densities plotted on a log scale. (c) Log densities for 0.1 μW received power. (d) Bit-error-rate plot.

Following the same procedure as we used for the *p-i-n* receiver, we can express the probability of error for the APD receiver as

$$
P_e \cong \frac{1}{2} \int_{-\infty}^{V_t} \frac{1}{\sqrt{2\pi}\sigma_{one}} \exp\left\{ -\frac{(v - v_{one})^2}{2\sigma_{one}^2} \right\} dv
$$
$$
+ \frac{1}{2} \int_{V_t}^{\infty} \frac{1}{\sqrt{2\pi}\sigma_{zero}} \exp\left\{ -\frac{(v - v_{zero})^2}{2\sigma_{zero}^2} \right\} dv. \tag{7.38}
$$

Using the Q-function to evaluate the integrals, Eq. 7.38 can be written as

$$
P_e \cong \frac{1}{2} Q\left(\frac{-V_t + v_{one}}{\sigma_{one}} \right) + \frac{1}{2} Q\left(\frac{V_t - v_{zero}}{\sigma_{zero}} \right). \tag{7.39}
$$

Since the variances in the two distributions are not equal, the optimum threshold does not lie halfway between the distributions as it did with the *p-i-n* receiver. The minimum probability of error is obtained using the same procedure we used for the *p-i-n* receiver. Applying Eq. 7.21 to Eq. 7.39, we obtain a solution for the threshold that is quadratic in V_t. The two solutions for the quadratic, corresponding to the two locations at which the distributions intersect, are given by

$$
V_t = \frac{\left(v_{zero}\sigma_{one}^2 - v_{one}\sigma_{zero}^2 \right) \pm \sqrt{(v_{one} - v_{zero})^2 + 2\left(\sigma_{one}^2 - \sigma_{zero}^2\right)\ln\left(\frac{\sigma_{one}}{\sigma_{zero}}\right)}}{\sigma_{one}^2 - \sigma_{zero}^2}. \tag{7.40}
$$

Although Eq. 7.40 is exact, it is complex to implement and a simpler criteria is often used. Instead of solving for the exact minimum probability of error receiver, we solve for an equal (or uniform) probability of error receiver. That is, we set the probability of mistaking a *zero* for a *one* to be equal to the probability of mistaking a *one* for a *zero*. This is also known as setting the probability of a false-alarm equal to the probability of a miss. This requires that the area under the tails of each distribution be equal. Even though the equal probability of error criterion is suboptimum in that it does not implement the exact minimum probability of error (i.e. a maximum-likelihood) receiver, the degradation is generally insignificant and a uniform probability of error criteria is adequate for virtually all practical optical communication receivers [6, 31].

An equal probability of error receiver is obtained by setting the arguments of the Q-function in Eq. 7.39 equal as

$$
\frac{-V_t + v_{one}}{\sigma_{one}} = \frac{V_t - v_{zero}}{\sigma_{zero}}. \tag{7.41}
$$

The expression for the optimum threshold is then

$$
V_{t_{opt}} \cong \frac{v_{zero}\sigma_{one} + v_{one}\sigma_{zero}}{\sigma_{one} + \sigma_{zero}}. \tag{7.42}
$$

As in the *p-i-n* receiver example, the optimum threshold is dependent on the amount of received signal and the noise in the receiver. This implies that the receiver's threshold must adapt to changing received power levels if a wide dynamic range is to be obtained. Obtaining the optimum threshold requires measuring both the signal levels and the noise variances for a received *zero* and *one*. This can often be a complex measurement to implement precisely and simpler algorithms are often desirable.

In addition to the optimum and near optimum thresholds given by Eqs. 7.40 and 7.42 respectively, we could also simply set the threshold to lie a fixed distance between the *zero* and *one* distributions, as we did for the *p-i-n* receiver. This could in fact be made optimum for one specific level of received signal power. We would have to accept the possibility of substantial performance degradation from optimum for other levels of received signal in exchange for a very simple, easily implemented algorithm for determining the threshold. In commercial receivers, the threshold setting is frequently fixed and the received powers must be confined to a relative narrow range.

Substituting the near-optimum threshold from Eq. 7.42 into Eq. 7.39 results in

$$P_e \cong Q(SNR_e), \tag{7.43}$$

where the electrical signal-to-noise ratio (the Q-factor) is given by

$$SNR_e = Q = \frac{v_{one} - v_{zero}}{\sigma_{zero} + \sigma_{one}}. \tag{7.44}$$

For the case of a high extinction ratio, $P_{rcvd_{zero}} \cong 0.0$. For reasonable levels of received signal power and multiplication gain the multiplied signal shot-noise dominates during a *one* and the multiplied dark-current-noise dominates the unmultiplied dark-current-noise. Eq. 7.44 can be then written as

$$Q \cong \frac{MR\left(P_{rcvd_{one}}\right)}{\sqrt{2qM^2F(M)RBI_1P_{rcvd_{one}}} + \sqrt{i_{elec}^2 + 2qM^2i_{dm}F(M)BI_2}}. \tag{7.45}$$

Figure 7.17(d) plots the probability of error obtained from Eqs. 7.43-7.45 as a function of the received signal power during a *one* for our example receiver.

Solving Eq. 7.45 for the average amount of received signal power needed to obtain a specified Q-factor assuming the receiver implements the near-optimum threshold from Eq. 7.42 yields

$$P_{rcvd_{one}} \cong 2Q\frac{h\upsilon}{\eta q}\left[QBI_1qF(M) + \sqrt{2qi_{dm}F(M)BI_2 + \frac{I_{elec}^2}{M^2}}\right]. \tag{7.46}$$

To obtain the 10^{-9} sensitivity of the APD receiver we use $Q = 6$. For our example APD, the sensitivity is 0.12 µW for a *one*, or an average power of -42 dBm. This is some 8 dB better than the -34.2 dBm of average received signal power needed with the *p-i-n* receiver. This improvement in sensitivity is a direct result of the internal avalanche multiplication mechanism. Note that the effect of the receiver electronics input current-noise in Eq. 7.46 is reduced by the multiplication gain of the APD.

When the multiplied dark-current noise is small enough that it can be neglected, Eq. 7.46 can be simplified as

$$P_{rcvd_{one}} \cong 2Q\frac{h\upsilon}{\eta q}\left[QBI_1qF(M)+\frac{I_{elec}}{M}\right]. \tag{7.47}$$

The effect of a finite extinction ratio defined as $P_{rcvd_{zero}} = \varepsilon P_{rcvd_{one}}$ can then be accounted for by modifying Eq. 7.47 as [63]

$$P_{rcvd_{one}} \cong 2Q\frac{h\upsilon}{\eta q}\left(\frac{1+\varepsilon}{1-\varepsilon}\right)\left[\frac{1+\varepsilon}{1-\varepsilon}QBI_1qF(M)+\frac{I_{elec}}{M}\right]. \tag{7.48}$$

When the gain is sufficiently large that the second term in brackets is negligible we can approximate the power loss penalty due to a small, $\varepsilon < 0.2$, finite extinction ratio as

$$P_{loss} \cong \frac{1+2\varepsilon}{1-2\varepsilon}. \tag{7.49}$$

The loss is approximately 1.8 dB for a 10 dB extinction ratio, $\varepsilon = 0.1$, and about 0.2 dB for a 20 dB extinction ratio, $\varepsilon = 0.01$. The impact of a finite extinction ratio is typically more severe with APD receivers than with p-i-n receivers. This can be seen by comparing Eq. 7.49 and Eq. 7.33. A finite extinction ratio will also cause the optimum multiplication gain to be lower than it would be for perfect extinction.

From Eq. 7.47 we see that the first term in the brackets increases with increasing multiplication gain because the excess-noise factor $F(M)$ increases with gain. The second term in the brackets accounts for the influence of the electronics current-noise and decreases with increasing gain. An optimum gain can therefore be defined. We can obtain a closed form estimate of the optimum multiplication gain by differentiating Eq. 7.47 with respect to the multiplication gain and solving for the point where the derivative is zero:

$$M_{opt} \cong \frac{1}{\sqrt{k_{eff}}}\sqrt{\frac{I_{elec}}{qBI_1Q}+k_{eff}-1}. \tag{7.50}$$

In practice, the Gaussian approximation that we used in Eq. 7.43 through Eq. 7.47 can result in a few dB of error when predicting the receiver sensitivity and several dB of error when predicting the value for the optimum gain. The accuracy of the Gaussian approximation can be significantly improved upon by taking into account the non-Gaussian nature of the avalanche multiplication gain. Following the approach taken by Conradi [86, 87] for convolving the avalanche multiplication statistics with the receiver electronics-noise, we can obtain more accurate estimates for the receiver sensitivity and optimum multiplication gain, assuming a high extinction ratio, as

$$P_{rcvd_{avg}} \cong 2Q\frac{hv}{\eta q}\left[QBI_1qF(M)\right]+\sqrt{\left[\left[F(M)-1\right]QBI_1q\right]^2+\frac{I_{elec}^2}{M^2}},$$

$$(7.51)$$

$$M_{opt} \cong \frac{M_G}{\left(3+\dfrac{2-4k_{eff}}{k_{eff}M_G}\right)^{1/4}},$$

where

$$M_G = \sqrt{\frac{I_{elec}}{kqQI_1B}}.$$

Figure 7.18 plots the receiver sensitivity from Eq. 7.51 as a function of avalanche multiplication for our example APD receiver. The sensitivity for two values of k_{eff} are shown. The lower value is $k_{eff}=0.01$. This is representative of a short-wavelength APD. The second value is $k_{eff}=0.25$, which is more representative of a long-wavelength device.

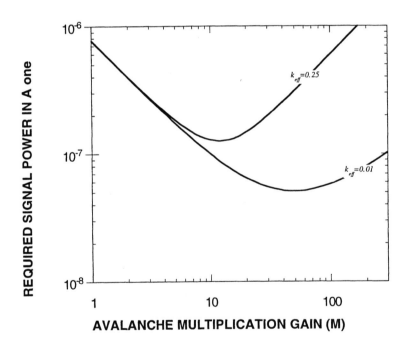

Figure 7.18 Receiver sensitivity for a 10^{-9} BER versus avalanche gain for k_{eff}=0.01 and 0.25.

The curve for the APD with the higher k_{eff} has an optimum gain near 10 and achieves a sensitivity of just over 120 nW. The lower k_{eff} device could operate with gains as high as 50 and would achieve a sensitivity of approximately 50 nW. This represents almost a 4 dB advantage in sensitivity.

7.3 Coherent Detection Digital Receivers

A simplified block diagram of a coherent detection receiver is shown in Fig. 7.19. A weak received signal of average power $P_{rcvd} = S^2$ at frequency v_s is combined with a strong local oscillator of average power $P_{lo} = L^2$ and frequency v_{lo} using a 50:50 optical coupler. Following the notation of Abbas $et\ al.$ [88], the inputs to the coupler are given by

$$S = \sqrt{2} \cdot \hat{S} \cdot e^{j\left(2\pi v_s t + \phi_s(t)\right)}, \quad L = \sqrt{2} \cdot \hat{L} \cdot e^{j\left(2\pi v_{lo}t + \phi_{lo}(t)\right)}. \tag{7.52}$$

Assuming that the coupler is linear, lossless, and symmetric, the fields at the output of the coupler are given by

$$\hat{E}_1 = \frac{1}{\sqrt{2}} e^{j\phi_c}\left[\hat{S} + \hat{L}e^{j\pi/2}\right] \text{ and } \hat{E}_2 = \frac{1}{\sqrt{2}} e^{j\phi_c}\left[\hat{S}e^{j\pi/2} + \hat{L}\right], \tag{7.53}$$

where ϕ_c is the phase shift in the optical coupler.

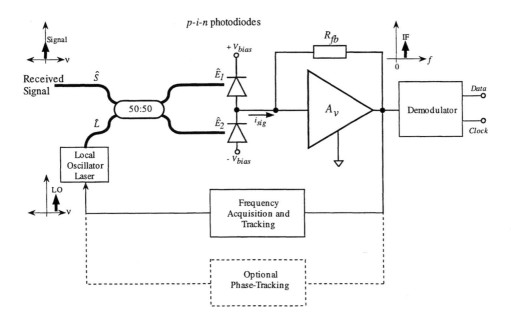

Figure 7.19 Coherent detection receiver.

The fields $\hat{E}_1$ and $\hat{E}_2$ illuminate a pair of p-i-n photodetectors configured in a balanced mixer configuration [88-92]. The photodetectors are assumed to be identical except for quantum efficiency. Balanced mixers have been implemented using discrete devices [92], using integrated photodetectors and waveguide couplers [93], and using complete monolithic photonic integrated circuits that include the LO laser as well as the coupler and photodetectors [94].

The photodetectors respond to the mean incident optical power and, assuming that the two fields are of identical polarization, generate photocurrents given by

$$
i_1(t) = R_1 \left\{ \frac{1}{2}\left(S^2 + L^2\right) + SL\cos\left[2\pi(v_s - v_{lo})t + \left(\phi_s(t) - \phi_{lo}(t)\right) - \frac{\pi}{2}\right]\right\} + n_1(t),
$$

$$(7.54)$$

$$
i_2(t) = R_2 \left\{ \frac{1}{2}\left(S^2 + L^2\right) + SL\cos\left[2\pi(v_s - v_{lo})t + \left(\phi_s(t) - \phi_{lo}(t)\right) + \frac{\pi}{2}\right]\right\} + n_2(t),
$$

where

$R_1 = \dfrac{\eta_1 q}{h v}$, $R_2 = \dfrac{\eta_2 q}{h v}$, and $n_1(t)$, $n_2(t)$ are the noise processes in each photodetector.

Because of the series balanced configuration of the photodetectors, current drawn from the middle node will contain the difference of the photocurrents as

$$
i_{sig}(t) = I_{DC_1} - I_{DC_2} + \frac{q}{h v}\left(\eta_1 + \eta_2\right)SL\cos\left(\omega_{if}t + \phi_{if}(t)\right) + n_1(t) - n_2(t) + n_e(t), \quad (7.55)
$$

where $n_e(t)$ is the noise from the receiver electronics and I_{DC_1} and I_{DC_2} are the DC photocurrents corresponding to the average optical power illuminating the photodetectors. The AC photocurrent appears at the difference frequency $\omega_{if} = 2\pi(v_s - v_{lo})$. The term $\phi_{if}(t)$ accounts for the phase-noise present in the difference signal.

The difference frequency is termed the intermediate frequency because the information bearing signal is now at a frequency that is intermediate between its initial optical frequency and its ultimate baseband form. The amplitude of the IF term is proportional to the product of the signal and the LO and so coherent systems are often described as exhibiting a mixing "gain." Excellent selectivity can also be obtained in a coherent receiver because the receiver passband is defined by comparatively narrow-band electrical filters rather than wide-bandwidth optical filters.

The IF phase-noise $\phi_{if}(t)$ is readily observed as a non-zero IF linewidth that can range from a few kHz in external cavity lasers to many tens of MHz in some DFB lasers [24, 25, 95, 96]. The spectrum of the IF phase-noise has two distinct regions. At low frequencies there is a 1/f region that corresponds to the slow frequency wander of the IF carrier frequency. At higher frequencies there is a broad white spectrum corresponding to fundamental quantum fluctuations in the laser [25, 95-97]. A coherent receiver makes use of a frequency acquisition system to accurately set the LO laser's frequency from a cold-start [98] and either a frequency tracking or phase tracking loop to stabilize the IF signal [99].

Heterodyne receivers utilize frequency tracking, or automatic frequency control (AFC), to remove the 1/f disturbances [99] and can be implemented with virtually any reasonable IF linewidth [100-103]. Homodyne receivers require the use of phase tracking in the form of a phase-locked-loop (PLL), which requires relatively narrow linewidths [102, 104, 105]. Phase-diversity techniques can also be used to advantage in constructing receivers that minimize the total bandwidth required, tolerate relatively large IF linewidths, and achieve acceptable sensitivities [106, 107].

Equation 7.55 reveals two advantages of the balanced-mixer configuration. The first advantage comes about because the balanced mixer generates the difference between the noise processes in the two photodetectors. This effectively cancels any correlated noise terms that appear at the two photodetectors. Intensity fluctuations due to RIN in the LO laser are correlated in the two photodetectors and are canceled [88]. The signal shot-noises are uncorrelated and do not cancel. Achieving an accurate and stable balance between the photocurrents and equal paths from the optical coupler to each photodetector is critical in obtaining large amounts of wideband excess-noise cancellation [88, 92]. Moderate bandwidth versions of the balanced mixer can be made to be autobalancing by using electrical feedback in a current-mirroring scheme [108]. Planar lightwave circuits are attractive for use in wideband optical balanced mixer receivers because the path lengths and splitting ratios are largely determined by the photolithography.

The total amount of signal shot-noise in a balanced mixer is the sum of the individual signal shot-noises generated in the two photodetectors. The signal-to-noise-ratio for the balanced mixer heterodyne receiver can then be expressed as

$$SNR_{het} = \frac{\frac{1}{2}\left(\frac{q}{h\nu}\right)^2 S^2 L^2 (\eta_1 + \eta_2)^2}{\left[2q\left(I_{DC_1} + I_{DC_2}\right) + \left(\sqrt{n_{corr_1}} - \sqrt{n_{corr_2}}\right)^2 + n_{rcvr}\right]B}, \quad (7.56)$$

where n_{corr_1} and n_{corr_2} are the spectral densities of any correlated noise processes in the two photodetectors, n_{rcvr} is the spectral density of the equivalent input noise from the receiver electronics noise, and B is the noise equivalent bandwidth of the receiver. Since the local oscillator is substantially stronger than the weak received signal, the DC photocurrents are dominated by the LO laser.

With a sufficiently large amount of DC photocurrent, the photocurrent shot-noise terms will dominate over the receiver electronics-noise and nearly shot-noise limited operation can be achieved [27, 88, 92, 109, 110]. This is often easy to accomplish. For example, if we use the same 2.0 pA per root-Hz receiver electronics current-noise density we used in the p-i-n and APD receiver analyses, we only need to generate a total of 1.25 mA of DC photocurrent for the LO shot-noise to be 10 times greater than the receiver electronics-noise. This requires a mW or so of LO power in high QE photodetectors [92].

The second advantage of the balanced mixer receiver is the efficient use of LO and signal power. Unlike a single detector receiver, which requires that a portion of both the received signal power and LO power be wasted, all of the available power can be used. If we substitute into Eq. 7.56 expressions for the DC photocurrents and we assume complete cancellation of any correlated noise terms, the SNR is

$$SNR_{het} = \frac{1}{2}\frac{(\eta_1 + \eta_2)}{h\nu B} P_{rcvd}. \quad (7.57)$$

In the limit of 100% quantum efficiency photodetectors the SNR becomes

$$SNR_{het} = \frac{P_{rcvd}}{h\nu B}, \quad (7.58)$$

which is the classic result for quantum limited heterodyne detection [90].

In a homodyne receiver, the IF is zero and the LO and signal are aligned in phase, causing the now baseband signal to have twice the power of a heterodyne system [3, 24]. The limiting case of the SNR for a balanced mixer homodyne receiver is then 3 dB greater than that of a heterodyne receiver, or [104, 111]

$$SNR_{hom} = \frac{2 P_{rcvd}}{h v B}.$$ (7.59)

Thus there is a factor of two difference in SNR between a quantum limited optical homodyne receiver and a quantum limited optical heterodyne receiver. This difference between heterodyne and homodyne does not occur in thermally limited RF receivers, where the noise is introduced *before* the mixing process, not *during* the mixing process as occurs in optical receivers [90, 104]. We note that an extensive quantum mechanical analysis of the balanced-mixer receiver has also been performed [112] that verifies the accuracy of the semi-classical formulation for virtually all practical receivers.

In deriving the SNR for the heterodyne receiver we assumed that the polarizations of the received signal and local oscillator were identical. This is often not the case [26, 106]. For example, signal polarization in virtually all currently installed fiber-optic systems is not controlled or predictable. The state of polarization is also generally time-varying, causing significant fading of the received signal.

We have also assumed that the spatial modes of the signal and local oscillator were well-matched. This is a good assumption in a single-mode fiber system but may not be true in a bulk-optics system. The general expression for the mode-matching efficiency in a coherent receiver is given by [113]

$$MME = \frac{\left| \iint A_s(x) \exp[j\phi_s(x)] dx \cdot A_{lo}(x) \exp[j\phi_{lo}(x)] dx \right|^2}{\iint A_s^2(x) dx \cdot \iint A_{lo}^2(x) dx},$$ (7.60)

where $A_s(x)$ and $\phi_s(x)$ are the amplitude and phase of the received signal field. $A_{lo}(x)$ and $\phi_{lo}(x)$ are the amplitude and phase of the local oscillator field and the integrals are performed over the photodetector's area. The use of high-quality optics and care in obtaining similar fields for both the signal and LO have resulted in demonstrated mode-matching efficiencies in excess of 75% (about a 1.25 dB loss) between two independent lasers in a bulk optic system [114]. Alternatively, photorefractives have been used as beamsplitters that automatically compensate for variations between the fields [115].

Several techniques have been proposed to deal with polarization control for coherent systems. One involves the use of PM fiber, which while demonstrated to be effective [116] is considered economically impractical except for specialized applications because it requires the installation of PM fiber. Other approaches utilize polarization switching [117] or polarization scrambling [118]. Naturally occurring polarization fluctuations usually are limited to frequencies below a few hundred Hz to a few kHz [26]. If we purposely induce a high-frequency polarization fluctuation so that we force the polarization to change during every bit-time, we will guarantee that some energy is received during each bit and polarization fading is avoided at the expense of a sensitivity penalty. Polarization tracking has also been demonstrated [26]. Its primary drawbacks are the need for an additional control loop and the use of either possibly fatigue inducing mechanical polarization controllers or lossy electro-optic devices.

A particularly attractive approach that allows a coherent receiver to operate with received signals of arbitrary polarization is the polarization diversity receiver [106, 119].

Several forms exist, one of which is illustrated in Fig. 7.20. The local oscillator laser's polarization is circularized and split into transverse electric (TE) and transverse magnetic (TM) components by a polarizing splitter. A similar polarizing split is performed on the received signal. The corresponding TE and TM polarizations are each photodetected by a balanced mixer receiver front-end. The outputs of the TE front-end and TM front-end are then electrically processed and combined to form the IF signal. An alternative implementation demodulates the TE and TM branches separately before combining.

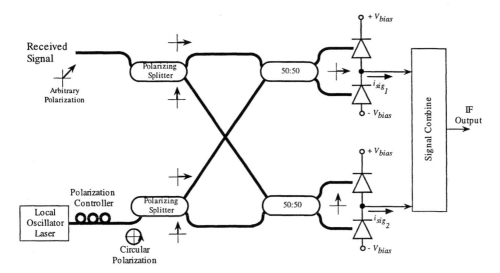

Figure 7.20 Polarization diversity receiver.

The principal drawbacks to the polarization diversity receiver are the need for higher LO laser powers if shot-noise limited operation is to be achieved in each photodetector, the additional complexity of the second balanced mixer front-end, the associated electrical combining, and a modest performance penalty that occurs when the received signal is split among two separate receivers, detected, and then recombined. The economic impact of this additional complexity can be reduced through monolithic integration. Complete balanced polarization diversity front-end modules have been demonstrated using both a combination of photonic integrated circuits [120] as well as a monolithic design [121]. It is also possible to realize the benefits of a balanced-receiver and an image reject mixer with a single photodetector by using an optical interferometer and highly birefringent optical fiber [122].

7.3.1 Modulation Formats and Demodulators for Coherent Receivers

Since the coherent detection process preserves information encoded into the frequency, amplitude, and phase of the received signal, coherent systems can utilize the amplitude-, frequency-, and phase-shift keyed modulation formats that were illustrated in Fig. 7.3. Each format has certain advantages and disadvantages from an overall system viewpoint. Examples of several popular demodulators are illustrated in Fig. 7.21. The design of these demodulators is very similar to that of a digital microwave receiver's demodulator. The primary difference is the phase-noise that is present in the IF signal due to the finite linewidths of the transmitter and LO lasers.

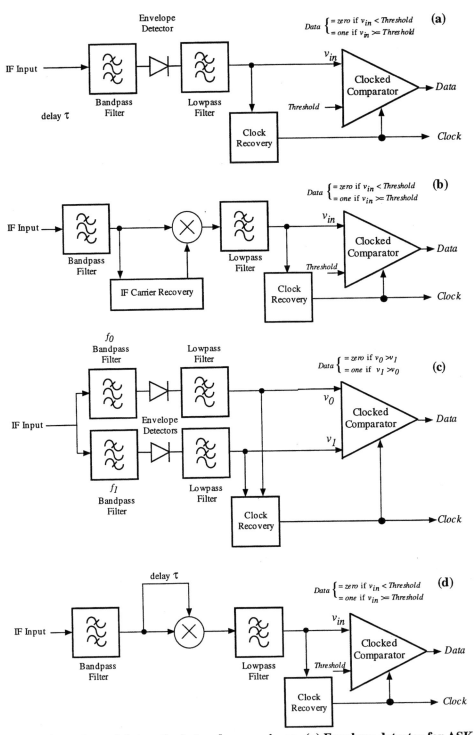

Figure 7.21 Demodulators for heterodyne receivers. (a) Envelope detector for ASK. (b) Synchronous detector for ASK. (c) Bandpass filter pair for binary FSK. (d) Delay line for either FSK, continuous phase FSK, or DPSK.

Amplitude Shift Keying

ASK is an extremely simple modulation format to generate. ASK is essentially the same as the on-off-keying format used with direct-detection receivers except that instead of appearing directly at baseband, the modulated signal appears at an IF as

$$v_{sig}(t) = Z_t m(t) 2 R \sqrt{P_{rcvd} P_{lo}} \cos\left(\omega_{if} t + \phi_{if}(t)\right) + noise, \qquad (7.61)$$

where $m(t)$ is the digital information we wish to recover, Z_t is the overall transimpedance of the receiver stages that precede the demodulator, and R is the responsivity of the balanced-mixer front-end. ϕ_{if} accounts for the phase-noise present in the signal. The additive noise term is ideally dominated by shot-noise from the local-oscillator laser.

For an ideal ASK signal, $m(t)$ will take on the value of 0 for a received *zero* and 1 for a received *one*. As in a conventional microwave system, ASK can be demodulated either asynchronously using an envelope detector, as shown in Fig. 7.21(a), or synchronously demodulated by using the scheme of Fig. 7.21(b) [35].

This illustrates an important point. In a coherent optical receiver it is necessary to further specify the demodulation as either coherent (synchronous) if phase knowledge is used or incoherent (asynchronous) if no phase knowledge is used. Thus an ASK receiver using the envelope detector demodulator of Fig. 7.21(a) would employ coherent detection with incoherent demodulation, while an ASK receiver using the synchronous demodulator of Fig. 7.21(b) would use coherent detection with coherent demodulation.

Since we are assuming a strong local oscillator so that LO shot-noise dominates the receiver noise performance, a great number of photons are received during each symbol-time, and we can safely assume that the voltage at the decision circuit's input will exhibit Gaussian statistics, as illustrated in Fig. 7.22(a). The dominant LO shot-noise also allows us to assume that the noise during a *zero* is the same as the noise during a *one* so that $\sigma_{one} = \sigma_{zero} = \sigma_n$.

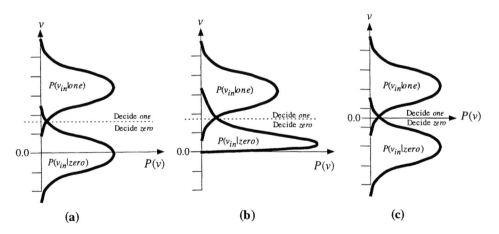

Figure 7.22 Example probability densities at the input to the decision circuit in a coherent receiver with various modulation formats and demodulators. (a) Coherent ASK. (b) Incoherent ASK. (c) Coherent FSK and PSK.

This makes the problem of determining the BER performance of an ideal zero linewidth optical heterodyne ASK receiver identical to the classic problem of the detection of a known pulse in Gaussian-noise [6]. The maximum likelihood demodulator is a matched filter [31]. Since in the demodulator in Fig. 7.21(b) the overall filter function is the cascade of the predetection bandpass and postdetection low-pass, either the bandpass-filter or the low-pass filter (or a combination) could be used to approximate the ideal sinc function matched-filter.

The voltage obtained during a *one* is proportional to the amplitude of the IF signal, while the noise is proportional to the DC photocurrent and receiver noise-equivalent-bandwidth as

$$v_{one} = Z_t 2R\sqrt{P_{rcvd_{one}}P_{lo}}, \quad \sigma_n = Z_t\sqrt{2qRP_{lo}B}. \tag{7.62}$$

The Gaussian statistics allow us to use the same analysis techniques we used with the *p-i-n* receiver. Starting with Eq. 7.19, we write the probability of error as the sum of two probabilities that correspond to integrals of Gaussian distributions. The first integral is the probability that the signal and noise during a *one* will fall below the decision threshold and be declared a *zero* . The second integral is the probability that the noise during a *zero* will exceed the threshold and be declared a *one*. We then use the same techniques employed to analyze the *p-i-n* receiver to determine an optimum threshold. The optimum threshold voltage is equal to half of the voltage generated by the receiver during a *one*. Once we have identified the optimum threshold, we substitute the Q-function for the integrals of the Gaussian distributions. This allows the BER for an ideal heterodyne ASK system with a coherent demodulator to be expressed as

$$P_e = Q\left(\frac{v_{one}}{2\sigma_n}\right) = Q\left(\frac{Z_t 2R\sqrt{P_{rcvd_{one}}P_{lo}}}{2Z_t\sqrt{2qRP_{lo}B}}\right) = Q\left(\sqrt{\frac{\eta P_{rcvd_{one}}}{2h\nu B}}\right) = Q\left(\sqrt{\frac{n_p}{2}}\right), \tag{7.63}$$

where $n_p = \dfrac{\eta P_{rcvd_{one}}}{h\nu B}$ is the peak number of photons required per bit of information.

Equation 7.63 assumes that the receiver noise bandwidth is equal to $1/T$, where T is the symbol time. This is the noise-bandwidth of a matched filter receiver. In order to achieve a BER of 10^{-9}, the argument of the Q-function must be 6. We must therefore receive 72 photons in each bit that is a *one*. This translates into an average of 36 photons per bit, assuming *ones* are equally as likely as *zeroes*.

When an incoherent demodulator is used, the bandpass filter used may still be an approximation to a matched filter, but the filter's output is now envelope detected instead of being coherently processed. The associated statistics for the *ones* and *zeroes* are no longer Gaussian because the envelope detector allows only positive voltages at its output. The probability density at the output of an envelope detector with a digitally modulated sinusoidal signal in narrow-band additive white Gaussian-noise at its input is well-known to be described by the Rician distribution when a *one* is received and by the Rayleigh distribution when a *zero* is received [123]. Examples of the probability densities observed at the decision circuit's input in an incoherent ASK demodulator are illustrated in Fig. 7.22(b).

To determine the BER performance for incoherent ASK demodulation, we again start with Eq. 7.19. The error probabilities used in Eq. 7.19 now correspond to integrals of the Rician and Rayleigh distributions. We can write the probability of error for an ideal zero-linewidth quantum-limited heterodyne ASK system with an incoherent demodulator as

$$P_e = \frac{1}{2}\left[1 - \int_{V_t}^{\infty} \frac{v}{\sigma_n^2} I_0\left(\frac{v \cdot v_{one}}{\sigma_n^2}\right)\exp\left(-\frac{v^2 + v_{one}^2}{2\sigma_n^2}\right)dv\right]$$
$$+ \frac{1}{2}\left[\int_{V_t}^{\infty} \frac{v}{\sigma_n^2}\exp\left(-\frac{v^2}{2\sigma_n^2}\right)dv\right].$$
(7.64)

The first term is the integral of the Rician distribution for the received *ones* and gives the probability we will mistake a *one* for a *zero*. The second term corresponds to the integral of Rayleigh distribution for the received *zeroes* and gives the probability we will mistake a *zero* for a *one*. $I_0(z)$ is the zero-order modified Bessel function of the first kind.

The BER from Eq. 7.64 can be expressed in a simpler form using Marcum's Q-function as

$$P_e = \frac{1}{2}\left[1 - Q\left(\frac{v_{one}}{\sigma_n}, \frac{V_t}{\sigma_n}\right)\right] + \frac{1}{2}\exp\left(-\frac{V_t^2}{2\sigma_n^2}\right),$$
(7.65)

where Marcum's Q-function is [124]

$$Q(a,b) = \int_b^{\infty} z I_0(az)\exp\left(-\frac{z^2 + a^2}{2}\right)dz.$$
(7.66)

This function is widely used in radar-receiver analysis and gives the probability that an envelope detected sinusoid of amplitude a plus additive noise of unity RMS amplitude will exceed the value b.

When the signal-to-noise ratio is high, the Marcum Q-function can be approximated using the standard Q-function as $Q(a,b) \cong 1 - Q(a - b)$. Using this approximation and assuming that the optimum threshold is at half of the v_{one} level, we can express the BER for the heterodyne ASK receiver using incoherent demodulation as

$$P_e \cong \frac{1}{2}Q\left(\frac{v_{one}}{2\sigma_n}\right) + \frac{1}{2}\exp\left(-\frac{V_{one}^2}{8\sigma_n^2}\right).$$
(7.67)

For all practical cases the second term in Eq. 7.67 dominates the probability of error, and we can further simplify the expression to

$$P_e \cong \frac{1}{2}\exp\left(-\frac{v_{one}^2}{8\sigma_n^2}\right) = \frac{1}{2}\exp\left(\frac{\left(Z_t 2R\sqrt{P_{rcvd_{one}}P_{lo}}\right)^2}{8\left(Z_t\sqrt{2qRP_{lo}B}\right)^2}\right) = \frac{1}{2}\exp\left(\frac{n_p}{4}\right).$$
(7.68)

In order to achieve a BER of 10^{-9} using an incoherent ASK demodulator, we must receive 80 photons in each bit that is a *one*. This translates into an average 40 photons per bit, assuming *ones* are equally as likely as *zeroes*. This is an average of 4 photons per bit more than required for coherent ASK demodulation. This is such a small degradation that incoherent demodulation is typically used because of its simplicity. Note that several approximations were made in obtaining Eq. 7.68. A more thorough numerical analysis of the Rayleigh-Rician distributions results in a value of 76 photons per received *one* [125].

Several examples of ASK system have been reported [29, 100, 126, 127], and analytical models that account for the characteristics of tuned receivers [128] and performance degradations due to finite linewidths, receiver-noise, and imperfect filter bandwidths have also been developed [129-133].

Frequency-Shift-Keying

In FSK the transmitted signal's envelope remains constant. This can be attractive in systems where amplitude changes in the signals can induce crosstalk effects. The form for the IF signal in an FSK system is

$$v_{sig}(t) = Z_t 2R\sqrt{P_{rcvd}P_{lo}} \cos\left(\left(\omega_{if} + m(t)\omega_d\right)t + \phi_{if}(t)\right) + noise, \qquad (7.69)$$

where ω_d is the deviation (or separation) between the IF frequency tones.

In the general *m*-ary FSK case, the digital information $m(t)$ takes on values from 0 to $m-1$, where m is the number of frequencies transmitted. Binary signaling with $m = 2$ is the most popular format [102]. The attraction to using $m > 2$ is a reduction in the number of photons required to transmit a bit [21, 134]. In an ideal zero-linewidth system, each FSK tone corresponds to a sinc function centered at the tone frequency, with nulls located at integer multiples of the symbol-rate.

For the case of binary FSK, it is common to define a modulation index similar to that used in analog-FM systems as [135]

$$\beta = \frac{\omega_d}{2\pi R} = \frac{tone\ spacing}{symbol\ rate}. \qquad (7.70)$$

For $\beta \approx 1$ or 2, the modulation is termed narrow-deviation FSK while for $\beta \gg 1$ it is termed wide-deviation FSK. If β is an integer, the modulation is termed orthogonal FSK and one tone is located in the nulls of the spectrum of the other tone.

A simple demodulator for FSK is illustrated in Fig. 7.21(c). A pair of bandpass filters is used. Ideally these are matched filters for the received pulse-shape. One is centered at the IF tone corresponding to a *zero*, while the other is centered at the IF tone corresponding to a *one*. The output from the bandpass filters are then envelope detected and filtered before appearing at the input to the digital decision circuitry. Note that unlike ASK and OOK receivers, the decision threshold is independent of the received signal power. Instead, it is determined by the FSK tone spacing.

A delay-line discriminator of the type shown in Fig. 7.21(d) can also be used to demodulate FSK. If the center of the discriminator is placed half-way between the FSK tones, the discriminator will convert the FSK signals into an amplitude modulation symmetrically centered around zero volts. The threshold can then be simply set at zero volts. For a receiver where the lower tone corresponds to a *zero* and is located below the discriminator center, any negative voltages are declared to be *zeroes*, while positive voltages are declared to be *ones*. The bandpass prefilter that precedes the delay-line discriminator plays a significant role in the overall performance of the demodulator. By properly selecting the prefilter, optimum matched-filter performance can be obtained [136]. An implementation using 2-pole Butterworth bandpass filters and a tapped delay-line has achieved performance within 0.5 dB of matched-filter performance [137].

Since a single laser oscillator is generally used to generate the FSK signal, the phase at the frequency transitions is actually continuous and the modulation format can be termed continuous-phase-FSK, or CPFSK [100]. The phase continuity between bits introduces memory into the system that can be exploited by a properly constructed demodulator [138]. For $\beta \approx 0.7$, it is possible to use a simple delay-line demodulator as a CPFSK demodulator [100, 127, 139].

To determine the BER performance for FSK, we first make the assumption that there is no crosstalk between the filters in the demodulator. This means that there is no energy appearing at the output of the filter located at the *zero* tone when a *one* is received, and vice-versa. This is relatively easy to guarantee in a narrow linewidth FSK system but can be difficult in systems with large IF linewidths unless a wide-deviation of several times the symbol rate is used.

For coherent demodulation the statistics at the decision circuit's input are Gaussian, as illustrated in Fig. 7.22(c). The output from the filter containing the tone is a pulse in additive Gaussian-noise, while the output from the other filter is Gaussian noise. The noise observed in the two filters will have the same variance since both noise processes originate from the LO shot-noise, which has a white frequency spectrum. The noises in the two filters are uncorrelated, however. The distribution for the voltage in the noise-only filter is a zero-mean Gaussian given by

$$P(v) = \frac{1}{\sqrt{2\pi}\sigma_n} \exp\left(-\frac{v^2}{2\sigma_n^2}\right), \tag{7.71}$$

while the distribution for the signal+noise filter is

$$P(v) = \frac{1}{\sqrt{2\pi}\sigma_n} \exp\left(-\frac{\left(v - v_{sig}\right)^2}{2\sigma_n^2}\right). \tag{7.72}$$

The probability of error is then the probability that the output voltage from the filter containing the noise is greater than the voltage from the filter with signal+noise. This is equivalent to computing the probability that the combination of the two noise terms exceeds the signal-voltage. Since the two noise terms are uncorrelated Gaussians, their variances add directly and the probability of error is then

$$P_e = \int_{v_{sig}}^{\infty} \frac{1}{\sqrt{2\pi}\sqrt{2}\sigma_n} \exp\left(-\frac{v^2}{4\sigma_n^2}\right) dv. \tag{7.73}$$

Using the Q-function instead of the integral we obtain [9, 140, 141]

$$P_e = Q\left(\frac{v_{sig}}{\sqrt{2}\sigma_n}\right) = Q\left(\frac{Z_t 2R\sqrt{P_{rcvd}P_{lo}}}{Z_t\sqrt{2qRP_{lo}B}}\right) = Q\left(\sqrt{n_p}\right) \tag{7.74}$$

as the probability of error for heterodyne FSK using a coherent demodulator. For the argument of the Q-function to be 6, we see that 36 photons per bit are required for a 10^{-9} BER. Since FSK is a constant envelope modulation format the peak and average number of photons-per-bit are identical.

For incoherent envelope detector demodulation of FSK, the statistics at the decision circuit's input are Rayleigh for the no-signal filter and Rician for the signal filter. Proceeding along lines similar to the coherent FSK case, the expression for the probability of error is then [9, 140, 141]

$$P_e \cong \frac{1}{2} \exp\left(-\frac{n_p}{2}\right). \tag{7.75}$$

Using incoherent demodulation, 40 photons-per-bit are required for a 10^{-9} BER.

Comparing FSK to ASK we see that FSK has a 3 dB advantage in sensitivity. The bandwidth required in the receiver is larger in FSK than in ASK, however, since we need to accommodate at least two tones instead of just one. This can cause difficulty when realizing very high data-rate receivers. There have been several demonstrations of FSK systems [21, 29, 100, 116, 126, 127], and extensive theoretical treatments that account for several of the commonly encountered degradation mechanisms are available [129, 130, 133, 134, 142].

Phase-Shift-Keying and Differential-Phase-Shift Keying

In PSK systems the transmitted signal's envelope is again held constant while the phase of the IF signal is varied by the information being transmitted. The form of the recovered IF signal is

$$v_{sig}(t) = Z_t 2R\sqrt{P_{rcvd}P_{lo}} \cos\left(\omega_{if}t + m(t)\pi + \phi_{if}(t)\right) + noise, \tag{7.76}$$

where $m(t)$ takes on the values of 0 or 1 depending on the information being transmitted.

In conventional PSK systems $m(t)$ corresponds directly to the data, that is $m = 0$ corresponds to a *zero* while $m = 1$ corresponds to a *one*. This type of a system requires absolute phase knowledge either through a homodyne receiver or a synchronous demodulator in a heterodyne receiver if we are to distinguish the symbols for *ones* from the symbols for *zeroes*.

In a homodyne receiver the LO is phase-locked to the received signal, the IF frequency is zero, and the recovered signal appears centered around DC as

$$v_{sig}(t) = \pm Z_t 2R\sqrt{P_{rcvd}P_{lo}} + noise. \tag{7.77}$$

The output of a homodyne receiver is a positive voltage if the received symbol is a *one* and a negative voltage if the received symbol is a *zero*. The decision circuit is then a simple comparison between the observed voltage and a threshold of zero volts. An error occurs if the noise causes the voltage to cross zero. Under the strong LO assumption, the statistics are Gaussian, and the probability that noise will cause the voltage during a received *one* to cross zero is

$$P_e = \int_0^\infty \frac{1}{\sqrt{2\pi}\sigma_n} \exp\left(-\frac{(v-v_{one})^2}{2\sigma_n^2}\right) dv. \tag{7.78}$$

Using a change of variable and the Q-function instead of the integral of the Gaussian, the probability of error is then [9, 140, 141]

$$P_e = Q\left(\frac{v_{sig}}{\sigma_n}\right) = Q\left(\frac{Z_t 4R\sqrt{P_{rcvd}P_{lo}}}{Z_t\sqrt{2qRP_{lo}\frac{B}{2}}}\right) = Q\left(\sqrt{4n_p}\right), \tag{7.79}$$

where, since the recovered signal appears directly at baseband, we have used the noise bandwidth of a baseband matched filter, or $B/2$. For the argument of the Q-function to be equal to 6, we see that 9 photons per bit are required for a 10^{-9} BER. This is 6 dB better than heterodyne FSK with coherent demodulation.

A simple and effective technique to realize an optical PLL is the use of a pilot-carrier [105, 143-145]. By purposely under-modulating the PSK transmitter by a slight amount, a small unmodulated carrier component is generated that is used as a pilot tone for the phase-locked-loop electronics. The DC and low-frequency output from the balanced mixer is proportional to the phase-error between the LO and signal and is used as the error signal for the PLL. The high-frequency output from the balanced mixer is the baseband signal that is fed to the digital decision circuit. Crosstalk between the data signal and the phase-lock signal can be significantly reduced by subtracting a scaled version of the recovered data from the phase-error signal [146, 147]. Other approaches to realize an optical PLL are to use a Costas-loop [148-151] or to transmit specific synchronization bits to aid in determining the phase-error [152]. One of the attractive features of the Costas-loop design is that it is more tolerant of linewidth than are balanced PLL designs [105, 153]. Homodyne receivers have also been demonstrated in QPSK systems [154].

In theory, even better performance than that obtained with homodyne receivers can be achieved if we invoke a local oscillator that *exactly* matches the received signal field in every way. It has the same amplitude, polarization, and frequency and is exactly in phase with the received signal during a *one* and exactly out of phase during a *zero*. The optical fields falling on the photodetectors combine destructively when the received symbol is out-of-phase with the LO and combine constructively when the received symbol is in-phase.

Since the fields are identical, the destructive interference results in no photons arriving at the photodetector, while the constructive case results in twice as many photons arriving. This is an example of the quantum-optimum receiver for binary signaling [155], whose BER performance is given by [156]

$$P_e = \frac{1}{2}\left\{1 - \sqrt{1 - \exp[-4n_p]}\right\}. \tag{7.80}$$

This is asymptotically equal to

$$P_e = \frac{1}{2} \exp\left[-4n_p\right].$$ (7.81)

In this case only 5 photons per bit are required for a 10^{-9} BER. This is 3 dB better than an optical homodyne receiver for binary PSK and is sometimes called the "super quantum limit" [102]. A near quantum-optimum receiver for the case of QPSK signaling has also been developed [157]. These quantum-optimum receivers are not readily realizable but are important theoretical constructs that provide bounds on ultimate receiver performance.

The BER for coherent demodulation in a heterodyne PSK system is also governed by Gaussian statistics. The analysis is similar to the homodyne case except that the signal appears at an IF frequency instead of at baseband and the noise-equivalent-bandwidth of the IF matched filter is now B instead of $B/2$. Under the assumption of a sufficiently high IF frequency [158], the BER is given by [9, 140, 141]

$$P_e = Q\left(\sqrt{2n_p}\right).$$ (7.82)

This requires 18 photons-per-bit for a 10^{-9} BER. Thus there is a 3 dB penalty for using heterodyne PSK with coherent demodulation instead of homodyne PSK.

A heterodyne PSK receiver may still make use of an optical PLL with an electrical offset equal to the IF center frequency [159] or it may use electrical carrier recovery techniques [160]. Heterodyne receivers with coherent demodulators can also be used in QPSK systems [23].

An alternative to PSK is to differentially encode the data being transmitted. This is termed differential PSK. In this scheme a *zero* is transmitted as a change in phase, while a *one* is sent as no change in phase. Since information is encoded into the relative phase from symbol to symbol, a simple heterodyne receiver with a one-bit delay and multiply of the form shown in Fig. 7.21(d) will demodulate the signal. The relative-phase comparison used in DPSK, as opposed to the absolute-phase comparison needed in PSK, implies that DPSK will have less exacting requirements on the laser phase-noise.

The analysis of DPSK follows along the lines of heterodyne FSK with incoherent demodulation except that the signal amplitude and noise power both double. Under the strong LO and zero-linewidths assumptions, the expression for the BER is [9, 140, 141]

$$P_e = \frac{1}{2} \exp\left[-n_p\right].$$ (7.83)

This requires 20 photons-per-bit for a 10^{-9} BER. This is 3 dB better than FSK with incoherent demodulation and is equal to the peak sensitivity of the OOK/photon-counting receiver. The DPSK signal is constant envelope, while the OOK signal is assumed to have a 50% duty-cycle. In terms of the average number of photons-per-bit, OOK has a 3 dB advantage. Several PSK and DPSK systems have been demonstrated [29, 100, 126, 127]. The sensitivity advantage of heterodyne PSK using coherent demodulation over heterodyne DPSK has been experimentally verified [161], and system models that account for common sources of degradation exist [142, 162].

7.3.2   Linewidth Effects

Linewidth is a major factor in determining the performance of a coherent optical receiver. The linewidth determines if a homodyne receiver is practical and whether coherent or incoherent demodulation can be used. The need to include a finite IF linewidth in a receiver's performance analysis is not usually encountered in microwave receiver analysis. It is not unusual for a microwave oscillator to have a linewidth of less than 1.0 Hz, and therefore most microwave receivers can safely assume that the received signal is nearly ideal from a phase-noise point-of-view. In contrast, a 1550 nm distributed-feedback-laser may have a linewidth of 10 MHz. This is a substantial percentage of the data-rate even for Gbps data rates.

The most easily observed effect of a finite IF linewidth is an increase in the observed BER for a given amount of received signal power. This is effectively a reduction in receiver sensitivity and is illustrated in Fig. 7.23. For a given modulation format, the ideal BER curve is obtained using an assumption that the IF linewidth is zero. As the linewidth is increased the slope of the BER curve dramatically changes and a BER floor may be observed [27, 134, 163].

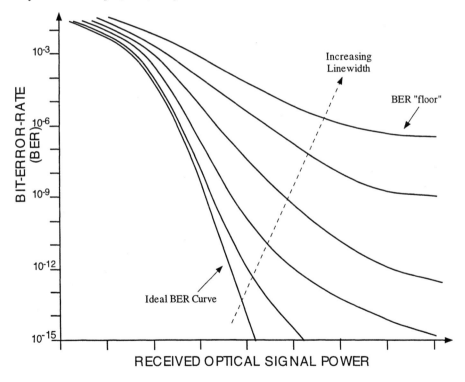

Figure 7.23 Effect of IF linewidth on receiver performance.

A non-zero IF linewidth spreads the IF signal energy over a broader bandwidth than is expected just from the modulation of the signal. This causes fluctuations in signal power and introduces cross-talk. The precise impact of linewidth on communications performance is difficult to predict in analytic form, and although the exact statistics are known for certain receiver architectures [164], numerical techniques [129] or simplified approximations [133] are usually used to estimate the performance.

The tolerance to linewidth varies depending on the modulation format used. Table 7.1 lists the ideal sensitivities and the approximate amount of IF linewidth that results in a 1 dB degradation in receiver sensitivity for conventional demodulators. The linewidths are expressed as a percentage of the data-rate, and since the demodulators can be implemented in a variety of ways, are only estimates of the required linewidth.

In general, since the more sensitive detection schemes require phase-locking, they will have the most stringent linewidth requirements. The BER in a phase-locked system is heavily dependent on the phase-noise present in the receiver and on the PLL bandwidth. Since wide-bandwidths introduce additional receiver electronics-noise into the loop, further corrupting the phase estimate, additional received signal power is required to maintain the SNR within the PLL at a sufficiently high level. In practice, even modest linewidths will cause severe degradations in a homodyne system's performance.

Table 7.1   Quantum limited receiver sensitivities and approximate linewidth requirements for conventional coherent receivers.

Format	*Homodyne*		*Heterodyne*			
			Coherent Demod.		*Incoherent Demod.*	
	n_p	Δf	n_p	Δf	n_p	Δf
PSK	9	~0.03%	18	~0.5%	-	-
DPSK	-	-	-	-	20	~1%
FSK	-	-	36	~0.5%	40	~5%
ASK	36	~0.03%	72	~0.5%	80	~10%

As an example, we see from the table that a 1 Gbps system that used PSK signaling and a homodyne receiver would require a linewidth of less than 300 kHz if we are to keep the receiver degradation due to phase noise to under 1 dB [153]. If we used a DPSK format and a heterodyne receiver, the phase need only stay coherent from symbol to symbol and we could tolerate a 1 MHz linewidth [162]. By using ASK signaling and a heterodyne receiver we could tolerate linewidths as large as 100 MHz [129].

The method used to combat the effects of a large IF linewidth depends on the type of signaling used. With PSK we can make use of a phase-diversity homodyne receiver [106]. ASK uses a wideband IF filter, envelope detector, and narrowband postdetection low-pass filter. A wideband predetection filter is used instead of a matched filter so that we capture most of the signal energy, which was spread in frequency due to the phase-noise. The postdetector filter substantially reduces the effects of the additional noise-bandwidth of the wide predetection filter. The sum of the two filter time-constants is usually equal to the symbol-time [129]. With binary FSK we use a wide tone-spacing and a pair of wideband predetection filters followed by envelope detectors and postdetection filtering. These techniques can allow heterodyne receivers to operate with linewidths on the same order as the data-rate while suffering only 3-4 dB penalties [106, 129].

7.4 Optically Preamplified Digital Receivers

We have seen how the use of avalanche multiplication in an APD or mixing gain in a heterodyne receiver helps overcome the electronics-noise in the receiver and improve the SNR at the input to the decision circuit. Optical amplification can also be used to improve receiver performance. A simplified diagram of an optically preamplified receiver is shown in Fig 7.24.

The received signal is processed by an optical amplifier and a bandpass filter. The amplifier usually consists of either a doped fiber amplifier or a semiconductor laser amplifier. In the 1550 nm region the erbium-doped fiber amplifier (EDFA) is a relatively mature technology [165]. At 1300 nm both the praseodymium doped fiber amplifier (PDFA) [166] and the fiber based Raman amplifier [167] have been demonstrated. The semiconductor laser amplifier can be fabricated at a variety of wavelengths including 1550, 1310, and 830 nm [168-171]. The SLA is particularly attractive since it lends itself to monolithic integration with other receiver components.

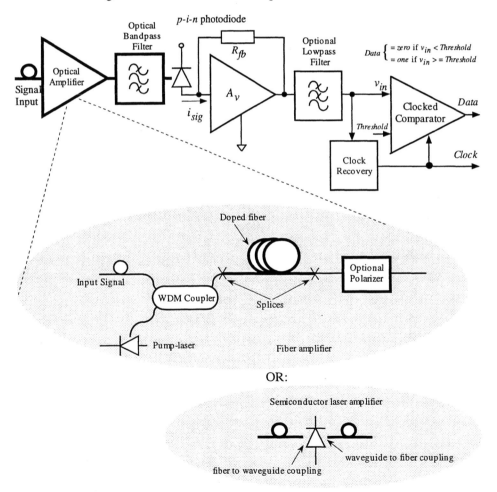

Figure 7.24 Optically preamplified receiver.

Once the optical signal is amplified it passes through an optical bandpass filter before being photodetected. The optical filter sets the optical noise-bandwidth of the receiver. Once the signal is detected it is processed using conventional electronics that may include an electrical lowpass filter before being sent to the digital decision circuit.

The effect of optical preamplification on the signal and noise levels in a receiver are illustrated in Fig 7.25. As with any receiver we anticipate the need for a certain amount of dynamic range so that we can accommodate a range of received signal levels. In some instances the optical preamplifier amplifier is configured to have a second stage that implements an optical AGC amplifier [165].

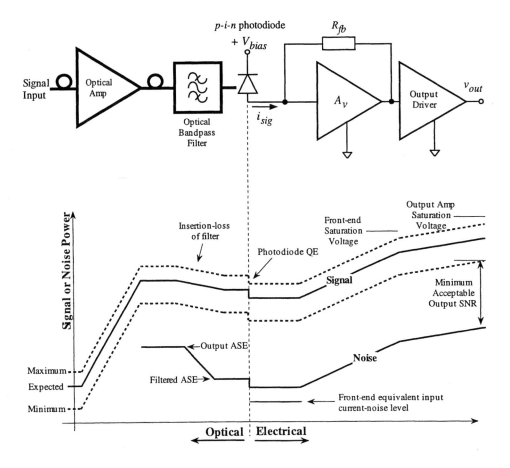

Figure 7.25 Signal and noise levels in a preamplified receiver.

A conventional optical amplifier has an associated amplified spontaneous emission (ASE) noise that is inherent in the optical amplification process [172]. An optical amplifier is usually modeled as a noise-free linear amplifier combined with an additive white Gaussian noise source that extends over the amplifier's bandwidth. In fact, the spectrum of the ASE is not flat over the entire several THz optical bandwidth of an optical amplifier and some structure is usually observed [165, 173].

The single-sided spectral density of the amplified spontaneous emission noise at the optical amplifier's output is given by [173]

$$n^2(f) = p \cdot n_{sp}(G-1)h\nu \quad \frac{\text{watts}}{\text{Hz}}, \tag{7.84}$$

where $p = 1$ if only a single polarization is present at the amplifier's output or $p = 2$ if both the TE and TM polarizations are present, n_{sp} is the spontaneous emission factor, and G is the linear gain of the amplifier. The spontaneous emission factor for an ideal amplifier is unity. In practice spontaneous emission factors between ~1.0 and 4 can be obtained, depending on the amplifier configuration [165, 168, 173, 174]. The amplifier's noise figure can be related to the spontaneous emission factor via

$$NF = 10\log_{10}\left[\frac{2n_{sp}(G-1)}{G} + \frac{1}{G}\right] \cong 10\log_{10}\left[2n_{sp}\right] \quad \text{for large gain.} \tag{7.85}$$

Careful attention to optical isolation and the use of a two-stage architecture have allowed optical amplifiers with nearly ideal 3 dB noise figures and gains in excess of 50 dB to be demonstrated [175].

For an optical amplifier with an optical noise-bandwidth B_o, the total amount of ASE noise power that would reach the photodetector is then

$$N = p \cdot n_{sp}(G-1)h\nu B_o \quad \text{watts.} \tag{7.86}$$

For a received signal of power P_{rcvd} watts at the input to the amplifier, the amount of signal power at the photodetector is simply $G \cdot P_{rcvd}$.

The photodetector in our semi-classical formulation is essentially a square-law detector for electric fields. This causes the received signal field to mix with the components of the ASE noise field that are polarized in the same direction as the signal. This is called the signal-cross-ASE-noise, and we will write this as $S \times ASE$. The various components in the ASE noise field will also mix with themselves. This is called the ASE-cross-ASE-noise; we will write this as $ASE \times ASE$.

The DC component in the photocurrent after the optically amplified signal is photodetected is given by [173]

$$I_{DC} = I_{sig} + I_{ASE},$$

$$I_{sig} = P_{rcvd}\eta_{in} \cdot G\eta_{out}\frac{q}{h\nu}, \qquad I_{ASE} = p \cdot n_{sp}q\eta_{out}(G-1)B_o, \tag{7.87}$$

where η_{in} represents any losses that occur at the preamplifier's input before the signal is optically amplified and η_{out} represents any losses that occur at the amplifier's output after optical amplification. The output loss factor can account for photodetector QE or bandpass filter insertion loss. Associated with the DC photocurrent will be a shot-noise with a single-sided current-noise spectral-density given by

$$i_{shot}^2(f) = 2qI_{DC}. \tag{7.88}$$

The single-sided current-noise spectral density that is associated with the $S \times ASE$ term is given by [173]

$$i_{S \times ASE}^2(f) = \begin{cases} 4q^2 n_{sp} \eta_{in} \eta_{out}^2 G(G-1)\dfrac{P_{rcvd}}{hv}; & f \le \dfrac{B_o}{2} \\ \\ 0; & f > \dfrac{B_o}{2} \end{cases}, \qquad (7.89)$$

where B_o is the optical noise-bandwidth.

The spectral-density for the $ASE \times ASE$ term is then [173]

$$i_{ASE \times ASE}^2(f) = \begin{cases} p \cdot 2\big(n_{sp}\eta_{out}(G-1)q\big)^2 B_o\left(1 - \dfrac{f}{B_o}\right); & f \le B_o \\ \\ 0; & f > B_o \end{cases}. \qquad (7.90)$$

Note that the $S \times ASE$ component is flat out to half the optical noise-bandwidth, while the $ASE \times ASE$ term has a triangular shape extending from a maximum at DC and reaching zero at the optical noise-bandwidth. The shapes of the spectral densities are illustrated in Fig. 7.26 for an idealized rectangular optical bandpass filter transfer function. Also shown are representative levels of the receiver electronics-noise and the photocurrent shot-noise.

From Eq. 7.89 and Eq. 7.90, we see that the height of the $S \times ASE$ spectral-density is proportional to the received signal power, while the $ASE \times ASE$ spectral-density is proportional to the optical noise-bandwidth. Figure 7.26 illustrates the case where the $S \times ASE$ component is larger than the $ASE \times ASE$ component. This is the desired operating point for a preamplified receiver since the performance will then be directly determined by the received signal power and amplifier noise figure.

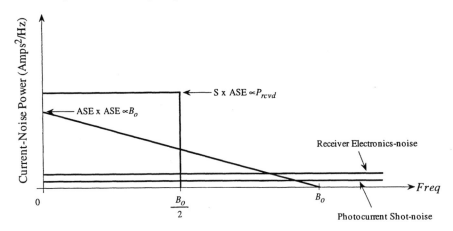

Figure 7.26 Example current-noise spectral densities in a preamplified receiver.

The total equivalent input current-noise spectral-density in the receiver is the sum of each of the individual noise processes as

$$i^2_{n_{eq}}(f) = i^2_{S \times ASE}(f) + i^2_{ASE \times ASE}(f) + i^2_{shot}(f) + i^2_{rcvr}(f). \tag{7.91}$$

The total current-noise in the receiver is obtained by integrating the current-noise spectral-density over the receiver's electrical noise-bandwidth. The individual current-noise components are then given by

$$I^2_{S \times ASE} = \begin{cases} 4q^2 n_{sp}\eta_{in}\eta^2_{out}G(G-1)\dfrac{P_{rcvd}}{hv}B_e; & B_e < \dfrac{B_o}{2} \\[3mm] 4q^2 n_{sp}\eta_{in}\eta^2_{out}G(G-1)\dfrac{P_{rcvd}}{hv}\dfrac{B_o}{2}; & B_e \geq \dfrac{B_o}{2} \end{cases},$$

$$I^2_{ASE \times ASE} = \begin{cases} p \cdot \left(n_{sp}\eta_{out}(G-1)q\right)^2 B_e(2B_o - B_e); & B_e < B_o \\[3mm] p \cdot \left(n_{sp}\eta_{out}(G-1)q\right)^2 B_o^2; & B_e \geq B_o \end{cases}, \tag{7.92}$$

$$I^2_{shot} = 2qI_{DC}B_e, \qquad I^2_{rcvr} = \int_0^{B_e} i^2_{rcvr}(f)df,$$

where B_e is the electrical noise-bandwidth of the receiver. In most cases, $B_e \ll B_o$.

We mentioned that the desirable operating condition is for the $S \times ASE$ noise to dominate the $ASE \times ASE$ noise. The point at which these noises are equal can be found using Eq. 7.92 as

$$4\eta_{in}G(G-1)\frac{P_{rcvd}}{hv} = p \cdot n_{sp}(G-1)^2\left(2B_o - B_e\right). \tag{7.93}$$

If we assume that $G \gg 1$, $n_{sp} \approx 1$, $\eta_{in} \approx 1$, $B_o \gg B_e$, and $p = 2$ we obtain

$$B_o = \frac{P_{rcvd}}{hv}. \tag{7.94}$$

The terms are equal when the optical bandwidth is equal to the photon arrival-rate (photons per second) at the input to the preamplifier. If we want the $S \times ASE$ noise to dominate the $ASE \times ASE$ noise by say about 10:1, we would like the optical bandwidth to be less than approximately 10% of the anticipated photon arrival-rate.

Using Eqs. 7.87, 7.91, and 7.92, we can write the electrical SNR for the preamplified receiver as

$$SNR = \frac{\left(P_{rcvd}\eta_{in}G\eta_{out}\dfrac{q}{hv}\right)^2}{I^2_{S \times ASE} + I^2_{ASE \times ASE} + I^2_{shot} + I^2_{rcvr}}. \tag{7.95}$$

For a high amplifier gain and an optical bandwidth that is narrow enough for the ASE × ASE noise to be negligible, the S × ASE term dominates all of the other receiver noise terms and we can write the postdetection SNR for the preamplified receiver as

$$SNR = \frac{\left(P_{rcvd} \eta_{in} G \eta_{out} \frac{q}{hv} \right)^2}{4 q^2 n_{sp} \eta_{in} \eta_{out}^2 G(G-1) \frac{P_{rcvd}}{hv} B_e}.$$

(7.96)

In the limit of large gain we obtain the S × ASE noise limited (or beat-noise limited) SNR as

$$SNR_e = \frac{\eta_{in} P_{rcvd}}{4 h v B_e}.$$

(7.97)

Note that the η_{out} losses that occurred after amplification do not appear. Preamplification is therefore a technique that overcomes both receiver electronics-noise and the effects of a photodetector with a relatively low quantum efficiency. The input losses do appear, however. This implies that whatever signal power is present at the input of the receiver should be efficiently coupled into the amplifier. Losses in an input optical isolator, wavelength division multiplexer that couples pump-energy into a fiber amplifier, or fiber to waveguide coupling losses in an semiconductor amplifier will directly effect the receiver's performance.

Eq. 7.97 should be compared to Eq. 3.35, where we obtained the shot-noise limited SNR for a direct detection system as

$$SNR_{dd} = \frac{\eta_{in} P_{rcvd}}{2 h v B_e}.$$

(7.98)

Taking the ratio of Eqs. 7.97 and 7.98 we see that

$$SNR_{pa} = \frac{1}{2} SNR_{dd}.$$

(7.99)

A preamplified direct detection receiver thus has a 3 dB lower SNR than a conventional direct detection receiver. This is a direct consequence of the fundamental fact that the minimum noise figure of a conventional optical amplifier is 3 dB.

Even though preamplification reduces the ultimately achievable SNR by at least 3 dB, in receivers that would otherwise be limited by electronics noise, the use of preamplification can yield a dramatic improvement in sensitivity. It is possible to improve an electronics-noise limited receiver's sensitivity by some 10-30 dB when using an optical preamplifier [170, 176, 177].

Preamplification can also be used in PPM systems [178] and in coherent systems. In coherent receivers, the local oscillator beats with the ASE noise, causing the ASE noise to appear at the IF. A minimum of a 3 dB penalty in SNR due to the preamplifier's noise figure is observed [179]. This implies that preamplification will be beneficial only if the coherent receiver was not operating near the shot-noise limit due to either lack of sufficient local oscillator power, large amounts of receiver electronics-noise, or low quantum-efficiency photodetectors [180]. Optical amplifiers have also been proposed for use in specialized homodyne PSK systems that can asymptotically approach the performance of the quantum-optimum receiver [181].

7.4.1 ASK or OOK with an Optically Preamplified Receiver

The performance analyses of a preamplified ASK or OOK receiver proceeds along similar lines to those we used with an APD based receiver. Both the APD and the preamplified receiver have statistics that are dependent on the received signal level. In the preamplified receiver the statistics for the received *zero* and received *one* are markedly different. In the limit of a high-gain preamplifier with an optical bandpass filter, the *zero* corresponds to detecting the $ASE \times ASE$ noise, while the *one* corresponds to detecting $S \times ASE$ This is illustrated in Fig. 7.27 for various locations in the receiver.

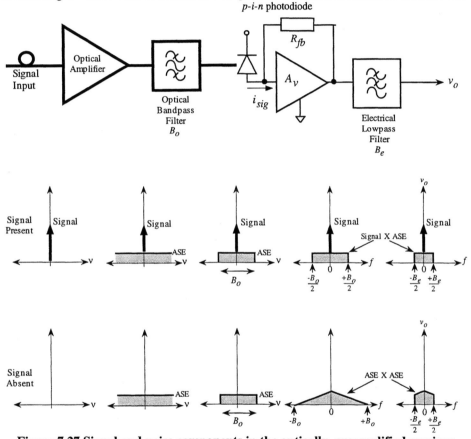

Figure 7.27 Signal and noise components in the optically preamplified receiver.

Strictly speaking, the photodetection statistics of amplified light are not described by the Poisson distribution. The detection statistics follow the non-central-negative binomial distribution [182]. In practice, because of the large number of photons being detected at the amplifier's output, the photodetector's output can be reasonably well modeled as a baseband signal in additive white-Gaussian noise and the conventional theory for incoherent demodulation of ASK can be used to describe the optically preamplified receiver's performance [35, 125, 183]. Whether the receiver is actually better described as using ASK or OOK is somewhat a matter of semantics. Since we need to consider the characteristics of the optical carrier, especially during optical filtering, we will describe this type of receiver as using ASK.

This is essentially the same theory that we used to analyze coherent detection of ASK with incoherent demodulation. The probability density is a Rician distribution when a *one* is received and a Rayleigh distribution when a *zero* is received [123]. To determine the BER performance we again start with Eq. 7.19. The error probabilities correspond to integrals of the Rician and Rayleigh distributions. Assuming a high extinction ratio transmitter, the probability of error for a preamplified ASK receiver is

$$P_e = \frac{1}{2}\left[1 - \int_{V_t}^{\infty} \frac{v}{\sigma_{one}^2} I_0\left(\frac{v \cdot v_{one}}{\sigma_{one}^2} \right) \exp\left(-\frac{v^2 + v_{one}^2}{2\sigma_{one}^2} \right) dv \right]$$
$$+ \frac{1}{2}\left[\int_{V_t}^{\infty} \frac{v}{\sigma_{zero}^2} \exp\left(-\frac{v^2}{2\sigma_{zero}^2} \right) dv \right],$$

(7.100)

where

$$\sigma_{zero}^2 = Z_t^2\left(I_{ASE\times ASE}^2 + I_{shot}^2 + I_{rcvr}^2 \right),$$

$$\sigma_{one}^2 = Z_t^2\left(I_{S\times ASE}^2 + I_{ASE\times ASE}^2 + I_{shot}^2 + I_{rcvr}^2 \right),$$

$$v_{one} = Z_t P_{rcvd_{one}} \eta_{in} G \eta_{out} \frac{q}{hv},$$

and Z_t is the overall receiver transimpedance from photodetector to decision circuit.

Following the same procedure involving the use of the Marcum Q-function and the associated simplifying assumptions that we used with coherent ASK, we obtain

$$P_e \cong \frac{1}{2}\exp\left(\frac{n_p}{4} \right).$$

(7.101)

In order to achieve a BER of 10^{-9} using a preamplified OOK direct detection receiver requires approximately 80 photons in each bit that is a *one*. This translates into an average 40 photons per bit, assuming *ones* are equally as likely as *zeroes*. As with the case for coherent ASK, more thorough numerical analysis of the Rayleigh-Rician distributions associated with an optically preamplified receiver results in a value of 76 photons per received *one* [125].

Because of the large number of photons being detected every symbol, we could also have simply used Gaussian statistics for both the *zero* and the *one* cases. It has been shown that the use of Gaussian statistics in predicting the ultimately achievable sensitivity in a preamplified receiver is accurate to within about 0.5 dB [182]. Of course, 0.5 dB may translate into an order of magnitude error in predicting the actual BER. This is not necessarily a problem in practice since the Gaussian approximation tends to underestimate performance. This is similar to what we found in the APD receiver analysis. Gaussian statistics are known to be slightly conservative and not particularly accurate for determining the optimum decision circuit threshold value. The Gaussian approximation remains important, however, because of the insight it provides into a receiver's operation.

Using the Gaussian approximation, we can write the probability of error as

$$P_e \cong Q(SNR_e),$$

(7.102)

where the electrical SNR (Q-factor) is given by Eq. 7.44 and we are using the approximation to the optimum threshold given by Eq. 7.42. For a 10^{-9} BER, we want the average SNR to be 6. For a 10^{-12} BER, we want the average SNR to be 7.

Assuming a perfect extinction ratio and an ideal high gain amplifier where the $S \times ASE$ noise dominates during a *one* and the $ASE \times ASE$ dominates during a *zero*, we can write the average SNR needed for a 10^{-9} BER as

$$\overline{SNR_e} = Q = \frac{v_{one} - v_{zero}}{\sigma_{zero} + \sigma_{one}} = \frac{\eta P_{rcvd_{one}}}{\sqrt{4 n_{sp} \eta_{in} P_{rcvd_{one}} h v B_e} + h v \sqrt{p n_{sp} B_e (2 B_o - B_e)}}. \quad (7.103)$$

Solving Eq. 7.103 to obtain a simple expression for the required amount of received power during a *one* involves a substitution of variables to obtain a quadratic, the solution of the quadratic, and the use of the approximation $\sqrt{1 + x} \cong 1 + x/2$ for small x.

In the ideal case, we would utilize an optical matched filter with a noise bandwidth equal to the inverse of the symbol-time [184]. In this case the receiver noise performance is determined by the optical filter noise-bandwidth, and we would generally use an electrical filter where $B_e > B_o/2$ so that we minimize any intersymbol interference that the electrical filter may introduce. In effect, the electrical filter is not needed in this case except to limit excessive receiver electronics-noise or other out-of-band noise sources.

Solving Eq. 7.103 for the power needed to obtain a specified Q-factor yields

$$P_{rcvd_{one}} \cong 2 Q^2 \frac{h v}{\eta_{in}} B_o n_{sp} \left(1 + \frac{1}{Q} \sqrt{\frac{p}{n_{sp}}} \right). \quad (7.104)$$

Since we are using an optical matched filter, $B_o = 1/T$, and we can obtain an expression for the number of photons required in a *one* as

$$n_p \cong 2 Q^2 \frac{n_{sp}}{\eta_{in}} \left(1 + \frac{1}{Q} \sqrt{\frac{p}{n_{sp}}} \right). \quad (7.105)$$

In the limit of an ideal amplifier $n_{sp} \approx 1$, $\eta_{in} \approx 1$. For a 10^{-9} BER, Q=6 and we obtain the sensitivity for an unpolarized amplifier with $p = 2$ under the Gaussian approximation as 89 photons per *one*, or an average of 44.5 photons per bit. The limiting sensitivity for a polarized amplifier with $p = 1$ is 84 photons per *one*, or an average of 42 photons per bit. Our Gaussian approximation has therefore overestimated the more exact sensitivity of 38 photons per bit by about 0.4 dB. We also see that the degradation for allowing both polarizations of ASE noise to mix is small, only about 0.3 dB.

In practice, an optical matched filter is difficult to realize and a simple Fabry-Perot filter is usually used [185]. In order to avoid intersymbol interference, this filter is typically several times wider than the symbol-rate. The additional noise that is introduced by the excessive optical noise-bandwidth is mitigated by the use of a lowpass electrical filter. The optimum electrical filter is now a matched filter for the baseband waveform. For a symbol time T, the noise bandwidth of the baseband matched filter is $1/2T$. The required number of photons per *one* is then given by

$$n_p \cong 2 Q^2 \frac{n_{sp}}{\eta_{in}} \left(1 + \frac{1}{Q} \sqrt{\frac{p}{n_{sp}} B_o T} \right). \quad (7.106)$$

The receiver sensitivity increases as the square-root of the product of the optical bandwidth and the symbol duration. This is a relatively slow rate of increase so that even with optical bandwidths as large as 10 times the symbol-rate, the use of an electrical matched-filter will allow the receiver sensitivity to be within about 1.5 dB of the theoretical maximum.

In the case where the transmitter laser has a finite extinction ratio, defined by $P_{rcvd_{zero}} = \varepsilon P_{rcvd_{one}}$, we can define an approximate power loss penalty from ideal as

$$P_{loss} \cong \frac{1 + 2\varepsilon + 2\sqrt{\varepsilon}}{1 - 2\varepsilon}. \tag{7.107}$$

The loss, in an ideal receiver, amounts to about 1 dB for a 20 dB extinction ratio with $\varepsilon = 0.01$ and about 3.6 dB for a 10 dB extinction ratio with $\varepsilon = 0.1$.

In addition to the limiting cases described above, there are many combinations of filters that can be used. Figure 7.28 illustrates the approximate degradation in sensitivity that a binary OOK receiver experiences for various combinations of optical and electrical noise-bandwidths. Gaussian statistics are used, and we are assuming that the $S \times ASE$ noise dominates during a *one* and the $ASE \times ASE$ dominates during a *zero*. Note that the electrical filter has little effect when an optical matched filter is used. In fact, we would not want much additional electrical filtering because of the likelihood of introducing unwanted intersymbol interference. The degradation is also not severe even when relatively wide bandwidth optical filters are used, as long as the electrical filter is nearly matched to the received symbols.

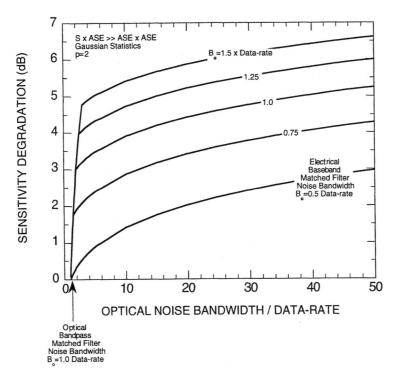

Figure 7.28 Effect of filtering on preamplified binary OOK receiver performance.

7.4.2 Preamplified Receivers for FSK and DPSK

In addition to the conventional OOK (ASK) format, a direct-detection receiver can be built to incorporate optical demodulation before the photodetectors [9]. This would allow it to demodulate either FSK or DPSK. Preamplification can be incorporated into the receivers, as illustrated in Fig. 7.29. Fig. 7.29(a) illustrates a delay and multiply demodulator that can be used to demodulate either FSK or DPSK [186]. A two filter demodulator for binary FSK is illustrated in Fig. 7.29(b) [187]. Single-filter FSK demodulators have also been demonstrated [188]. For either demodulator structure we would ideally like the optical filters to be matched filters for the received symbols [184].

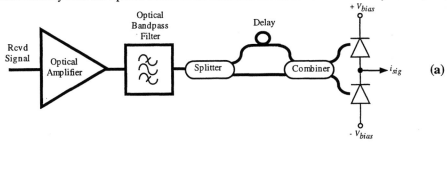

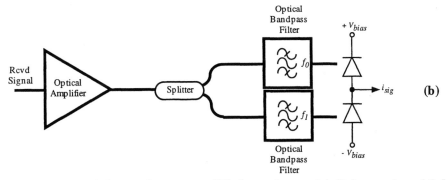

Figure 7.29 Demodulators for preamplified receivers. (a) Delay and multiply demodulator for FSK or DPSK. (b) Two-filter detection of FSK.

That we obtained the same fundamental sensitivity for both optically preamplified direct detection of OOK (i.e. ASK) and coherent detection of ASK with incoherent demodulation is an important point. This occurs because both the ideal amplifier and an ideal heterodyne receiver are both quantum-limited receiver structures [90, 189]. That an ideal optically preamplified receiver and an ideal heterodyne receiver are capable of achieving the same sensitivity had been known for some time [190-192] but was largely considered a theoretical curiosity because of the lack of a readily available, high-gain, low noise-figure optical amplifier. The relatively recent advent of low-noise erbium-doped fiber amplifiers combined with additional theoretical understanding of the properties of preamplified receivers [165, 184, 193] has enabled practical, optically preamplified direct-detection receivers to equal, and in many cases exceed, the demonstrated sensitivity of coherent receivers [170, 176, 177].

7.5   Clock Recovery

Another key subsystem in the digital receiver is the clock recovery system. Fig. 7.30 illustrates several circuits that can be used to implement clock recovery for NRZ signals. The exact configuration used for clock recovery depends on the type of modulation format used. Non-return-to-zero coding does not produce a component at the clock frequency, while return-to-zero coding will have some components at the clock-rate [32]. Manchester coded waveforms have their energy centered around the clock frequency, so it is relatively easy to recover the clock information [19, 32].

The basic idea of all of the schemes illustrated is to use some form of nonlinearity to generate a component at the clock frequency [194]. Fig. 7.30(a) illustrates a simple circuit that implements the nonlinearity using a squarer. The recovered signal spectrum is first high-pass (or bandpass) filtered to emphasize components near $1/2T$. The filtered signal is then squared to extract a component at the clock rate. This is then filtered and sent through a zero crossing detector to generate a digital clock waveform. The performance of this type of clock recovery scheme is very dependent on the filter bandwidths. High-Q filters realized using either dielectric resonators [195] or surface-acoustic-wave (SAW) filter are usually required [196]. A PLL is also frequently used to help achieve a low-jitter clock. The use of a differentiator instead of a filter is illustrated in Fig. 7.30(b) [197], while a half-bit delay and multiply is shown in Fig. 7.30(c).

Clock recovery systems are usually specified in terms of the clock waveform's phase-jitter and phase-deviation. Phase-jitter refers to the phase-noise associated with the clock signal. Some random phase-jitter is inevitable due to the presence of noise in the waveform that is being used to generate the clock. The transmitted information that is modulating the recovered waveform also introduces a systematic jitter into the clock waveform. This type of data-induced jitter is often termed pattern dependent jitter. The two jitter terms are uncorrelated, so the total jitter is the root-sum-square of the individual systematic and random jitters. It is typical for high quality clock recovery systems to be able to keep the phase-jitter to less than a few degrees RMS [198].

Phase-deviation refers to changes in the absolute phase between the input signal and the output signal. We would ideally like this to be a fixed value. In practice, as the signal level varies, the absolute phase will also vary somewhat. This can be particularly pronounced in circuits with strong nonlinearities [197]. Another parameter is the time required for the clock recovery system to accurately recover the clock. A circuit switched system may require millisecond acquisition times. A fast packet switching system may utilize a burst-mode receiver that requires rapid clock acquisition times [199-201].

The phase-jitter and phase-deviation associated with the recovered clock waveform directly impact the performance of the receiver's digital decision circuitry. Decision circuits are usually specified in terms of their sensitivity and phase-margin. Sensitivity is usually defined as the minimum peak-to-peak input voltage needed at the time a decision is made to maintain the bit-error-rate below a certain level, typically 10^{-9}. This value is usually obtained under the assumption of perfect decision clock timing. The decision circuit's phase-margin is the maximum phase variation that can occur in the clock waveform, assuming that there is sufficient signal level to satisfy the decision circuit's sensitivity requirements. In practice the actual level the decision circuit operates on is usually maintained by an AGC or limiting amplifier to be much larger than its minimum sensitivity specification. The clock recovery circuit must then guarantee that the worst case combination of phase-jitter and phase-variation do not exceed the decision circuit's own phase-margin.

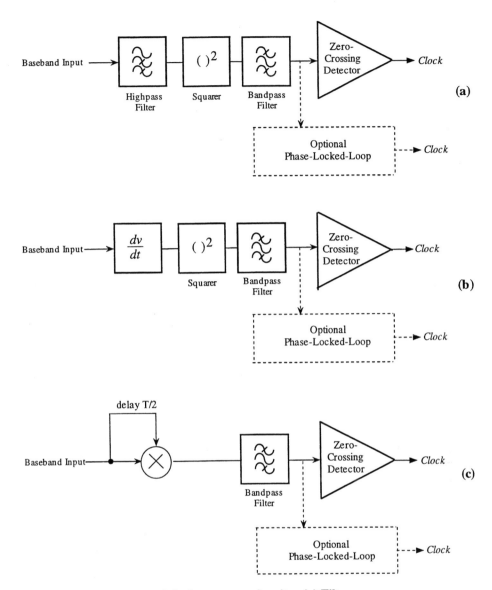

**Figure 7.30 Examples of clock recovery circuits. (a) Filter-squarer.
(b) Differentiate-and-square. (c) Delay-and-multiply.**

Transmitted data that contain long strings of contiguous *ones* or *zeroes* are particularly problematic for clock recovery systems because there are no symbol transitions present for comparatively long periods of time. The clock recovery system must then "flywheel" through the period in which transitions are absent. This often forces unacceptably complex circuit implementations. A simpler solution is to line-code the transmitted information to purposely limit the number of contiguous *ones* or *zeroes* (called the run-length) that can be transmitted [8].

7.6   Analog Receivers

Lasers can be amplitude (intensity), phase, frequency, and polarization modulated. Any of these techniques can be used to transmit analog information. The most commonly used format is intensity modulation using either internal modulation of the laser bias current or external modulation via either an absorptive or interferometric modulator. The information to be transmitted can either directly modulate the optical field or it can be used to first modulate a subcarrier that is then used to modulate the laser. The subcarrier technique is of particular interest in local-access systems and in CATV applications, where each TV channel can be carried on an individual subcarrier [202-205]. Figure 7.31 illustrates the functional components usually found in an analog receiver.

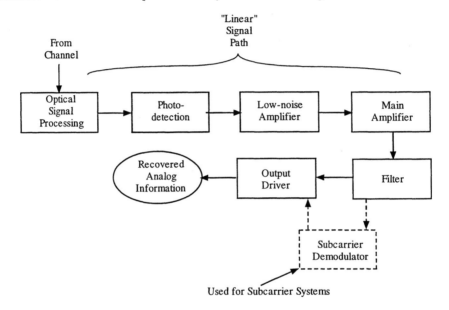

Figure 7.31 Analog receiver subsystems.

The received optical signal from the channel can be optically processed by filters, optical amplifiers, or interferometers before being detected. The photodetector converts the processed optical field into an electrical signal and introduces loss due to its finite quantum efficiency. The photodetector will also have an optical bandwidth and an electrical bandwidth and may have a multiplication gain associated with it.

The photodetector is followed by a low-noise amplifier. The combination of optical processing, photodetector quantum efficiency and quantum shot-noise, and amplifier-noise will typically determine the sensitivity of the receiver. After the front-end there are often additional stages of gain in the main amplifier that boost the low level output of the front-end to levels that are sufficient to drive the demodulator. Within the linear signal-path there will be a filter that determines the amount of noise that passes to the final stages of the analog receiver. The filter may be contained within the amplifier transfer function or it may be a separate stage. The output of this filter will typically drive a final amplifier stage that functions as an output driver. Most analog systems are direct-detection systems or preamplified systems, although there are examples of coherent analog systems [206].

In a subcarrier system an additional electrical demodulation stage is required to recover the transmitted information. For example, if the information to be transmitted was a video signal, it may have been encoded as frequency modulation (FM) of an electrical subcarrier that was then combined with several other modulated subcarriers before being transmitted. At the receiver, the desired subcarrier is selected, the signal is passed to a subcarrier FM demodulator, and the transmitted video signal is recovered.

Let us examine the simple example of an analog receiver, illustrated in Fig. 7.32. The receiver is intended for use in a 3 km long 1300 nm analog system that carries a single modulated RF carrier signal from an antenna to a remote receiver. The carrier is at a frequency of 500 MHz and the type of modulation is not specified, other than that it is band-limited to less than 10 MHz. The link is required to provide a 40 dB carrier-to-noise ratio at the receiver output with intermod products at least 50 dB below the carrier level. The transmitter laser is known to have a RIN level below -150 dB/Hz. The receiver uses an 80% efficient p-i-n photodiode and a transimpedance amplifier that has a noise density of $5.0 \text{pA}/\sqrt{\text{Hz}}$. The front-end is followed by a bandpass filter centered on the 500 MHz carrier and an output driver stage that combines a modest voltage gain with high current gain, allowing the output to swing several volts across low-impedance loads. The bandpass filter is selected to be third-order and to follow a Bessel function response in order to minimize the group-delay distortion experienced by the recovered modulated microwave signal [207, 208]. The noise-bandwidth of the bandpass filter is 15.0 MHz.

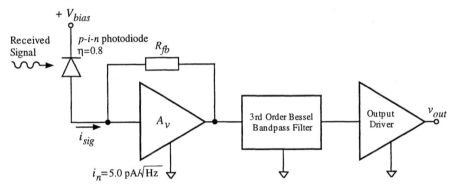

Figure 7.32 Example analog receiver.

In order to determine the CNR present at the receiver's output, we model the power in the received optical signal as

$$P_{rcvd}(t) = \overline{P}_{rcvd}\left(1 + m \cos \omega_m t\right), \tag{7.108}$$

where ω_m is the modulation frequency (500 MHz) and m is the modulation index. The electrical photocurrent signal at the output of the photodetector consists of two components, an AC carrier signal component given by

$$i_{sig}(t) = m\frac{\eta q}{h\nu}\overline{P}_{rcvd} \cos \omega_m t \tag{7.109}$$

and a DC component given by

$$i_{DC} = \frac{\eta q}{h v} \overline{P}_{rcvd}. \tag{7.110}$$

The carrier-to-noise ratio at the output of the receiver from Fig. 7.32 is then given by

$$CNR = \frac{\frac{1}{2}\left(m\frac{\eta q}{h v}\overline{P}_{rcvd}\right)^2}{\left[2q\frac{\eta q}{h v}\overline{P}_{rcvd} + \left(25.0 \times 10^{-24}\right)\right] \times 15\text{MHz}}. \tag{7.111}$$

The depth of modulation m that can be used is a strong function of the linearity in the system. In an analog system, nonlinearities in the transmitter, channel, and receiver all play a role in determining the overall system performance. Nonlinearities can sometimes be difficult to predict analytically, and commercial CAD circuit analysis packages that support nonlinear analysis can often be used to advantage.

To understand the types of effects that arise from receiver nonlinearities, we will use a simple technique that allows us to obtain an estimate for the distortion products present in the receiver output. We can approximate the receiver output voltage as a simple truncated power series of the form

$$v_{out}(t) = Z_t i_{sig}(t) \cong k_1 i_{sig}(t) + k_2 i_{sig}^2(t) - k_3 i_{sig}^3(t). \tag{7.112}$$

Using this power-series model, the third-order term is responsible for the gain compression (or output saturation) observed in most practical amplifiers [209].

In this particular receiver analysis we will simplify our analysis by ignoring the second order term and assume $k_2 = 0$. We make this simplification because this is a single channel system and the relatively narrow bandpass response of the receiver will effectively filter out any second-order terms. In actual practice, even when filtering is used, it may be necessary to verify that any second-order terms that are generated do not cause DC offsets or mix with other signals, causing the generation of additional signals that do ultimately appear at the receiver output.

The effective transimpedance gain seen at the fundamental frequency of the carrier signal is given by

$$Z_t = k_1 - \frac{3}{4}k_3\left(m\frac{\eta q}{h v}\overline{P}_{rcvd}\right)^2. \tag{7.113}$$

The 1 dB output compression point occurs when the electrical power in the output signal is 1 dB below what would be expected if the receiver was perfectly linear. The 1 dB compression point occurs when

$$k_1 - \frac{3}{4}k_3\left(m\frac{\eta q}{h v}\overline{P}_{rcvd}\right)^2 = 0.891 k_1. \tag{7.114}$$

The example receiver illustrated in Fig. 7.32 is designed to have a linear midband transimpedance with k_1 equal to $10\,\text{k}\Omega$ and is known to have a 1 dB output compression point of approximately 5.0 volts. Using the known output saturation voltage, a relatively large modulation depth of $m = 0.5$, measured quantum efficiency of the photodetector of $\eta = 0.8$, and linear transimpedance of $k_1 = 10,000$, we can solve Eqs. 7.113 and 7.114 for the value of the cubic coefficient, obtaining $k_3 = 10^{10}$. Our complete expression for the receiver response is then

$$v_{out}(t) \cong 10^4 i_{sig}(t) - 10^{10} i_{sig}^3(t). \tag{7.115}$$

A nonlinearity such as that of Eq. 7.115 results in the production of three principal types of distortion: harmonic distortion, intermodulation distortion, and crossmodulation distortion. In many cases the intermodulation performance is of particular interest and will be the type of distortion investigated here.

Second and third order intermods are particularly important in multi-channel systems such as those used with analog amplitude modulated CATV distribution. The composite second order (CSO) and composite triple beat (CTB) are widely used measures of the second-order and third order intermodulation products, respectively [8]. To evaluate this receiver's intermodulation performance, we use an AC signal photocurrent of the form

$$i_{sig}(t) = m \frac{\eta q}{h\nu} \overline{P}_{rcvd}[\cos \omega_1 t + \cos \omega_2 t], \tag{7.116}$$

where both ω_1 and ω_2 are contained within the 10.0 MHz modulation spectrum around the 500 MHz carrier frequency.

In general, passing a pair of sinusoids through a nonlinearity will result in the generation of intermods at the frequencies $n\omega_1 \pm p\omega_2$, where n and p are any integer. The sum of the coefficients $n + p$ is known as the *order* of the intermodulation product. In many practical cases, only a few of the intermods that are produced will be significant.

The cubic portion of Eq. 7.115 will result in the creation of third-order intermods within the receiver at the frequencies $3\omega_1$, $3\omega_2$, $2\omega_1 \pm \omega_2$, and $2\omega_2 \pm \omega_1$. Of these third-order intermod products, those at $2\omega_1 - \omega_2$ and $2\omega_2 - \omega_1$ fall within the receiver passband and will contribute to distortion in the receiver's output.

Figure 7.33 plots the various output voltages, the carrier-to-noise ratio, and the carrier-to-intermod ratio for our example receiver. The curves shown were obtained using $m = 0.5$, From Fig. 7.33(a) we see that for received signal-powers below about -10 dBm the output-noise is constant at a level that is equal to the receiver's equivalent input current-noise density times the transimpedance times the noise bandwidth of the bandpass filter. At higher levels of received signal, the signal quantum shot-noise becomes noticeable.

As the received signal power increases, the output signal increases linearly until power levels of approximately -8 dBm are reached. Above this point the receiver enters into compression, the output saturates, and there is no increase in output signal even for an increase in received signal. What does increase rather dramatically is the distortion present in the output. For received power levels above about -15 dBm the intermodulation products present in the output exceed the noise present in the output, and by the point at which the receiver is illuminated by -3 dBm, the output intermodulation products are the same level as the output carrier.

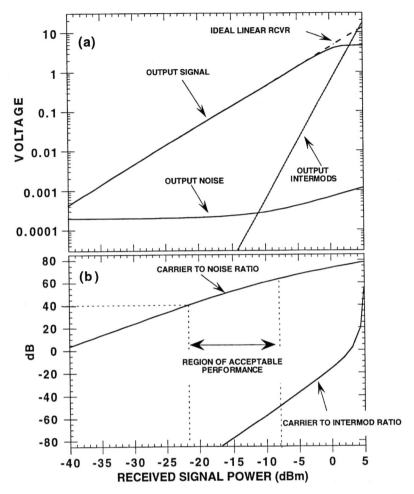

Figure 7.33 Calculated performance of the example analog receiver from Fig. 7.32.
(a) Output voltages. (b) CNR and carrier to intermod ratio.

The output CNR and carrier to intermod ratio are shown in Fig 7.33(b). For the levels of received power shown the CNR is always positive. The change in CNR slope with received power that corresponds to the receiver changing from a receiver-noise limited regime to a quantum shot-noise limited regime is seen for received power levels near -10 dBm. Using this receiver we could obtain CNRs in excess of our design goal of 40 dB for received power greater than about -22 dBm. We can maintain the intermodulation distortion products at the desired 50 dB below the output carrier for received optical signal powers less than about -8 dBm. Our region of acceptable operation for this intensity modulated subcarrier system is therefore from -22 to -8 dBm.

Analog systems are not limited to just intensity modulation. The laser can be frequency, phase, or polarization modulated. An example of a receiver for an optical FM analog system is shown in Fig. 7.34(a). The frequency modulated laser is optically demodulated by the optical interferometer and photodetector. The interferometer forms an optical discriminator that converts frequency modulation into intensity modulation.

Analog systems can also utilize frequency modulated subcarriers [203, 210].

Subcarrier FM is considered attractive because high output signal-to-noise ratios (>50 dB) can be obtained with only modest carrier-to-noise ratios (10-20 dB) at the input to the FM demodulator. This improvement occurs if the FM deviation is relatively high and the FM demodulator is operated above threshold [211]. Demodulators for subcarrier FM systems are illustrated in Fig. 7.34(b) and (c).

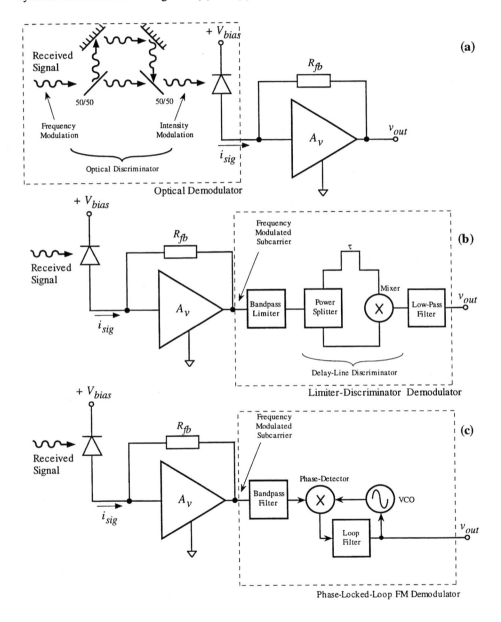

Figure 7.34 Demodulators for FM analog modulation. (a) Interferometer discriminator for optical FM. (b) Phase-locked-loop demodulator for subcarrier FM. (c) Delay-line demodulator for subcarrier FM.

7.7 Reported Receiver Sensitivities

We conclude this tutorial text with a short survey of optical communication receivers that have been reported in the open literature through early 1994 and a table listing the quantum-limited sensitivities for a variety of popular modulation formats. The chart of reported sensitivities is intended to be representative, not all-inclusive, and some noteworthy results have undoubtedly been omitted. Lines of constant received power are also shown.

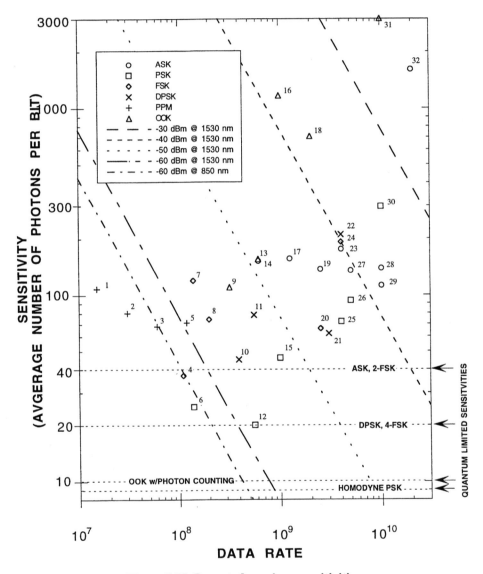

Figure 7.35 Reported receiver sensitivities.

Num.	Data Rate	Wavelength	Sensitivity	Modulation	Receiver	Ref.
1	15 Mbps	810 nm	108	4-PPM	APD-FET	[12]
2	30 Mbps	810 nm	80	4-PPM	APD-FET	[12]
3	60 Mbps	810 nm	68	4-PPM	APD-FET	[12]
4	110 Mbps	830 nm	37	4-FSK	Heterodyne	[21]
5	120 Mbps	810 nm	71	4-PPM	APD-FET	[12]
6	140 Mbps	1320 nm	25	PSK	Homodyne	[144, 212]
7	140 Mbps	1542 nm	119	FSK	Heterodyne	[213]
8	200 Mbps	1530 nm	74	FSK	Heterodyne	[214]
9	325 Mbps	820 nm	110	OOK	APD-trans-Z	[215]
10	400 Mbps	1530 nm	45	DPSK	Heterodyne	[216]
11	560 Mbps	1540 nm	78	DPSK	Heterodyne	[217]
12	565 Mbps	1064 nm	20	PSK	Homodyne	[152]
13	622 Mbps	1297 nm	155	OOK	APD-HEMT	[218]
14	622 Mbps	1537 nm	152	ASK	Preamp	[219]
15	1 Gbps	1500 nm	46	PSK	Homodyne	[220]
16	1 Gbps	1310 nm	1162	OOK	APD-FET	[221]
17	1.24 Gbps	1533 nm	156	ASK	Preamp	[222]
18	2 Gbps	1540 nm	705	OOK	APD-FET	[223]
19	2.5 Gbps	1530 nm	137	ASK	Preamp	[176]
20	2.5 Gbps	1554 nm	66	CPFSK	Heterodyne	[110, 224]
21	3 Gbps	1532 nm	62	DPSK	Preamp	[225]
22	4 Gbps	1528 nm	209	DPSK	Heterodyne	[127]
23	4 Gbps	1528 nm	175	ASK	Heterodyne	[127]
24	4 Gbps	1528 nm	191	FSK	Heterodyne	[127]
25	4 Gbps	1508 nm	72	PSK	Homodyne	[145]
26	5 Gbps	1550 nm	93	PSK	Homodyne	[150]
27	5 Gbps	1534 nm	135	ASK	Preamp	[226]
28	10 Gbps	1530 nm	139	ASK	Preamp	[176]
29	10 Gbps	1536 nm	112	ASK	Preamp	[177]
30	10 Gbps	1550 nm	297	PSK	Homodyne	[148]
31	10 Gbps	1530 nm	3000	OOK-RZ	APD-FET	[227]
32	20 Gbps	1552 nm	1600	ASK	Preamp	[228]

Shot-Noise Limited Sensitivities of Various Binary Receivers*
(listed in order of decreasing sensitivity)

Modulation Format	Receiver Structure	BER Expression	Sensitivity Peak Photons per bit n_p	Sensitivity Average Photons per bit
PSK	Quantum Optimum	$\cong \dfrac{1}{2} e^{-4n_p}$	5	5
PSK	Homodyne	$= Q\left(\sqrt{4n_p}\right)$	9	9
PSK	Heterodyne Synchronous	$= Q\left(\sqrt{2n_p}\right)$	18	18
OOK	Photon Counting	$= \dfrac{1}{2} e^{-n_p}$	20	10
DPSK	Heterodyne	$= \dfrac{1}{2} e^{-n_p}$	20	20
DPSK	Preamplified Direct detection	$= \dfrac{1}{2} e^{-n_p}$	20	20
ASK (OOK)	Homodyne	$= Q\left(\sqrt{n_p}\right)$	36	18
FSK	Heterodyne Synchronous	$= Q\left(\sqrt{n_p}\right)$	36	36
FSK	Heterodyne Asynchronous	$= \dfrac{1}{2} e^{-n_p/2}$	40	40
FSK	Preamplified Direct detection	$= \dfrac{1}{2} e^{-n_p/2}$	40	40
ASK (OOK)	Heterodyne Synchronous	$= Q\left(\sqrt{n_p/2}\right)$	72	36
ASK (OOK)	Preamplified Direct detection	$= \dfrac{1}{2} e^{-n_p/4}$	76	38
ASK (OOK)	Heterodyne Asynchronous	$= \dfrac{1}{2} e^{-n_p/4}$	80	40

* Sensitivity is specified for a 10^{-9} BER. Equally likely *ones* and *zeroes* are used in computing the average number of photons per bit. Preamplifier includes a polarizer to eliminate ASE in the polarization orthogonal with the signal.

7.8 References

1.   M. Nakamura, *et al.* "15-GHz AlGaAs/GaAs HBT Limiting Amplifier with Low Phase Deviation," in *IEEE GaAs IC Symposium*. 1991.

2.   M. Moller, *et al.*, "13 Gb/s Si-Bipolar AGC Amplifier IC with High Gain and Wide Dynamic Range for Optical-Fiber Receivers," IEEE Journal of Solid-State Circuits, 1994, vol. 29, no. 7, pp. 815-822.

3.   D.W. Faulkner, "A Wideband Limiting Amplifier for Optical-Fiber Receivers," IEEE Journal of Solid-State Circuits, 1983, vol. SC-18, no. 3, pp. 333-340.

4.   R.M. Gagliardi and S. Karp, *Chap. 7 - Digital Communications - Binary Systems* in *Optical Communications,* 1976, John Wiley & Sons: New York.

5.   R.E. Ziemer and W.H. Tranter, *Chap. 8 - Optimum Receivers and Signal Space Concepts* in *Principles of Communications,* 1976, Houghton Mifflin Company: Boston.

6.   H. Van Trees, *Chap. 2 - Classical Detection and Estimation Theory* in *Detection Estimation and Modulation Theory - Part 1,* 1968, John Wiley and Sons, Inc.: New York.

7.   G.R. Cooper and C.D. McGillem, *Chap. 3 - Several Random Variables* in *Probabilistic Methods of Signal and Systems Analysis,* 1971, Holt, Rinehart and Winston, Inc.: New York.

8.   Y. Takasaki, *et al.*, "Optical Pulse Formats for Fiber Optic Digital Communications," IEEE Transactions on Communications, 1976, vol. COM-24, no. 4, pp. 404-413.

9.   Y. Yamamoto, "Receiver Performance Evaluation of Various Digital Optical Modulation-Demodulation Systems in the 0.5-10 micron Wavelength Region," IEEE Journal of Quantum Electronics, 1980, vol. QE-16, no. 11, pp. 1251-1259.

10.   W.S. Holden, "An Optical Frequency Pulse Position Modulation Experiment," Bell System Technical Journal, 1975, vol. 54, pp. 285-296.

11.   T.V. Muoi and J.L. Hullett, "Receiver Design for Optical PPM Systems," IEEE Transactions on Communications, 1978, vol. COM-26, no. 2, pp. 295-300.

12.   A. MacGregor and B. Dion. "39 Photons/bit Direct Detection Receiver at 810 nm, BER=1E-6, 60 Mb/s, QPPM," in *Free-Space Laser Communication Technologies III.* 1991, SPIE, Vol. 1417.

13.   X. Sun, *et al.* "50 Mbps Free-Space Direct Detection Laser Diode Optical Communication System with Q=4 PPM Signaling," in *Free-Space Laser Communication Technologies II.* 1990, SPIE, Vol. 1218.

14.   J.R. Lesh. "Power Efficient Communications for Space Applications," in *International Telemetering Conference.* 1982, International Foundation for Telemetering, Vol. XVIII.

15. X. Sun and F. Davidson, "Word Timing Recovery in Direct Detection Optical PPM Communication Systems with Avalanche Photodiodes Using a Phase Lock Loop," IEEE Transactions on Communications, 1990, vol. 38, no. 5, pp. 666-673.

16. F. Davidson and X. Sun, "Slot Clock Recovery in Optical PPM Communication Systems with Avalanche Photodiode Photodetectors," IEEE Transactions on Communications, 1989, vol. 37, no. 11, pp. 1164-1172.

17. J. Katz, "Average Power Constraints in AlGaAs Semiconductor Lasers Under Pulse-Position-Modulation Conditions," Optics Communications, 1986, vol. 56, no. 5, pp. 330-333.

18. F. Davidson and X. Sun. "Bandwidth Requirements for Direct Detection Optical Communication Receivers with PPM Signaling," in *Free-Space Laser Communication Technologies III*. 1991, SPIE, Vol. 1417.

19. T.V. Muoi, "Receiver Design for Digital Fiber Optic Transmission Systems Using Manchester (Biphase) Coding," IEEE Transactions on Communications, 1983, vol. COM-31, no. 5, pp. 608-619.

20. E.E. Basch and T. Brown, *Chap. 16 - Introduction to Coherent Fiber-Optic Communication* in *Optical-Fiber Transmission,* E.E. Basch, Editor. 1987, Howard Sams Division of Macmillan: Indianapolis.

21. S.B. Alexander, *et al.*, "4-ary FSK Coherent Optical Communication System," Electronics Letters, 1990, vol. 26, no. 17, pp. 1346-1348.

22. B. Glance, "Performance of Homodyne Detection of Binary PSK Optical Signals," Journal of Lightwave Technology, 1986, vol. LT-4, no. 2, pp. 228-235.

23. J.M. Kahn, *et al.*, "Heterodyne Detection of 310-Mb/s Quadriphase-Shift Keying Using Fourth-Power Optical Phase-Locked-Loop," IEEE Photonics Technology Letters, 1992, vol. 4, no. 12, pp. 1397-1400.

24. T. Okoshi and K. Kikuchi, *Chap. 3 - Semiconductor-Laser Noise and Its Effects* in *Coherent Optical Fiber Communications,* 1988, Kluwer Academic Publishers: Boston.

25. K. Petermann, *Chap. 7 - Noise Characteristics of Solitary Laser Diodes* in *Laser Diode Modulation and Noise,* 1988, Kluwer Academic Publishers: Boston.

26. T. Okoshi and K. Kikuchi, *Chap. 8 - Fluctuation and Control of State-of-Polarization in Single-Mode Fibers* in *Coherent Optical Fiber Communications,* 1988, Kluwer Academic Publishers: Boston.

27. P.S. Henry and S.D. Personick, *Part 6 - Receivers*, in *Coherent Lightwave Communications,* Progress in Lasers and Electro-Optics, P. Smith ed. 1990, IEEE Press: New York.

28. I. Kaminow, "Balanced Optical Discriminator," Applied Optics, 1964, vol. 3, no. 4, pp. 507-510.

29. R. Vodhanel, *et al.*, "Performance of Directly Modulated DFB Lasers in 10 Gb/s ASK, FSK and DPSK Lightwave Systems," Journal of Lightwave Technology, 1990, vol. LT-8, no. 9, pp. 1379-1386.

30. K. Ishii, *et al.*, "High-Bit-Rate, High-Input Sensitivity Decision Circuit Using Si Bipolar Technology," IEEE Journal of Solid-State Circuits, 1994, vol. 29, no. 5, pp. 546-550.

31. J.M. Wozencraft and I.M. Jacobs, *Chap. 4 - Optimum Receiver Principles* in *Principles of Communication Engineering,* 1965, John Wiley & Sons: New York.

32. R.E. Ziemer and W.H. Tranter, *Chap. 7 - Digital Data Transmission* in *Principles of Communications,* 1976, Houghton Mifflin Company: Boston.

33. R. Lucky, *et al.*, *Chap. 4 - Baseband Pulse Transmission* in *Principles of Data Communication,* 1968, McGraw-Hill: New York.

34. W.B. Davenport and W.L. Root, *Chap. 7 - Shot Noise* in *An Introduction to the Theory of Random Signals and Noise,* 1958, McGraw-Hill: New York.

35. I. Korn, *Chap. 4 - ASK Baseband Systems* 1985, Van Nostrand Reinhold Company: New York.

36. H. Nyquist, "Certain Topics in Telegraph Transmission Theory," AIEE Transactions, 1928, vol. 47, pp. 617-643.

37. R. Gibby, "Some Extensions of Nyquist's Telegraph Transmission Theory," Bell System Technical Journal, 1965, vol. 44, pp. 1487-1510.

38. J. Scanlon, "Pulses Satisfying the Nyquist Criterion," Electronics Letters, 1992, vol. 28, no. 1, pp. 50-52.

39. G. May, *et al.*, "Extended 10 Gb/s Fiber Transmission Distance at 1538 nm Using a Duobinary Receiver," IEEE Photonics Technology Letters, 1994, vol. 6, no. 5, pp. 648-650.

40. R.E. Ziemer and R.L. Peterson, *Chap 3* in *Digital Communications and Spread Spectrum Systems,* 1985, Macmillan: New York.

41. R. Lucky, *et al.*, *Chap. 5 - Optimization of the Baseband System* in *Principles of Data Communication,* 1968, McGraw-Hill: New York.

42. J.M. Wozencraft and I.M. Jacobs, *Chap. 7 - Important Channel Models* in *Principles of Communication Engineering,* 1965, John Wiley & Sons: New York.

43. J. Winters and R. Gitlin, "Electrical Signal Processing Techniques in Long-Haul Fiber-Optic Systems," IEEE Transactions on Communications, 1990, vol. 38, no. 9, pp. 1439-1453.

44. R. Rugemalira, "Optimum Linear Equalization of a Digital Fibre Optic Communication System," Optical and Quantum Electronics, 1981, vol. 13, pp. 153-163.

45. R. Dogliotti, *et al.*, "Baseband Equalization in Fibre Optic Digital Transmission," Optical and Quantum Electronics, 1976, vol. 8, pp. 343-353.

46. I. Bar-David, "Communication Under the Poisson Regime," IEEE Transaction on Information Theory, 1969, vol. IT-15, pp. 31-37.

47. K. House, "Filters for the Detection of Binary Signaling: Optimization Using the Chernoff Bound," IEEE Transactions on Communications, 1980, vol. COM-28, no. 2, pp. 257-259.

48. C. Gokcek and Z. Unver, "Design Method for Optimal Data Transmission Filters," Electronics Letters, 1990, vol. 26, no. 6, pp. 373-374.

49. S. Nader and L. Lind, "Optimal Data Transmission Filters," IEEE Transactions on Circuits and Systems, 1979, vol. CAS-26, no. 1, pp. 36-45.

50. P. Monteiro, *et al.* "Pulse Shaping Microstrip Filters for 10 and 15 Gbit/s Optical Receivers," in *High-Speed Fiber Networks and Channels II.* 1992, SPIE, Vol. 1784.

51. I. Korn, "Probability of Error in Binary Communication Systems with Causal Bandlimiting Filters," IEEE Transactions on Communications, 1973, vol. COM-21, no. 8, pp. 878-898.

52. J. O'Reilly, *et al.*, "New Strategy for the Design and Realisation of Optimised Receiver Filters for Optical Telecommunications," IEE Proceedings Part J, 1990, vol. 137, no. 3, pp. 181-185.

53. B. Enning, "Matched Filter for Gbit/s Applications," Electronics Letters, 1992, vol. 28, no. 1, pp. 96-97.

54. J. Craig, "Optimum Approximation to a Matched Filter Response," IRE Transactions on Information Theory, 1960, pp. 409-410.

55. J. Watson and W. Stanley, "Matched-Filter Lumped Approximations for Pulse Applications," IEEE Transactions on Circuits and Systems, 1974, vol. CAS-21, no. 3, pp. 364-368.

56. L. Lind and T. Labarre, "Realisable 100% Raised Cosine Filters," Electronics Letters, 1981, vol. 17, no. 5, pp. 197-198.

57. C. Medeiros, *et al.*, "Multigigabit Optically Preamplified Receiver Designs with Improved Timing Tolerance," IEE Proceedings Part J, 1992, vol. 139, no. 2, pp. 143-147.

58. J. O'Reilly, *et al.*, "Optical Fiber Direct Detection Receivers Optimally Tolerant to Jitter," IEEE Transactions on Communications, 1986, vol. COM-34, no. 11, pp. 1141-1147.

59. J. O'Reilly and R. Fyath, "New APD-Based Receivers Providing Tolerance to Alignment Jitter for Binary Optical Communications," IEE Proceedings Part J, 1988, vol. 135, no. 2, pp. 119-125.

60. R.C. Hooper, "Optimum Equalisers for Binary Digital Transmission Systems and Their Application to Optical Receivers," IEE Proceedings Part J, 1989, vol. 136, no. 2, pp. 137-140.

61. M. El-Hadidi and B. Hirosaki, "The Bayes' Optimal Receiver for Digital Fibre-Optic Communication Systems," Optical and Quantum Electronics, 1981, vol. 13, pp. 469-486.

62. S.D. Personick, "Receiver Design for Digital Fiber Optic Communication Systems, I & II," Bell System Technical Journal, 1973, vol. 52, no. 6, pp. 843-886.

63. R.G. Smith and S.D. Personick, *Receiver Design for Optical Fiber Communication Systems*, in *Semiconductor Devices for Optical Communication*, 2nd, H. Kressel ed. Vol. 39, 1982, Springer Verlag: New York, pp. 89-160.

64. J. Proakis, *Chap. 6 - Digital Signaling over a Bandwidth Constrained Linear Filter Channel* in *Digital Communications*, 1989, McGraw-Hill: New York.

65. T.V. Muoi, "Receiver Design for Optical-Fiber Systems," Journal of Lightwave Technology, 1984, vol. LT-2, no. 3, pp. 243-265.

66. K. Wilson. "Prediction of the Required Received Power for Bipolar Raised Cosine Modulation," in *Free-Space Laser Communication Technologies II*. 1990, SPIE, Vol. 1218.

67. R. Durrant and W. Kennedy. "Experimental Results of an Optical 20 Mbps Bipolar Coded Raised Cosine Digital Communication Link," in *Free-Space Laser Communication Technologies II*. 1990, SPIE, Vol. 1218.

68. H. Nishimoto, *et al.*, "New Method for Analyzing Eye Patterns and Its Application to High-Speed Optical Transmission Systems," Journal of Lightwave Technology, 1988, vol. 6, no. 5, pp. 678-685.

69. W. Davenport and W. Root, *Chap. 8 - The Gaussian Process* in *An Introduction to the Theory of Random Signals and Noise*, 1958, McGraw-Hill: New York.

70. P. Balaban, "Statistical Evaluation of the Error Rate of the Fiberguide Repeater Using Importance Sampling," Bell System Technical Journal, 1976, vol. 55, no. 6, pp. 745-766.

71. S.D. Personick, *et al.*, "A Detailed Comparison of Four Approaches to the Calculation of the Sensitivity of Optical Fiber System Receivers," IEEE Transactions on Communications, 1977, vol. COM-25, no. 5, pp. 541-548.

72. W.M. Hubbard, "The Approximation of a Poisson Distribution by a Gaussian Distribution," Proceedings of the IEEE, 1970, vol. 58, no. 9, pp. 1374-1375.

73. J.B. Kennedy and A.M. Neville, *Chap. 10 - Normal Distribution* in *Basic Statistical Methods for Engineers and Scientists*, 1986, Harper & Row: New York.

74. N. Bergano, *et al.*, "Margin Measurements in Optical Amplifier Systems," IEEE Photonics Technology Letters, 1993, vol. 5, no. 3, pp. 304-306.

75. R.M. Gagliardi and G. Prati, "On Gaussian Error Probabilities in Optical Receivers," IEEE Transactions on Communications, 1980, vol. COM-28, no. 9, pp. 1742-1747.

76. D.R. Smith and I. Garrett, "A Simplified Approach to Digital Optical Receiver Design," Optical and Quantum Electronics, 1978, vol. 10, pp. 211-221.

77. R. McIntyre, "The Distribution of Gains in Uniformly Multiplying Avalanche Photodiodes: Theory," IEEE Transactions on Electron Devices, 1972, vol. ED-19, no. 6, pp. 703-712.

78.   J. Conradi, "The Distribution of Gains in Uniformly Multiplying Avalanche Photodiodes: Experimental," IEEE Transactions on Electron Devices, 1972, vol. ED-19, no. 6, pp. 703-712.

79.   N. Sorensen and R. Gagliardi, "Performance of Optical Receivers with Avalanche Photodetection," IEEE Transactions on Communications, 1979, vol. COM-27, no. 9, pp. 1315-1321.

80.   R. Rugemalira, "Calculation of Average Error Probability in an Optical Communication Channel in the Presence of Intersymbol Interference Using a Characteristic Function Method," Optical and Quantum Electronics, 1980, vol. 12, pp. 119-129.

81.   M. Masuripur, *et al.*, "Fiber-Optics Receiver Error Rate Prediction Using the Gram-Charlier Series," IEEE Transactions on Communications, 1980, vol. COM-28, no. 3, pp. 402-407.

82.   W. Hauk, *et al.*, "Calculation of Error Rates for Optical Fiber Systems," IEEE Transactions on Communications, 1978, vol. COM-26, no. 7, pp. 1119-1126.

83.   P.P. Webb, *et al.*, "Properties of Avalanche Photodiodes," RCA Review, 1974, vol. 35, pp. 234-278.

84.   F. Davidson and X. Sun, "Gaussian Approximation Versus Nearly Exact Performance Analysis of Optical Communication Systems with PPM Signaling and APD Receivers," IEEE Transactions on Communications, 1988, vol. 36, no. 11, pp. 1185-1192.

85.   X. Sun and F. Davidson, "Photon Counting with Avalanche Photodiodes," Journal of Lightwave Technology, 1992, vol. 10, no. 8, pp. 1023-1032.

86.   J. Conradi, "A Simplified Non-Gaussian Approach to Digital Optical Receiver Design with Avalanche Photodiodes: Theory," Journal of Lightwave Technology, 1991, vol. 9, no. 8, pp. 1019-1026.

87.   J. Conradi, "A Simplified Non-Gaussian Approach to Digital Optical Receiver Design with Avalanche Photodiodes: Experimental," Journal of Lightwave Technology, 1991, vol. 9, no. 8, pp. 1027-1030.

88.   G.L. Abbas, *et al.*, "A Dual Detector Optical Heterodyne Receiver for Local Oscillator Noise Suppression," Journal of Lightwave Technology, 1985, vol. LT-3, no. 5, pp. 1110-1122.

89.   T. Waite and R.A. Gudmundsen, "A Balanced Mixer for Optical Heterodyning: The ANN Detector," Proceedings of the IEEE, 1966, vol. 54, pp. 297-299.

90.   B.M. Oliver, "Signal-to-Noise Ratios in Photoelectric Mixing," Proceedings of the IRE, 1961, vol. 49, pp. 1960-1961.

91.   B.L. Kasper, *et al.*, "Balanced Dual-Detector Receiver for Optical Heterodyne Communication at Gigabit per Second Rates," Electronics Letters, 1986, vol. 22, no. 8.

92. S.B. Alexander, "Design of Wide-Band Optical Heterodyne Balanced Mixer Receivers," Journal of Lightwave Technology, 1987, vol. LT-5, no. 4, pp. 523-537.

93. R.J. Deri, *et al.*, "Low-Loss Monolithic Integration of Balanced Twin-Photodetectors with a 3 dB Waveguide Coupler for Coherent Lightwave Receivers," IEEE Photonics Technology Letters, 1990, vol. 2, no. 8, pp. 581-584.

94. T.L. Koch, *et al.*, "Balanced Operation of a GaInAs/GaInAsP Multiple-Quantum-Well Integrated Heterodyne Receiver," IEEE Photonics Technology Letters, 1990, vol. 2, no. 8, pp. 577-580.

95. Y. Yamamoto, "AM and FM Quantum Noise in Semiconductor Lasers - Part I: Theoretical Analysis," IEEE Journal of Quantum Electronics, 1983, vol. QE-19, no. 1, pp. 34-46.

96. Y. Yamamoto, "AM and FM Quantum Noise in Semiconductor Lasers - Part II: Comparison of Theoretical and Experimental Results for AlGaAs Lasers," IEEE Journal of Quantum Electronics, 1983, vol. QE-19, no. 1, pp. 47-58.

97. F.G. Walther and J.E. Kaufmann. "Characterization of GaAlAs Laser Diode Frequency Noise," in *Sixth Topical Meeting on Optical Fiber Communication, OFC*. 1983, New Orleans, Optical Society of America,

98. Y.C. Chung, *et al.*, "A Cold-Start 1.7 Gb/s FSK Heterodyne Detection System Using Single Electrode DFB Lasers with No Equalization," IEEE Photonics Technology Letters, 1990, vol. 2, no. 2, pp. 132-134.

99. J.E. Kaufmann. "Phase and Frequency Tracking Considerations for Heterodyne Optical Communication," in *Proceeding of the International Telemetering Conference*. 1982, International Foundation for Telemetering.

100. P.S. Henry and S.D. Personick, *Part 7 - Systems* in *Coherent Lightwave Communications,* P. Smith, Editor 1990, IEEE Press: New York.

101. T. Okoshi and K. Kikuchi, *Chap. 4 -Improvement of Semiconductor-Laser Performance* in *Coherent Optical Fiber Communications,* 1988, Kluwer Academic Publishers: Boston.

102. J. Salz, "Coherent Lightwave Communications," AT&T Technical Journal, 1985, vol. 64, no. 10, pp. 2153-2209.

103. S.K. Nielsen, *et al.*, "Universal AFC for Use in Optical DPSK Systems," Electronics Letters, 1993, vol. 29, no. 16, pp. 1445-1446.

104. L.G. Kazovsky, "Optical Heterodyning versus Optical Homodyning: A Comparison," Journal of Lightwave Technology, 1988, vol. LT-6, no. 1, pp. 18-24.

105. L.G. Kazovsky, "Balanced Phase-Locked Loops for Optical Homodyne Receivers," Journal of Lightwave Technology, 1986, vol. LT-4, no. 2, pp. 182-185.

106. L.G. Kazovsky, "Phase- and Polarization-Diversity Coherent Optical Techniques," Journal of Lightwave Technology, 1989, vol. 7, no. 2, pp. 279-292.

107. R. Corvaja and G. Pierobon, "Performance Evaluation of ASK Phase-Diversity Lightwave Systems," Journal of Lightwave Technology, 1994, vol. 12, no. 2, pp. 519-525.

108. P. Hobbs, "Reaching the Shot-Noise Limit for $10," Optics and Photonics News, 1991, vol. 2, no. 4, pp. 17-23.

109. T. Okoshi and K. Kikuchi, *Chap. 5 -Receivers for Coherent Optical Fiber Communications* in *Coherent Optical Fiber Communications,* 1988, Kluwer Academic Publishers: Boston.

110. N. Ohkawa, *et al.*, "A Highly Sensitive Balanced Receiver for 2.5 Gb/s Heterodyne Detection Systems," IEEE Photonics Technology Letters, 1991, vol. 3, no. 4, pp. 375-377.

111. S.D. Personick, "An Image Band Interpretation of Optical Heterodyne Noise," Bell System Technical Journal, 1971, vol. 49, no. 1, pp. 213-217.

112. J. Shapiro, "Quantum Noise and Excess Noise in Optical Homodyne and Heterodyne Receivers," IEEE Journal of Quantum Electronics, 1985, vol. QE-21, no. 3, pp. 237-250.

113. D. Fink, "Coherent Detection Signal-to-Noise," Applied Optics, 1975, vol. 14, no. 3, pp. 689-691.

114. K.A. Winick and P. Kumar, "Spatial Mode-Matching Efficiencies for Heterodyned GaAlAs Semiconductor Lasers," Journal of Lightwave Technology, 1988, vol. 6, no. 4, pp. 513-520.

115. F. Davidson, *et al.*, "50 Mbps Optical Homodyne Communications Receiver with a Photorefractive Optical Beam Combiner," IEEE Photonics Technology Letters, 1992, vol. 4, no. 1, pp. 1295-1298.

116. K. Iwashita and H. Kano, "FSK Transmission Experiment Using 10.5 km Polarisation Maintaining Fibre," Electronics Letters, 1986, vol. 22, no. 4, pp. 214-215.

117. G. Marone, *et al.* "Polarization Independent Detection by Synchronous Intra-bit Polarization Switching in Optical Coherent Systems," in *Proceedings of the International Conference on Communications (ICC).* 1990, IEEE: New York.

118. T.K. Hodgkinson, *et al.*, "Polarisation Insensitive Heterodyne Detection Using Polarisation Scrambling," Electronics Letters, 1987, vol. 23, no. 10, pp. 513-514.

119. B. Glance, "Polarization Independent Optical Receiver," Journal of Lightwave Technology, 1987, vol. LT-5, no. 2, pp. 274-276.

120. M. Hamacher, *et al.*, "Coherent Receiver Front-End Module Including a Polarization Diversity Waveguide OIC and a High-Speed InGaAs Twin-Dual p-i-n Photodiode OEIC Based on InP," IEEE Journal of Photonics Technology Letters, 1992, vol. 4, no. 11, pp. 1234-1237.

121. R. Deri, *et al.*, "Ultracompact Monolithic Integration of Balanced, Polarization Diversity Photodetectors for Coherent Lightwave Receivers," IEEE Photonics Technology Letters, 1992, vol. 4, no. 11, pp. 1238-1240.

122. R.J. Pedersen and F. Ebskamp, "New All-Optical RIN Suppressing, Image Rejection Receiver with Efficient Use of LO- and Signal-Power," IEEE Photonics Technology Letters, 1993, vol. 5, no. 12, pp. 1462-1464.

123. C.W. Helstrom, *Section 3.4 - Functions of Two Random Variables* in *Probability and Stochastic Processes for Engineers,* 1984, Macmillan Publishing Company: New York.

124. S. Stein and J. Jones, *Chap. 10* in *Modern Communication Principles,* 1966, McGraw-Hill: New York.

125. P.S. Henry. "Error-Rate Performance of Optical Amplifiers," in *Conference on Optical Fiber Communication, OFC.* 1989, Houston, OSA, Paper ThK3.

126. T. Okoshi and K. Kikuchi, *Chap. 7 - System Experiments* in *Coherent Optical Fiber Communications,* 1988, Kluwer Academic Publishers: Boston.

127. A. Gnauck, *et al.,* "4 Gb/s Heterodyne Transmission Experiments using ASK, FSK and DPSK Modulation," IEEE Photonics Technology Letters, 1990, vol. 2, no. 12, pp. 908-910.

128. I. Garrett and G. Jacobsen, "Theoretical Analysis of ASK Heterodyne Optical Receivers with Tuned Front Ends," IEE Proceedings Part J, 1988, vol. 135, no. 3, pp. 255-259.

129. G. Foschini, *et al.,* "Noncoherent Detection of Coherent Lightwave Signals Corrupted by Phase Noise," IEEE Transactions on Communications, 1988, vol. 36, no. 3, pp. 306-314.

130. I. Garrett and G. Jacobsen, "Theory for Optical Heterodyne Narrow-Deviation FSK Receivers with Delay Demodulation," Journal of Lightwave Technology, 1988, vol. 6, no. 9, pp. 1415-1423.

131. G. Jacobsen and I. Garrett, "Theory for Heterodyne Optical ASK Receivers Using Square-Law Detection and Postdetection Filtering," IEE Proceedings Part J, 1987, vol. 134, no. 5, pp. 303-312.

132. J. Franz, "Receiver Analysis for Incoherent Optical ASK Heterodyne Systems," Journal of Optical Communications, 1987, vol. 8, no. 2, pp. 57-66.

133. L.G. Kazovsky and O. Tonguz, "ASK and FSK Coherent Lightwave Systems: A Simplified Approximate Analysis," Journal of Lightwave Technology, 1990, vol. 8, no. 3, pp. 338-351.

134. L. Jeromin and V. Chan, "M-ary FSK Performance for Coherent Optical Communication Systems Using Semiconductor Lasers," IEEE Transactions on Communications, 1986, vol. COM-34, no. 4, pp. 375-381.

135. M. Schwartz, *Chap. 4 - Modulation Techniques* in *Information Transmission, Modulation, and Noise,* 1970, McGraw-Hill: New York.

136. D. Boroson, "Optimal Design of Prefilters for Binary FSK Discriminators," Electronics Letters, 1992, vol. 28, no. 3, pp. 284-286.

137. M.L. Stevens, *et al.*, "A Near-Optimum Discriminator Demodulator for Binary FSK with Wide Tone Spacing," IEEE Microwave and Guided Wave Letters, 1993, vol. 3, no. 7, pp. 227-229.

138. T. Imai, *et al.*, "A High Sensitivity Receiver for Multigigabit-per-Second Optical CPFSK Transmission Systems," Journal of Lightwave Technology, 1991, vol. 9, no. 9, pp. 1136-1144.

139. K. Iwashita and N. Takachio, "2 Gbit/s Optical CPFSK Heterodyne Transmission Through 200 km Single-Mode Fiber," Electronics Letters, 1987, vol. 23, no. 7, pp. 341-342.

140. T. Okoshi and K. Kikuchi, *Chap. 2 - Improvement of Receiver Sensitivity* in *Coherent Optical Fiber Communications,* 1988, Kluwer Academic Publishers: Boston.

141. G. Jacobsen, "Sensitivity Limits for Digital Optical Communication Systems," Journal of Optical Communications, 1993, vol. 14, no. 2, pp. 52-64.

142. G. Jacobsen, "Performance of DPSK and CPFSK Systems with Significant Post-Detection Filtering," Journal of Lightwave Technology, 1993, vol. 11, no. 10, pp. 1622-1631.

143. J. Kahn, *et al.*, "Optical Phaselock Receiver with Multigigahertz Bandwidth," Electronics Letters, 1989, vol. 25, no. 10, pp. 626-628.

144. D.A. Atlas and L.G. Kazovsky, "An Optical PSK Homodyne Transmission Experiment using 1320 nm Diode-Pumped Nd:YAG Lasers," IEEE Photonics Technology Letters, 1990, vol. 2, no. 5, pp. 367-370.

145. J.M. Kahn, *et al.*, "4 Gb/s PSK Homodyne Transmission System using Phase-Locked Semiconductor Lasers," IEEE Photonics Technology Letters, 1990, vol. 2, no. 4, pp. 285-287.

146. L. Sun and P. Ye, "Optical Homodyne Receiver Based on an Improved Balanced Phase-Locked Loop with Data-to-Phaselock Crosstalk Suppression," IEEE Photonics Technology Letters, 1990, vol. 2, no. 9, pp. 678-679.

147. J. Kahn, "BPSK Homodyne Detection Experiment Using Balanced Optical Phase-Locked Loop with Quantized Feedback," IEEE Photonics Technology Letters, 1990, vol. 2, no. 11, pp. 840-843.

148. S. Norimatsu, *et al.*, "10 Gbit/s Optical PSK Homodyne Transmission Experiments Using External Cavity DFB LDs," Electronics Letters, 1990, vol. 26, no. 10, pp. 648-649.

149. W.R. Leeb, "Optical 90-Degree Hybrid for Costas Type Receivers," Electronics Letters, 1990, vol. 26, no. 18, pp. 1431-1432.

150. S. Norimatsu, *et al.*, "PSK Optical Homodyne Detection Using External Cavity Laser Diodes in a Costas Loop," IEEE Photonics Technology Letters, 1991, vol. 2, no. 5, pp. 374-376.

151. S. Norimatsu, *et al.*, "An Optical 90 Degree Hybrid Balanced Receiver Module Using a Planar Lightwave Circuit," IEEE Photonics Technology Letters, 1994, vol. 6, no. 6, pp. 737-740.

152. B. Wandernoth, "20 Photon/Bit 565 Mbit/s PSK Homodyne Receiver Using Synchronisation Bits," Electronics Letters, 1992, vol. 28, no. 4, pp. 387-388.

153. L.G. Kazovsky, "Decision-Driven Phase-Locked Loop for Optical Homodyne Receivers," Journal of Lightwave Technology, 1985, vol. LT-3, no. 6, pp. 1238-1247.

154. F. Derr, "Optical QPSK Homodyne Transmission of 280 Mbit/s," Electronics Letters, 1990, vol. 26, no. 6, pp. 401-403.

155. S. Dolinar, "Quantum Optimum Receivers," 1973, Report Number 111, MIT Research Laboratory for Electronics.

156. C.W. Helstrom, *Chapter IV* in *Quantum Detection and Estimation Theory,* 1976, Academic Press: New York.

157. R.S. Bondurant, "Near-Quantum Optimum Receivers for the Phase-Quadrature Coherent-State Channel," Optics Letters, 1993, vol. 18, no. 22, pp. 1896-1898.

158. D.A. Atlas and D.G. Daut, "Finite Intermediate Frequency Penalty in Optical PSK Synchronous Heterodyne Receivers," IEEE Photonics Technology Letters, 1991, vol. 3, no. 8, pp. 769-772.

159. L.G. Kazovsky, *et al.*, "Optical Phase-Locked PSK Heterodyne Experiment at 4 Gb/s," IEEE Photonics Technology Letters, 1990, vol. 2, no. 8, pp. 588-590.

160. S. Watanabe, *et al.*, "560 Mbit/s Optical PSK Heterodyne Detection Using Carrier Recovery," Electronics Letters, 1989, vol. 25, no. 9, pp. 588-589.

161. A. Schoepflin, "565 Mbit/s Optical DPSK and PSK Heterodyne Transmission Experiment using Nd:YAG Lasers at $1.3\,\mu$," Electronics Letters, 1990, vol. 26, no. 4, pp. 255-256.

162. G. Nicholson, "Probability of Error for Optical Heterodyne DPSK System with Quantum Phase Noise," Electronics Letters, 1984, vol. 20, no. 24, pp. 1005-1007.

163. I. Garrett and G. Jacobsen, "Theoretical Analysis of Heterodyne Optical Receivers for Transmission Systems Using (Semiconductor) Lasers with Nonnegligible Linewidth," Journal of Lightwave Technology, 1986, vol. LT-4, no. 3, pp. 323-334.

164. X. Zhang, "Exact Analytical Solution for Statistical Characterisation of Phase Noise in Envelope Detection," Electronics Letters, 1993, vol. 29, no. 17, pp. 1511-1513.

165. E. Desurvire, *Erbium-Doped Fiber Amplifiers: Principles and Applications*, 1994, New York: John Wiley & Sons.

166. T. Whitley, *et al.*, "Towards a Practical 1.3 micron Optical Fiber Amplifier," British Telecom Technical Journal, 1993, vol. 11, no. 2.

167. S. Grubb, *et al.* "1.3 μm Cascaded Raman Amplifier in Germanosilicate Fibers," in *Topical Meeting on Optical Amplifiers and Their Applications.* 1994, Breckenridge, Colorado, Optical Society of America, Postdeadline paper PD3.

168. T. Durhuus, *et al.* "High Performance Semiconductor Optical Preamplifier," in *Topical Meeting on Optical Amplifiers and Their Applications.* 1994, Breckenridge, Colorado, Optical Society of America, WB-4.

169. J. Crowe and W. Ahearn, "Semiconductor Laser Amplifier," IEEE Journal of Quantum Electronics, 1966, vol. QE-2, no. 8, pp. 283-289.

170. T. Saito, *et al.*, "High Receiver Sensitivity at 10 Gb/s Using an Er-Doped Fiber Preamplifier Pumped with a 0.98 μm Laser Diode," IEEE Photonics Technology Letters, 1991, vol. 3, no. 6, pp. 530-533.

171. C. Mikkelsen, *et al.*, "High-Performance Semiconductor Optical Preamplifier Receiver at 10 Gb/s," IEEE Photonics Technology Letters, 1993, vol. 5, no. 9, pp. 1096-1097.

172. A. Yariv, *Quantum Electronics,* 1967, John Wiley & Sons: New York.

173. N.A. Olsson, "Lightwave Systems with Optical Amplifiers," Journal of Lightwave Technology, 1989, vol. 7, no. 7, pp. 1071-1082.

174. "Special Issue on Optical Amplifiers," Journal of Lightwave Technology, 1991, vol. 9, no. 1.

175. R. Laming, *et al.*, "Erbium-Doped Fiber Amplifier with 54 dB Gain and 3.1 dB Noise Figure," IEEE Photonics Technology Letters, 1992, vol. 4, no. 12, pp. 1345-1347.

176. A.H. Gnauck and C.R. Giles, "2.5 and 10 Gb/s Transmission Experiments Using a 137 Photon/Bit Erbium-Fiber Preamplifier Receiver," IEEE Photonics Technology Letters, 1992, vol. 4, no. 1, pp. 80-82.

177. R. Laming, *et al.*, "High-Sensitivity Two-Stage Erbium-Doped Fiber Preamplifier at 10 Gb/s," IEEE Photonics Technology Letters, 1992, vol. 4, no. 12, pp. 1348-1350.

178. R. Cryan, *et al.*, "An Optically Preamplified PPM Threshold Detection Receiver for Satellite Communication," Microwave and Optical Technology Letters, 1994, vol. 7, no. 7, pp. 324-327.

179. B. Glance, *et al.*, "Sensitivity of an Optical Heterodyne Receiver in Presence of an Optical Preamplifier," Electronics Letters, 1988, vol. 24, no. 19, pp. 1229-1230.

180. S. Ryu and Y. Horiuchi, "Use of an Optical Amplifier in a Coherent Receiver," IEEE Photonics Technology Letters, 1991, vol. 3, no. 7, pp. 663-665.

181. S. Miller, "Performance of a Coherent Optical PSK Receiver Employing Optical Amplification," IEE Proceedings - Part J, 1992, vol. 139, no. 3, pp. 215-222.

182. T. Li and M. Teich, "Bit-Error Rate for a Lightwave Communication System Incorporating an Erbium-Doped Fibre Amplifier," Electronics Letters, 1991, vol. 27, no. 7, pp. 598-599.

183. P. Humblet and M. Azizoglu, "On the Bit Error Rate of Lightwave Systems with Optical Amplifiers," Journal of Lightwave Technology, 1991, vol. 9, no. 11, pp. 1576-1582.

184. P. Humblet. "Design of Optical Matched Filters," in *IEEE Globecom*. 1991, Phoenix, IEEE Press, 35.1.1.

185. S.R. Chinn, "Error-Rate Performance of Optical Amplifiers with Fabry-Perot Filters," Electronics Letters, 1995, vol. 31, no. 9, pp. 756-757.

186. T.J. Paul, *et al.*, "3 Gbit/s Optically Preamplified Direct Detection DPSK Receiver with 116 photon/bit Sensitivity," Electronics Letters, 1993, vol. 29, no. 7.

187. G. Jacobsen, "Heterodyne and Optically Implemented Dual-Filter FSK Systems: Comparison Based on Rigorous Theory," IEEE Photonics Technology Letters, 1992, vol. 4, no. 7, pp. 771-774.

188. R. Steele and G. Walker, "High-Sensitivity FSK Signal Detection with an Erbium-Doped Fiber Preamplifier and Fabry-Perot Demodulation," IEEE Photonics Technology Letters, 1990, vol. 2, no. 10, pp. 753-755.

189. A. Yariv, *Appendix D - Noise in Laser Amplifiers* in *Optical Electronics,* 1985, Holt, Rinehart and Winston: New York.

190. F. Arams and M. Wang, "Infrared Laser Preamplifier System," Proceedings of the IEEE, 1965, vol. 53, pp. 329.

191. E.V. Hoversten, *Chap. F8 - Section 3.2 - Preamplifier and Homodyne Systems* in *Laser Handbook,* F.T. Arecchi and E.O. Schulz-DuBois, Editors. 1972, American Elsevier Publishing Company: New York.

192. R.H. Kingston, *Chap. 8 - Laser Preamplification* in *Detection of Optical and Infrared Radiation,* 1978, Springer-Verlag: New York.

193. O. Tonguz and R. Wagner, "Equivalence Between Preamplified Direct Detection and Heterodyne Receivers," IEEE Photonics Technology Letters, 1991, vol. 3, no. 9, pp. 835-837.

194. J.J. Spilker, *Chap. 14 - Bit Synchronizers for Digital Communication* in *Digital Communications by Satellite,* 1977, Prentice-Hall Information Theory Series: Englewood Cliffs, New Jersey.

195. D. Millicker and R. Standley, "2 Gb/s Timing Recovery Circuit Using Dielectric Resonator Filter," Electronics Letters, 1987, vol. 23, pp. 738-739.

196. S. Rosenberg, *et al.*, "Optical Fiber Repeatered Transmission Systems Using SAW Filters," IEEE Transactions on Ultrasonics, 1983, pp. 119-126.

197. Y. Imai, *et al.*, "Design and Performance of Clock-Recovery GaAs ICs for High-Speed Optical Communication Systems," IEEE Transactions on Microwave Theory and Techniques, 1993, vol. 41, no. 5, pp. 745-751.

198. P. Trischitta and E. Varma, *Jitter in Digital Transmission Systems*, 1989, Norwood, MA: Artech House.

199. C. Eldering, *et al.* "Transmitter and Receiver Requirements for TDMA Passive Optical Networks," in *Third IEEE Conference on Local Optical Networks.* 1991, TokyoJapan.

200. Y. Ota, *et al.*, "DC-1 Gb/s Burst-Mode Compatible Receiver for Optical Bus Applications," Journal of Lightwave Technology, 1992, vol. 10, no. 2, pp. 244-249.

201. M. Banu and A. Dunlop, "Clock Recovery Circuits with Instantaneous Locking," Electronics Letters, 1992, vol. 28, no. 23.

202. W.I. Way, "Subcarrier Multiplexed Lightwave System Design Considerations for Subscriber Loop Applications," Journal of Lightwave Technology, 1989, vol. 7, no. 11, pp. 1806-1818.

203. R. Olshansky, *et al.*, "Subcarrier Multiplexed Lightwave Systems for Broadband Distribution," Journal of Lightwave Technology, 1989, vol. 7, no. 9, pp. 1329-1340.

204. R. Olshansky, *et al.* "Design and Performance of Wideband Subcarrier Multiplexed Lightwave Systems," in *European Conference on Optical Communication.* 1988, Brighton U. K., IEE: London.

205. "Special Issue on Broad-Band Lightwave Video Transmission," Journal of Lightwave Technology, 1993, vol. 11, no. 1.

206. R. Kalman, *et al.*, "Dynamic Range of Coherent Analog Fiber-Optic Links," Journal of Lightwave Technology, 1994, vol. 12, no. 7, pp. 1263-1277.

207. G.C. Temes and J.W. LaPatra, *Chap. 12 - Approximation Theory* in *Introduction to Circuit Synthesis and Design,* 1977, McGraw-Hill Book Company: New York.

208. L. Weinberg, *Network Analysis and Synthesis,* 1962, McGraw-Hill, Inc.: New York.

209. T.T. Ha, *Chap. 6 - Signal Distortion Characterizations* in *Solid-State Microwave Amplifier Design,* 1981, John Wiley & Sons: New York.

210. T. Darcie, "Subcarrier Multiplexing for Multiple-Access Lightwave Networks," Journal of Lightwave Technology, 1987, vol. LT-5, pp. 1103-1110.

211. R.E. Ziemer and W.H. Tranter, *Chap. 6 - Noise in Modulation Systems* in *Principles of Communications,* 1976, Houghton Mifflin Company: Boston.

212. L.G. Kazovsky and D.A. Atlas. "PSK Synchronous Heterodyne and Homodyne Experiments Using Optical Phase-Locked Loops," in *Conference on Optical Fiber Communication, OFC.* 1990, Optical Society of America, Postdeadline Paper PD11.

213. R. Noe, *et al.*, "Optical FSK Transmission with Pattern-Independent 119 Photoelectrons/Bit Receiver Sensitivity with Endless Polarisation Control," Electronics Letters, 1989, vol. 25, no. 12, pp. 757-758.

214. B. Glance, *et al.* "Densely Spaced FDM Optical Coherent System with Near Quantum-Limited Sensitivity and Computer Controlled Random Access Channel Selection," in *Conference on Optical Fiber Communication, OFC.* 1989, Optical Society of America, Postdeadline Paper PD11.

215. T.V. Muoi. "Extremely Sensitive Direct Detection Receiver for Laser Communications," in *Conference on Lasers and Electro-Optics (CLEO).* 1987, Optical Society of America, Paper THS4.

216. R.A. Linke, *et al.*, "Coherent Lightwave Transmission Over 150 km Fibre Lengths at 400 Mbit/s and 1 Gbit/s Data Rates Using Phase Modulation," Electronics Letters, 1985, vol. 22, no. 1, pp. 30-31.

217. T. Chikama, *et al.*, "Modulation and Demodulation Techniques in Optical Heterodyne PSK Transmission Systems," Journal of Lightwave Technology, 1990, vol. 8, no. 3, pp. 309-321.

218. L.D. Tzeng, *et al.*, "A High Performance Optical Receiver for 622 Mb/s Direct Detection Systems," IEEE Photonics Technology Letters, 1990, vol. 2, no. 10, pp. 759-761.

219. P.P. Smythe, *et al.*, "152 Photons per Bit Detection at 622 Mbit/s to 2.5 Gbit/s Using an Erbium Fibre Preamplifier," Electronics Letters, 1990, vol. 26, no. 19, pp. 1604-1605.

220. J.M. Kahn, "1 Gbit/s PSK Homodyne Transmission System Using Phase-Locked Semiconductor Lasers," IEEE Photonics Technology Letters, 1990, vol. 1, no. 5, pp. 340-342.

221. J.C. Campbell, *et al.*, "High Performance Avalanche Photodiode with Separate Absorption, Grading, and Multiplication Regions," Electronics Letters, 1983, vol. 19, pp. 818-820.

222. K. Kannan, *et al.*, "High-Sensitivity Receiver Optical Preamplifiers," IEEE Photonics Technology Letters, 1992, vol. 4, no. 3, pp. 272-275.

223. M. Shikada, *et al.* "1.5 μ High Bit Rate Long Span Transmission Experiments Employing a High Power DFB-DC-PBH Laser Diode," in *European Conference on Optical Communication.* 1985, Istituto Internazionale delle Communicazioni: Geneva. Post Deadline Paper.

224. T. Imai, *et al.*, "Over 300 km CPFSK Transmission Experiment Using 67 Photon/bit Sensitivity Receiver," Electronics Letters, 1990, vol. 26, no. 6, pp. 1357-1358.

225. E.A. Swanson, *et al.*, "High Sensitivity Optically Preamplified Direct Detection DPSK Receiver with Active Delay-Line Stabilization," IEEE Photonics Technology Letters, 1994, vol. 6, no. 2, pp. 263-265.

226. Y.K. Park, *et al.* "5 Gbit/s Optical Preamplifier Receiver with 135-Photons/bit Usable Receiver Sensitivity," in *Conference on Optical Fiber Communication, OFC.* 1993, Optical Society of America, Paper Number TuD4.

227. Y. Miyamoto, *et al.*, "A 10 Gb/s High Sensitivity Optical Receiver Using an InGaAs-InAlAs Superlattice APD at 1.3 μm / 1.5 μm," IEEE Photonics Technology Letters, 1991, vol. 3, no. 4, pp. 372-374.

228. K. Hagimoto, *et al.* "Twenty-Gbit/s Signal Transmission Using a Simple High-Sensitivity Optical Receiver," in *Conference on Optical Fiber Communication, OFC*. 1992, Optical Society of America, Paper Number Tul3.

Index

Stephen B. Alexander is Vice President and Chief Technology Officer of CIENA Corporation. He served as Vice President, Transport Products during 1996 and 1997 and as the Director of Lightwave Systems at the company since joining it in 1994. From 1982 until joining CIENA, he was employed at MIT Lincoln Laboratory, where he last held the position of Assistant Leader of the Optical Communications Technology Group. While at Lincoln Laboratory he initiated the development of a 20-channel WDM test-bed for the AT&T, DEC, MIT Consortium on Wide-Band All-Optical Networks (AON). He also worked on high-capacity intersatellite optical communication systems, opto-electronic device characterization, and high-speed coherent detection systems. Mr. Alexander is an Associate Editor for the Journal of Lightwave Technology and is a member of the steering committee for the Conference on Optical Fiber Communication (OFC). He was a general chair of the 1997 conference and a technical program chair for OFC'95. He chaired the systems subcommittee for OFC 92–93 and was a member of the program committee for 90–91. He has also served on various technical committees for the Conference on Lasers and Electro-Optics (CLEO), the International Conference on Integrated Optics and Optical Fiber Communication (IOOC), the Lasers and Electro-Optics Society (LEOS) annual meeting, the Conference on Optical Amplifiers and their Applications, and various topical meetings on optical communications. Mr. Alexander received both his BS and MS degrees in electrical engineering from the Georgia Institute of Technology.